Synthetic DNA and RNA Programming

Synthetic DNA and RNA Programming

Special Issue Editors

Patrick O'Donoghue
Ilka Heinemann

MDPI • Basel • Beijing • Wuhan • Barcelona • Belgrade

MDPI

Special Issue Editors

Patrick O'Donoghue
The University of
Western Ontario
Canada

Ilka Heinemann
Western University
Canada

Editorial Office
MDPI
St. Alban-Anlage 66
4052 Basel, Switzerland

This is a reprint of articles from the Special Issue published online in the open access journal *Genes* (ISSN 2073-4425) from 2018 to 2019 (available at: https://www.mdpi.com/journal/genes/special_issues/Synthetic_Programming).

For citation purposes, cite each article independently as indicated on the article page online and as indicated below:

LastName, A.A.; LastName, B.B.; LastName, C.C. Article Title. *Journal Name* **Year**, *Article Number*, Page Range.

ISBN 978-3-03921-734-2 (Pbk)
ISBN 978-3-03921-735-9 (PDF)

Cover image courtesy of Yuka Naraki, Space-Time Inc.

Contents

About the Special Issue Editors

Patrick O'Donoghue is Canada Research Chair in Chemical Biology and Associate Professor of Chemistry and Biochemistry at the University of Western Ontario (London, Ontario, Canada). He received his bachelor's degree in Biophysics and Ph.D. in Chemistry, supervised by Dr. Zan Luthey-Schulten, from the University of Illinois Urbana-Champaign. He was a Postdoctoral Fellow first with Dr. Carl Woese at Illinois and then at Yale University with Dr. Dieter Söll. In 2013, he was appointed Assistant Professor at Western where he was promoted to Associate Professor with tenure in 2019. His research focuses on genetic code evolution and engineering. The O'Donoghue lab develops methods for site-specific insertion of post-translational modifications into proteins and applies these methods toward elucidating the role of protein modifications in signaling networks linked to cancer and neurodegenerative diseases. The O'Donoghue lab also investigates the role of mistranslation in health and disease.

Ilka Heinemann is Assistant Professor of Biochemistry at the University of Western Ontario (London, Ontario, Canada). She received her master's degree and Ph.D. in Microbiology, supervised by Dr. Dieter Jahn at the Technical University of Braunschweig, Germany. She was a Postdoctoral Fellow at Yale University with Dr. Dieter Söll. In 2013, she was recruited as Assistant Professor to Western. Her research focuses on the regulation of microRNA metabolism by terminal nucleotidyltransferases. The Heinemann lab also uses synthetic biology approaches to engineer a small RNA editing protein towards a high fidelity reverse 3'–5' RNA polymerase.

genes

MDPI

Editorial

Synthetic DNA and RNA Programming

Patrick O'Donoghue [1,2,*] and Ilka U. Heinemann [1,*]

[1] Department of Biochemistry, The University of Western Ontario, London, ON N6A 5C1, Canada
[2] Department of Chemistry, The University of Western Ontario, London, ON N6A 5C1, Canada
* Correspondence: patrick.odonoghue@uwo.ca (P.O.); ilka.heinemann@uwo.ca (I.U.H.)

Received: 26 June 2019; Accepted: 2 July 2019; Published: 11 July 2019

Abstract: Synthetic biology is a broad and emerging discipline that capitalizes on recent advances in molecular biology, genetics, protein and RNA engineering as well as omics technologies. Together these technologies have transformed our ability to reveal the biology of the cell and the molecular basis of disease. This Special Issue on "Synthetic RNA and DNA Programming" features original research articles and reviews, highlighting novel aspects of basic molecular biology and the molecular mechanisms of disease that were uncovered by the application and development of novel synthetic biology-driven approaches.

Keywords: genetic code expansion; genome synthesis; genome editing; microRNA; protein modification; RNA metabolism; tRNA; synthetic biology; unnatural amino acids; unnatural nucleotides

---

Synthetic biology is a broad and emerging discipline that capitalizes on recent advances in molecular biology, genetics, protein and RNA engineering, as well as omics technologies. Together these biotechnologies have transformed our ability to reveal the biology of the cell and the molecular basis of disease. This special issue of *Genes* on "Synthetic DNA and RNA Programming" features 12 original research articles and reviews that highlight novel aspects of basic molecular and cellular biology, uncovered by the application and development of synthetic biology-driven approaches.

The approaches highlighted here involve programming genes, RNAs, and proteins, both in the test tube and in living cells, to explore areas of molecular biology related to genetic code evolution [1,2] and genetic code expansion [3–5], including applications in programming protein modifications [6,7], RNA metabolism [8,9], and genetic systems for genome engineering [10,11] and biocontainment of modified microorganisms [12]. These contributions showcase both the diversity of research approaches emerging as synthetic biology tools and the extreme level of detail with which these tools enable studies and the manipulation of individual protein modifications or cellular pathways. The contributed articles cluster into four thematic categories: *genetic code expansion, genetic code evolution, novel genetic systems and molecular tools*, and *RNA programming*.

1. Expanding the Genetic Code

Genetic code expansion studies focus on methods that enable the production of proteins with amino acids beyond the canonical or standard 20 genetically encoded amino acids. Cells with engineered and expanded genetic codes can produce proteins with 21 [13] or even 22 [14,15] different genetically encoded amino acids. These additional amino acids allow the introduction of new and specific chemically functionalized side chains.

Genetic code expansion normally requires the addition of a new aminoacyl-tRNA synthetase and tRNA pair. Efficient and high-fidelity genetic encoding requires that the AARS tRNA pair is mutually orthogonal to the endogenous AARSs and tRNAs in the cell. These methods normally reassign the meaning of a stop codon, usually UAG, to the direct incorporation of an additional, non-canonical amino acid (ncAA). The most commonly used orthogonal pairs include the archaeal enzymes tyrosyl-tRNA syntetase (TyrRS) [13], the pyrrolysyl-tRNA synthetase (PylRS) [16,17], and the phosphoseryl-tRNA synthetase (SepRS) [18]. Balasuriya et al. used the phosphoserine system to produce highly active human kinases with programmed phosphorylation from facile *Escherichia coli* expression systems [7]. The authors used these reagents to identify a specific phosphorylation site on protein kinase B (Akt1) that interferes with a clinically relevant kinase inhibitor.

Genetic code expansion also enables the incorporation of other posttranslational modifications at specific or programmed sites in proteins. Umehara et al. [6] used a mutant of the PylRS tRNA^{Pyl} orthogonal pair to genetically encode N^ε-acetylated lysine into the *E. coli* alanyl-tRNA synthetase. The resulting acetylated AlaRS was catalytically deficient. The authors next used in vivo assays to determine that two Cob deacetylases were able to remove the K73 acetylation, identifying a novel regulatory pathway to control AlaRS activity.

The study from Hoffmann et al. focused on the central role of tRNAs in genetic code expansion [3]. Specifically, this review article addressed two 'fundamentally different translation systems' that evolved in natural organisms. These include the systems that genetically encode selenocysteine and pyrrolysine, the so-called 21st and 22nd amino acids. Design and engineering of tRNA variants for optimal genetic code expansion were examined, as were synthetic biology applications of genetic code expansion.

Finally, Gang et al. focused on the application of peptides with ncAAs as antibiotics. Antibiotic development is a matter of urgent clinical need and one of the most promising areas is related to the development of synthetic cyclic peptides. These compounds are often inspired by natural products, but through the use of genetic code expansion, ncAAs can be included in these antibiotic peptides [4]. The review article from Gang et al. focused on methods for peptide cyclization and applications of cyclic peptides.

2. Genetic Code Evolution

Although Crick's frozen accident hypothesis once envisioned a perfectly interpreted genetic code [19], recent studies have found that cells tolerate surprisingly high levels of amino acid misincorporation [20]. The next pair of featured articles [1,2] investigated biochemical mechanisms that regulate translation fidelity.

Berg et al. [1] provided the first detailed characterization of human tRNA^{Ser} identity elements. These are the critical nucleotides that are essential for tRNA recognition by the cognate aminoacyl-tRNA synthetase and determine the serine accepting identity of this tRNA. The tRNA identity elements define the correspondence of amino acids with anticodons and thus are fundamental to the accurate decoding of the genetic code. Using a unique in vivo assay, relying on the misincorporation of serine at proline codons in yeast, this work identified the evolution of new identity elements in the human tRNA^{Ser}.

Similarly, the article by Rathnayake et al. [2] also highlighted the fact that the mechanisms and molecular entities tasked with ensuring the faithful interpretation of genetic code have evolved over time into idiosyncratic variants. The aspartyl-tRNA synthetase (AspRS) enzymes occur as two variants, the discriminating AspRS, which exclusively charges Asp to tRNA^{Asp}, as well as a non-discriminating variant (ND-AspRS). ND-AspRS charges Asp onto tRNA^{Asn} in a pathway generating Asn-tRNA^{Asn} via Asp-tRNA^{Asn}. In some bacteria, both AspRS and ND-AspRS are capable of not only charging Asp to tRNA^{Asp} and tRNA^{Asn}, but also of have an unexpected glutamyl-tRNA synthetase (GluRS)-like activity. In these cases, AspRS acts as a GluRS, and is able to ligate Glu to tRNA^{Glu} but not to tRNA^{Asp}. Although the AspRS enzyme targets a non-cognate tRNA, amino acid misincorporation does not occur, thus preserving translation fidelity.

3. Novel Genetic Systems and Molecular Tools

Gordon et al. opened the door to studying the unusual biology of yeast of the genus *Metschnikowia* [11]. These yeasts have a genetic code variation in which the leucine CUG codons are instead decoded as serine. The authors developed a new and efficient transformation system for

Metschnikowia and an additional 21 yeast species, providing new platforms for genome synthesis and engineering efforts.

Protein expression is an essential methodology for biochemical studies, including those mentioned above, as well as for the production of biopharmaceuticals such as antibodies and therapeutic peptides. Chen et al. [10] used a high-throughput approach to screen for the impact of the expression of any *E. coli* gene on cell growth. The most significant impact on *E. coli* growth was observed upon the overexpression of membrane proteins. Interestingly, the authors found that for certain proteins, the use of lower copy number plasmids stabilized cell growth rate and increased overall recombinant protein yields. The impact on cell growth could also often be remedied by amino acid supplementation. Using a related methodology, Schwark et al. used a fluorescent protein reporter system and investigated an *E. coli* release factor deletion strain with a highly efficient archaeal orthogonal AARS tRNA pair. Their data suggested that the engineered strain will be a highly useful tool for applications requiring the incorporation of multiple ncAAs [5].

A major ethical responsibility for synthetic biologists is the ability to retain control or confinement of genetically modified microorganisms. This drove the field to develop fascinating and effective methods for containing microorganisms in the lab that are commonly referred to as biocontainment. Some of these approaches included the use of genetic code expansion to create synthetic auxotrophs, such as an *E. coli* strain dependent on ncAAs for growth [21]. Diwo et al. further extended this idea with their novel formulation of an 'alien genetic code' [12]. The authors presented the concept that an ideal biocontainment system would involve the creation of microbes with genetic codes that are totally incompatible or 'alien' to the natural genetic code. The construction of such an organism is considered to be achieved via the directed evolution of an existing cell or through de novo construction of 'synthetic' genomes and cells.

4. RNA Programming

RNAs play important roles in controlling translation, yet there are still limited tools to study and engineer RNAs and their activity in cells. Turk et al. developed a novel reporter system to quantify the activity of microRNAs (miRNAs) in living cells [9]. While microRNAs can be quantified by methods such as quantitative PCR, a variable and unknown portion of cellular miRNAs is inactive. To assess the amount of active miRNA in the cell, the authors developed a GFP-based reporter system that allows for time- and space-resolved monitoring of active miRNAs in single cells.

Another limitation of analyzing RNAs is the limitation of commercially available polymerases to extend RNAs in the forward 5' to 3' direction. Chen et al. [8] reviewed recent progress made towards the characterization and engineering of tRNA^{His} guanylyltransferase homologs capable of reverse 3' to 5' nucleotide polymerization. Tools and enzymes with broadened RNA substrates and extended template-dependent reverse polymerization activity were discussed and will be invaluable for RNA labeling, engineering, and analysis efforts.

In summary, this collection of articles represents new directions in multiple areas of interest to synthetic biologists. From expanding the number of genetically encoded amino acids to creating new genetic tools in diverse microbes, these articles are each exemplars of the sophisticated synthetic biology approaches that produce cells with new capabilities, enabling the production of designer proteins and RNAs. Some of these tools reveal molecular events in living cells that were previously inaccessible.

Funding: This research was funded by grants from the Natural Sciences and Engineering Research Council of Canada (04776-2014 to I.U.H.; 04282-2014 to P.O.; 530175-2018 to P.O.); J.P. Bickell Foundation to I.U.H.; Ontario Early Researcher Award (ER18-14-183 to I.U.H); Canada Foundation for Innovation (229917 to P.O.); the Ontario Research Fund (229917 to P.O.); Canada Research Chairs (950-229917 to P.O.); Ontario Centres of Excellence (28922 to P.O.).

Acknowledgments: We are deeply grateful to each of the contributors to this special issue, which would not have been possible without their continuous and dedicated efforts in scientific discovery and advancing our field.

Conflicts of Interest: The authors declare no conflict of interest.

References

1. Berg, M.D.; Genereaux, J.; Zhu, Y.; Mian, S.; Gloor, G.B.; Brandl, C.J. Acceptor stem differences contribute to species-specific use of yeast and human tRNA^{Ser}. *Genes* **2018**, *9*, 612. [CrossRef] [PubMed]
2. Rathnayake, U.M.; Hendrickson, T.L. Bacterial aspartyl-tRNA synthetase has glutamyl-tRNA synthetase activity. *Genes* **2019**, *10*, 262. [CrossRef] [PubMed]
3. Hoffman, K.S.; Crnkovic, A.; Söll, D. Versatility of synthetic tRNAs in genetic code expansion. *Genes* **2018**, *9*, 537. [CrossRef] [PubMed]
4. Gang, D.; Kim, D.W.; Park, H.S. Cyclic peptides: Promising scaffolds for biopharmaceuticals. *Genes* **2018**, *9*, 557. [CrossRef] [PubMed]
5. Schwark, D.G.; Schmitt, M.A.; Fisk, J.D. Dissecting the contribution of release factor interactions to amber stop codon reassignment efficiencies of the *Methanocaldococcus jannaschii* orthogonal pair. *Genes* **2018**, *9*, 546. [CrossRef] [PubMed]
6. Umehara, T.; Kosono, S.; Söll, D.; Tamura, K. Lysine acetylation regulates alanyl-tRNA synthetase activity in *Escherichia coli*. *Genes* **2018**, *9*, 473. [CrossRef] [PubMed]
7. Balasuriya, N.; McKenna, M.; Liu, X.; Li, S.S.C.; O'Donoghue, P. Phosphorylation-dependent inhibition of Akt1. *Genes* **2018**, *9*, 450. [CrossRef]
8. Chen, A.W.; Jayasinghe, M.I.; Chung, C.Z.; Rao, B.S.; Kenana, R.; Heinemann, I.U.; Jackman, J.E. The role of 3′ to 5′ reverse RNA polymerization in tRNA fidelity and repair. *Genes* **2019**, *10*, 250. [CrossRef] [PubMed]
9. Turk, M.A.; Chung, C.Z.; Manni, E.; Zukowski, S.A.; Engineer, A.; Badakhshi, Y.; Bi, Y.; Heinemann, I.U. MiRAR-miRNA activity reporter for living cells. *Genes* **2018**, *9*, 305. [CrossRef]
10. Chen, H.; Venkat, S.; Wilson, J.; McGuire, P.; Chang, A.L.; Gan, Q.; Fan, C. Genome-wide quantification of the effect of gene overexpression on *Escherichia coli* growth. *Genes* **2018**, *9*, 414. [CrossRef]
11. Gordon, Z.B.; Soltysiak, M.P.M.; Leichthammer, C.; Therrien, J.A.; Meaney, R.S.; Lauzon, C.; Adams, M.; Lee, D.K.; Janakirama, P.; Lachance, M.A.; et al. Development of a transformation method for *Metschnikowia borealis* and other CUG-serine yeasts. *Genes* **2019**, *10*, 78. [CrossRef]
12. Diwo, C.; Budisa, N. Alternative biochemistries for alien life: Basic concepts and requirements for the design of a robust biocontainment system in genetic isolation. *Genes* **2018**, *10*, 17. [CrossRef] [PubMed]
13. Wang, L.; Brock, A.; Herberich, B.; Schultz, P.G. Expanding the genetic code of *Escherichia coli*. *Science* **2001**, *292*, 498–500. [CrossRef] [PubMed]
14. Wright, D.E.; Altaany, Z.; Bi, Y.; Alperstein, Z.; O'Donoghue, P. Acetylation regulates thioredoxin reductase oligomerization and activity. *Antioxid. Redox Signal.* **2018**, *29*, 377–388. [CrossRef]
15. Wan, W.; Huang, Y.; Wang, Z.; Russell, W.K.; Pai, P.J.; Russell, D.H.; Liu, W.R. A facile system for genetic incorporation of two different noncanonical amino acids into one protein in *Escherichia coli*. *Angew. Chem. Int. Ed.* **2010**, *49*, 3211–3214. [CrossRef] [PubMed]
16. Blight, S.K.; Larue, R.C.; Mahapatra, A.; Longstaff, D.G.; Chang, E.; Zhao, G.; Kang, P.T.; Green-Church, K.B.; Chan, M.K.; Krzycki, J.A. Direct charging of tRNA$_{CUA}$ with pyrrolysine in vitro and in vivo. *Nature* **2004**, *431*, 333–335. [CrossRef] [PubMed]
17. Polycarpo, C.; Ambrogelly, A.; Berube, A.; Winbush, S.M.; McCloskey, J.A.; Crain, P.F.; Wood, J.L.; Söll, D. An aminoacyl-tRNA synthetase that specifically activates pyrrolysine. *Proc. Natl. Acad. Sci. USA* **2004**, *101*, 12450–12454. [CrossRef] [PubMed]
18. Park, H.S.; Hohn, M.J.; Umehara, T.; Guo, L.T.; Osborne, E.M.; Benner, J.; Noren, C.J.; Rinehart, J.; Söll, D. Expanding the genetic code of *Escherichia coli* with phosphoserine. *Science* **2011**, *333*, 1151–1154. [CrossRef]
19. Crick, F.H. The origin of the genetic code. *J. Mol. Biol.* **1968**, *38*, 367–379. [CrossRef]
20. Lant, J.T.; Berg, M.D.; Heinemann, I.U.; Brandl, C.J.; O'Donoghue, P. Pathways to disease from natural variations in human cytoplasmic tRNAs. *J. Biol. Chem.* **2019**, *294*, 5294–5308. [CrossRef]
21. Mandell, D.J.; Lajoie, M.J.; Mee, M.T.; Takeuchi, R.; Kuznetsov, G.; Norville, J.E.; Gregg, C.J.; Stoddard, B.L.; Church, G.M. Biocontainment of genetically modified organisms by synthetic protein design. *Nature* **2015**, *518*, 55–60. [CrossRef] [PubMed]

![genes](GCAT TACG GCAT) *genes*

MDPI

Article

MiRAR—miRNA Activity Reporter for Living Cells

Matthew A. Turk [1], Christina Z. Chung [1], Emad Manni [1], Stephanie A. Zukowski [1], Anish Engineer [2], Yasaman Badakhshi [1], Yumin Bi [1] and Ilka U. Heinemann [1,*]

[1] Department of Biochemistry, The University of Western Ontario, 1151 Richmond Street, London, ON N6A 5C1, Canada; mturk5@uwo.ca (M.A.T.); cchung88@uwo.ca (C.Z.C.); emanni@uwo.ca (E.M.); szukowsk@uwo.ca (S.A.Z.); ybadakhs@uwo.ca (Y.B.); ybi@uwo.ca (Y.B.)

[2] Department of Physiology and Pharmacology, The University of Western Ontario, 1151 Richmond Street, London, ON N6A 5C1, Canada; aengine@uwo.ca

* Correspondence: ilka.heinemann@uwo.ca.com; Tel.: +1-519-850-2949

Received: 6 June 2018; Accepted: 15 June 2018; Published: 19 June 2018

Abstract: microRNA (miRNA) activity and regulation are of increasing interest as new therapeutic targets. Traditional approaches to assess miRNA levels in cells rely on RNA sequencing or quantitative PCR. While useful, these approaches are based on RNA extraction and cannot be applied in real-time to observe miRNA activity with single-cell resolution. We developed a green fluorescence protein (GFP)-based reporter system that allows for a direct, real-time readout of changes in miRNA activity in live cells. The miRNA activity reporter (MiRAR) consists of GFP fused to a 3′ untranslated region containing specific miRNA binding sites, resulting in miRNA activity-dependent GFP expression. Using qPCR, we verified the inverse relationship of GFP fluorescence and miRNA levels. We demonstrated that this novel optogenetic reporter system quantifies cellular levels of the tumor suppressor miRNA let-7 in real-time in single Human embryonic kidney 293 (HEK 293) cells. Our data shows that the MiRAR can be applied to detect changes in miRNA levels upon disruption of miRNA degradation pathways. We further show that the reporter could be adapted to monitor another disease-relevant miRNA, miR-122. With trivial modifications, this approach could be applied across the miRNome for quantification of many specific miRNA in cell cultures, tissues, or transgenic animal models.

Keywords: fluorescent reporter; live cell imaging; microRNA quantification; optogenetics; small molecule drug screening

1. Introduction

Human gene expression and RNA transcript stability can be regulated before, during, and after transcription. MicroRNAs (miRNAs) regulate transcript stability by binding to messenger RNAs (mRNAs) in complementary regions, inducing endonuclease-mediated cleavage or inhibiting protein synthesis [1]. In mammals, miRNAs usually contain sequence homology to their target transcript in their 5′ and 3′ untranslated regions (UTR) [2,3]. Genes encoding miRNAs are transcribed as long precursor transcripts and processed to yield short 18–24-nucleotide-long miRNAs. These miRNAs are subsequently integrated into protein complexes to induce mRNA silencing. While translational aspects of miRNA research are primarily focused on cancer, deregulation of miRNA stability and activity has relevance to many diseases. Dysfunctional miRNA expression, processing, and degradation have been found in diseases including breast cancer [4], acute myeloid leukemia [5], ovarian cancer [6], and hepatocellular carcinoma [7], but links between miRNAs and Alzheimer's disease [8], diabetes [9], and schizophrenia [10] are also emerging.

The miRNA let-7 is often implicated in disease and the let-7 miRNA sequence and timing of expression during development are highly conserved amongst vertebrates [11]. In normal cells and tissues, let-7 suppresses tumor proliferation and cell survival by negatively regulating oncogenic signaling pathways [12]. Let-7 directly binds to complementary regions of mRNAs with protein products involved in cell cycle proliferation and apoptosis, such as e.g., Ras, high mobility group A2 (Hmga2), Caspase 3, and others [11,13–17]. Let-7 levels are significantly lower in cancer cells and stem cells compared to differentiated cell types, highlighting the role for let-7 in cell cycle regulation [18,19]. Similarly, let-7 is down-regulated in numerous cancers [20–22] and low let-7 levels are associated with shortened post-operative survival [23].

Recent work has begun to reveal the role of let-7 in maintaining cell differentiation and cancer proliferation [12,13,15]. In lung cancers, let-7 and the oncogene Kirsten rat sarcoma viral oncogene homolog (*kras*) have a reciprocal relationship [16]. High KRas levels and low let-7 levels generate a highly cancerous phenotype. Increasing let-7 levels, however, cause KRas levels to decrease and normal cell morphology to return. The KRas mRNA has seven predicted let-7 binding sites in its 3′-UTR [16]. Other genes regulated by let-7 include *Hmga2* and *Caspase 3*. *Hmga2* regulates the G2/M checkpoint in cell cycling and contains binding sites for let-7 miRNA in its 3′-UTR [24]. Let-7 also regulates apoptosis via let-7 binding sites in the 3′-UTR of *Caspase 3*. By interfering with *Caspase 3* expression, let-7 allows cells to escape apoptotic effector caspases [17].

KRas, *Hmga2*, *Caspase 3*, and other oncogenes are directly regulated by let-7 levels in the cell. Thus, let-7 biosynthesis and the regulation of let-7 levels are of increasing interest as new therapeutic targets [25]. Current treatments have focused on a let-7 replacement strategy [20], yet the delivery of RNA therapeutics has proven difficult [26]. Another approach focuses on inhibitors of let-7 degradative enzymes Lin28 [27,28] and the terminal uridylyltransferase Tut4 [29] as targets for small molecule chemotherapeutics. Lin28 binds to precursor miRNA (pre-miRNA) let-7 and recruits Tut4, which subsequently polyuridylates the pre-miRNA. Polyuridylated RNAs are degraded by the U-specific exonuclease Dis3L2 [1]. Screening for small molecule inhibitors of let-7 degradative enzymes currently relies on in vitro biochemical assays to screen for functional inhibition of the respective proteins. Unfortunately, identified small molecule inhibitors often fail to effectively alter miRNA metabolism in vivo due to off-target activities, unspecific side effects, and failure to efficiently enter the cell. Screens directly assessing miRNA levels and activity in the cell would circumvent these technical difficulties. Several methodologies are available to assess overall and specific expression levels of miRNAs in cells. Next-generation sequencing and miRNA arrays are used to identify changes in the overall miRNome. Real-time quantitative polymerase chain reaction (RT-qPCR) and northern blotting are tools to probe individual miRNAs [30,31], yet these assays are not amenable to studies in live cells and they fail to report the level of active miRNA.

The observed variability between tissues and even between single cells call for the development of methods to follow expression and activity of miRNAs in tissues or individual living cells [27–29]. Furthermore, miRNA quantity does not necessarily correspond to miRNA activity, as miRNAs can be silenced by single nucleotide additions without affecting miRNA prevalence in the cell [32]. We developed an optogenetic green fluorescence protein (GFP)-based reporter to assess the level of active let-7 in live cells. We fused a let-7 regulated 3′-UTR from the human *kras* gene to a GFP reporter, allowing for a direct readout of let-7 activity in vivo, thus generating a miRNA activity reporter (MiRAR). We further show that this MiRAR reporter system can be adapted as a reporter for other miRNAs. Our proof using principle experiments shows that genetically encoded sensors of miRNA activity will be highly useful tools in investigating the biological role of RNA-regulating enzymes in vivo.

2. Materials and Methods

2.1. Genomic DNA Extraction

Genomic DNA was prepared as a template for the amplification of 3′-UTRs as described previously [33]. Briefly, human embryonic kidney 293 (HEK 293) cells were grown to confluency in Dulbecco's modified eagle medium (DMEM) from Gibco (Thermo Fisher Scientific, Waltham, MA,

USA) and 10% fetal bovine serum (FBS). Cells from half of a 150 mm plate (~10^7 cells) were resuspended in 2.4 mL lysis buffer (0.6% sodium dodecyl sulfate (SDS), 10 mM ethylenediaminetetraacetic acid (EDTA), 10 mM Tris-HCl pH 8.0) and 25 units RNase I (NEB M0243S) and incubated at 37 °C for 1 h. This was followed by the addition of 240 μL 5 M NaCl (6 mmol) and 1 h of incubation on ice. The solution was centrifuged at 10,000× *g* for 30 min at 4 °C and the DNA was extracted from the supernatant via phenol chloroform extraction. The extracted DNA was stored at −20 °C until further use.

2.2. Reporter Gene Construct for Let-7

We generated a reporter system to link miRNA content in the cell to GFP fluorescence. GFP was cloned into pcDNA3.1. The 3′-UTR of human wildtype *kras* was amplified from pRL KRas 3′-UTR (plasmid 14804, Addgene, Cambridge, MA, USA) and mutant KRas 3′-UTR from pRL KRas 3′-UTR (plasmid 14805 Addgene) [34] using primers KRasfor (5′-TCTGGGTGTTGATGATGCCTTC-3′) and KRasrev (5′-CCTGGTAATGATTTAAATGTAGTTATAGAAATAAATAATATG-3′). The resulting PCR product was cloned downstream of GFP into pcDNA3.1-GFP using *Kpn*I and *Bam*HI restriction sites, yielding the plasmid pMiRAR-let-7 and pMiRAR-let-7-mutant. Successful cloning was verified by DNA sequencing at the London Regional Genomics Centre (London, ON, Canada). The construct sequences are supplied in the supplementary file S1.

2.3. Reporter Gene Construct for miR-122

As a reporter system for miR-122, we chose the cytoplasmic polyadenylation element binding protein (CPEB) 3′-UTR. The CPEB 3′-UTR (NM_001079533.1) was amplified from HEK 293 genomic DNA with primers CPEB-*Kpn*I-for (5′-ATCAGGTACCTAAAGGAGCTGGCCTTG-3′) and CPEB-*Bam*HI-rev (5′-TTAAGGATCCCTGCTGCAACGTGTT-3′). The amplified DNA was then inserted into pCDNA3.1 downstream of GFP, yielding pMiRAR-miR-122. Successful cloning was verified by DNA sequencing at the London Regional Genomics Centre. The construct sequence is supplied in the supplementary file S1.

2.4. Quantification of Green Fluorescent Protein Fluorescence in Live Cells

HEK 293 cells were grown to ~80% confluency as described above on a 6-well plate using DMEM supplemented with 1% v/v penicillin-streptomycin (Wisent Inc., Saint-Jean-Baptiste, QC, Canada). Cells were then co-transfected with pMiRAR plasmids, pCMV-tdTomato (632534, Clontech, Mountain View, CA, USA), and RNAs as indicated using Lipofectamine 2000 in Opti-MEM transfection media (11668019, Invitrogen, Carlsbad, CA, USA). Cells were harvested 48 h after transfection. Cell fluorescence was measured using the Synergy H1 microplate reader (BioTek, Winooski, VT, USA) at excitations of 480 nm and 554 nm, and emissions at 509 nm and 581 nm. In each well, a grid of 11 × 11 fields was scanned and fluorescence intensity was recorded. The data reported represent the average GFP and tdTomato fluorescence intensity per well. RNAs co-transfected were as follows: let-7 (5′-p-UGAGGUAGUAGGUUGUGUGGUU-3′) and anti-let-7 (5′-AACCACACAACCUACUACCUCA-3′) at 80 pM concentration; hsa-miR-122 (5′-p-UGGAGUGUGACAAUGGUGUUUG-3′) and anti-hsa-miR-122 (5′-CAAACACCAUUGUCA CACUCCA-3′) at 100 nM. Tut4 knockdown was carried out with anti-Tut4 small interfering RNA (siRNA) (Dharmacon OnTargetPlus System, L-021797-01-0005, Lafayette, CO, USA) according to manufacturer's instructions. Successful Tut4 knockdown was confirmed by separating 50 μg of total protein from HEK 293 cells treated with Tut siRNA or a scrambled control via SDS-PAGE (SDS-polyacrylamide gel electrophoresis). Proteins were transferred to a polyvinylidene difluoride (PVDF) membrane by western blotting and Tut4 and GAPDH were detected with protein specific antibodies 18980-1-AP (Proteintech, Chicago, IL, USA) and MAB374 (Sigma-Aldrich, St. Louis, MO, USA). All experiments were carried out at least in triplicate; representative cell images are shown.

2.5. MicroRNA Quantification by Real-Time Quantitative Polymerase Chain Reaction

RT-qPCR was performed as described previously [35]. Briefly, a primer with an internal stem loop structure was designed to target mature let-7 miRNA (5′-GTTGGCTCTGGTGCAGG GTCCGAGGTATTCGCACCAGAGCCAACAACTAT-3′) or miR-122 (5′-GTCGTATGCAGAGC AGGGTCCGAGGTATTCGCACTGCATACGACCAAACA-3′). This primer was unfolded for 5 min at 65 °C and then refolded for 2 min on ice to form a stem loop structure. The primer was then incubated with total purified cellular RNA. Then, 0.125 pmol RNA was synthesized into complementary DNA (cDNA) using SuperScript III RT (200 units/µL) and stem loop primers. cDNA synthesis was carried out for both RNA extracted from wildtype and Tut4-knockdown cell lines. The reaction was incubated in a thermocycler for 30 min at 16 °C, followed by pulsed RT of 60 cycles at 30 °C for 30 s, 42 °C for 30 s, and 50 °C for 1 min. The cDNA generated was later diluted 10-fold, and quantitative PCR was conducted using SYBR Green qPCR MasterMix (Thermo Fisher Scientific) and qPCR primers (300 nM). Forward primers were designed for miR-122 (5′-AGGCTGGAGTGTGACAATG-3′), let-7 (5′-TGAGGTAGTAGGTTGTATAGTTGTTGG-3′) and universal miR reverse (5-GAGCAGGGTCCGAGGT-3′). Samples were amplified for 35 cycles with Eppendorf Realplex (Eppendorf, Hamburg, Germany), and miRNA levels were extrapolated using a comparative C_T (Cycle Threshold)method described previously [35].

3. Results

3.1. Let-7 micro RNA Reduces Green Fluorescent Proteins Fluorescence in Live Cells

To assess miRNA levels in live cells, we generated a GFP-based reporter system. The 3′-UTR of KRas was cloned downstream of a *gfp* gene into pcDNA3.1 to generate a reporter system where GFP fluorescence is responsive to changes in let-7 concentration in the cell (Figure 1a). Cells transfected with the reporter construct pMiRAR-let-7 expressed GFP (Figure 2a,c), indicating that endogenous let-7 levels do not entirely silence *gfp* expression. Background fluorescence of cells without miRAR (1858 ± 44 RFU (relative fluorescence units)) was subtracted from the fluorescence intensities. To evaluate the responsiveness of GFP production, we co-transfected the reporter pMiRAR-let-7 with 80 pMlet-7 miRNA or separately with anti-miR RNA complementary to let-7 (anti-let-7). Supplementing cells with exogeneous let-7 effectively inhibited GFP translation, reducing fluorescence by more than 3-fold (Figure 2a,c). In contrast, supplementing cells with anti-let-7, which binds to and de-activates cellular let-7, led to a marked decrease in active let-7 molecules in the cell as reported by a 1.3-fold increase in GFP production and fluorescence (Figure 2a,c).

3.2. Visualizing Let-7 Accumulation due to Inhibition of Let-7 Degradative Enzymes

We further tested the reporter system by assessing miRNA levels in cells depleted for the let-7 degradative enzyme Tut4, which has been shown previously to affect miRNA degradation [1,29,36,37]. Tut4 polyuridylates let-7 miRNAs, marking them for degradation by the exonuclease Dis3L2 (Figure 1b) [1]. Tut4 was knocked down using siRNA, and partial knockdown of ~50% of Tut4 was confirmed by western blotting (Figure 2b). As expected, the depletion of Tut4 resulted in a decrease of GFP fluorescence by 2.4-fold, confirming an increase in cellular let-7 levels (Figure 2a,c). Thus, elevation of miRNA concentrations caused by inhibition of the uridylyltransferase Tut4, and the subsequent lack of U-dependent let-7 degradation can be measured using our GFP reporter system.

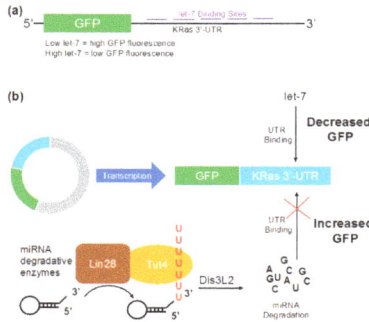

Figure 1. Schematic of the microRNA (miRNA) activity reporter (MiRAR) for let-7 levels in vivo. The KRas 3'-UTR was fused downstream of green fluorescence protein (GFP) to allow quantification of cellular let-7 levels. (**a**) Schematic of pMiRAR-let-7 construct; (**b**) miRNA degradative enzymes Lin28 and Tut4 collaborate to mark let-7 miRNA for degradation by the exonuclease Dis3L2. The RNA binding protein Lin28 recruits Tut4 to polyuridylate miRNA and pre-miRNAs, leading to degradation by the U-specific exonuclease Dis3L2. Lowered miRNA levels lead to an increase in GFP translation and fluorescence. KRas: Kirsten rat sarcoma viral oncogene homolog, UTR: untranslated region.

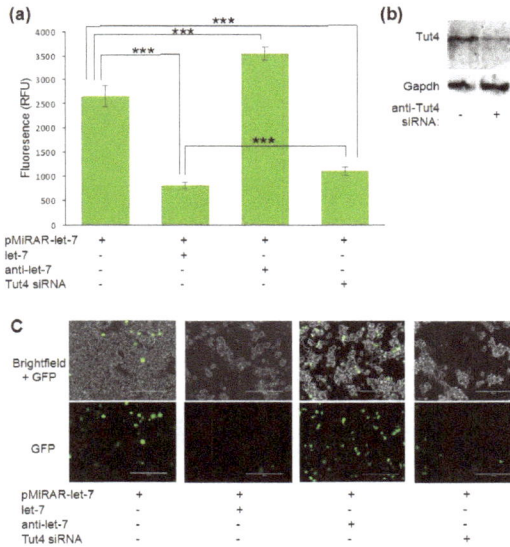

Figure 2. The MiRAR in live cells. Fluorescence intensity measurements and cell images for different treatments of MiRAR-transfected cells. Human embryonic kidney 293 (HEK 293) cells were grown to confluency, transfected with the MiRAR containing the KRas-3'-UTR, and treated as outlined. (**a**) Fluorescence intensities validate miRAR-let-7 as a miRNA reporter. Background fluorescence of untreated cells was subtracted from the experiments. Error bars are based on at least three biological replicates and represent one standard deviation; (**b**) Western blot of untreated HEK 293 cells and treated cells after a knockdown of Tut4, confirming partial depletion of Tut4; (**c**) Images of live cells co-transfected with MiRAR and indicated as small interfering RNAs (siRNAs) or miRNAs. Row 1: overlay of phase light microscopy and GFP UV microscopy; row 2: GFP UV microscopy alone. The white bar represents 200 μm. *p* values are *** < 0.001.

3.3. Mutation of Let-7 Binding Sites Abolishes the Sensitivity of the pMiRAR to Changes in Let-7 Levels

To confirm that the changes in fluorescence were indeed exclusively due to let-7 binding to the KRas-UTR in our reporter, we generated a variant of the reporter gene construct with mutated let-7 binding sites. Mutations in the let-7 binding sites abolish the regulatory effect of let-7 on gene expression. The let-7 un-responsive KRas-3′UTR mutant was described previously [34], and cloned downstream of the *gfp* coding sequence. As before, we observed a significant decrease in florescence in cells co-transfected with the wildtype pMiRAR-let-7 construct with Tut4 siRNA compared to untreated cells (Figure 3a,b). In contrast, GFP fluorescence in cells carrying a plasmid with a mutated KRas-UTR fused to GFP (pMiRAR-let-7-mutant) did not respond to a Tut4 knockdown (Figure 3a,b). These data confirm that the change in fluorescence is indeed due to binding of let-7 to the KRas-3′-UTR. As a control experiment, a second plasmid coding for tdTomato was co-transfected to probe for differences in transfection efficiency. No significant changes in transfection efficiency of tdTomato were observed (Figure 3a), indicating that variation in transfection efficiency does not account for the decrease or increase of GFP fluorescence. To further confirm that the Tut4 knockdown decreases let-7 levels in the cell, we quantified let-7 miRNA levels by qPCR (Figure 4). Indeed, in the Tut4 knockdown, a 2.7-fold increase in let-7 miRNA levels was observed, correlating with the 2.4-fold decrease in GFP fluorescence in the pMiRAR-let-7 reporter. These data further confirm that depletion of Tut4 leads to an increase of let-7 levels in the cells, as described previously [29]. We here show that a 2.4-fold decrease in GFP fluorescence corresponds to a 2.7-fold increase in cellular let-7 levels, showing direct inverse reporting of let-7 levels by pMiRAR-let-7.

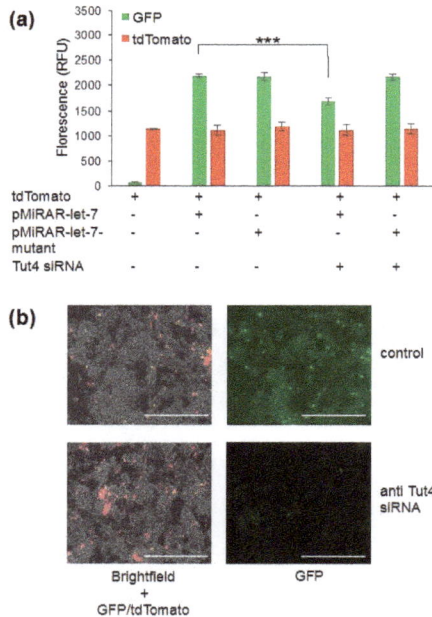

Figure 3. MiRAR–let-7 with KRas-3-UTR is let-7 specific. Fluorescence intensities (**a**) and cell images (**b**) for HEK 293 cells co-transfected with pMiRAR-let-7 and tdTomato. HEK 293 cells were grown to confluence and treated as outlined. Cells were either transfected with the original pMiRAR-let-7 construct (Wild Type (WT)-KRas) or a construct containing mutated let-7 binding sites (pMiRAR-let-7-mutant). Error bars are based on three biological replicates and show one standard deviation. White bars represent 200 μm. *p* values are *** < 0.001.

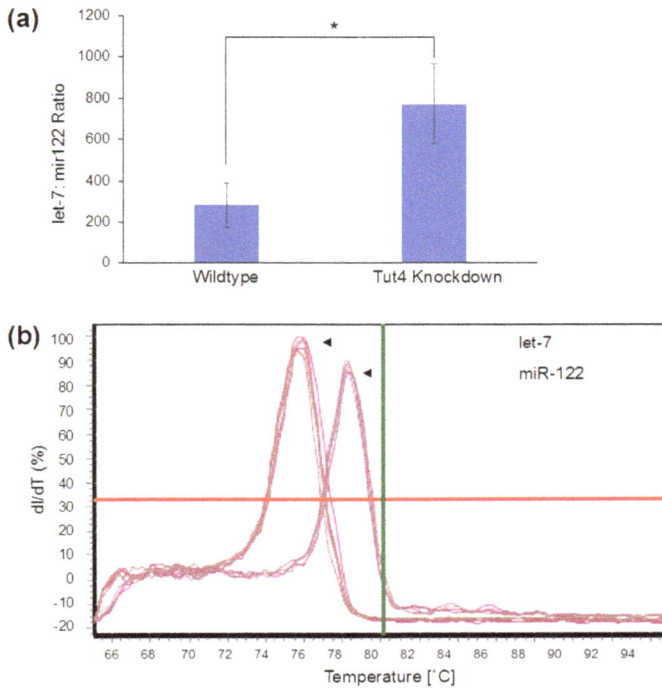

Figure 4. qPCR for let-7 miRNA in wildtype and Tut4-knockdown HEK293 cells. Cells were grown to confluence and total RNA was extracted. (**a**) Let-7 levels were quantified in relation to miRNA miR-122. miRNA levels in the wildtype and Tut4 knockdown are significantly different ($p < 0.03$). Error bars are based on three biological replicates and show one standard deviation. (**b**) Melting curves of let-7 and miR-122 primers are distinct and specific. p values are * < 0.05.

3.4. Adapting the Optogenetic Reporter for Quantifying miR-122

To investigate whether the reporter system is applicable and useful for the reporting of other miRNAs, we generated a similar reporter system for miR-122. The CPEB 3′-UTR encodes 2 miR-122 binding sites and CPEB translation is regulated by cellular miR-122 levels [38]. We fused the CPEB 3′-UTR to *gfp* to generate a reporter system for miR-122 (pMiRAR-miR-122). Low concentrations of co-transfected miR-122 and anti-miR-122 showed no significant change in fluorescence and transfected RNA concentrations were increased to 100 nM. This led to significant changes in GFP fluorescence (Figure 5a). Transfection with miR-122 at 100 nM led to a 1.2-fold decrease in fluorescence, while the anti-miR-122 led to a 1.3-fold increase in fluorescence (Figure 5a,b). Thus, a reporter with two miRNA binding sites is applicable to more pronounced changes in miRNA activity, while a reporter with five miRNA binding sites may be useful for subtle changes. This demonstrates that with minor modification the MiRAR can be adapted to likely any desired target miRNA or range of sensitivity.

(a)

(b)

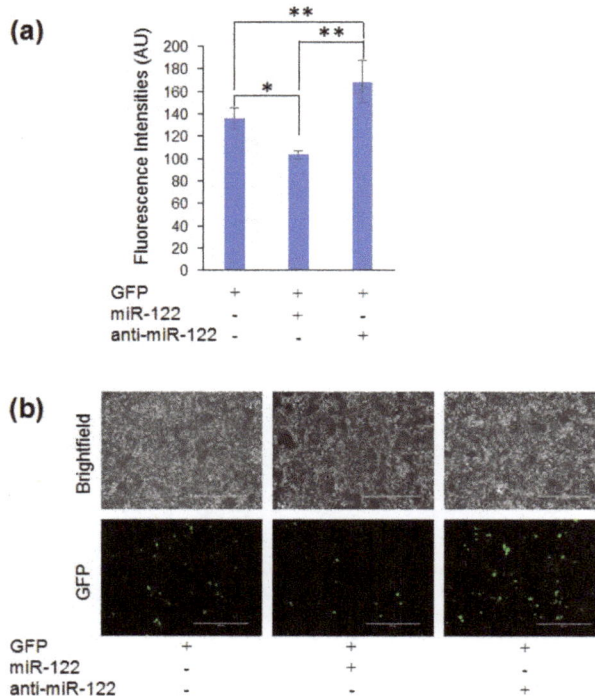

Figure 5. MiRAR-miR-122: Reporter gene construct for cellular miR-122 levels. Fluorescence intensities (**a**) for HEK 293 cells co-transfected MiRAR-miR-122 (CPEB-3′UTR) and miR-122 or anti-miR-122. HEK 293 cells were grown to 60–80% confluency and treated as indicated. Fluorescence intensities were measured by the Synergy H1 microplate reader at an excitation of 480 nm and emission of 509 nm. Error bars represent one standard deviation. (**a**) Cells transfected with 100 nM RNA; (**b**) MiRAR cell images of cells treated with RNAs as indicated. White bars show 400 μm. *p* values are * < 0.05 and ** < 0.01. AU: arbitrary units.

4. Discussion

The level of miRNA in cells is usually determined by next-generation sequencing or qPCR, which are methods that require cell lysis and subsequent RNA extraction. To observe and quantify miRNA activity in live cells and to capture real-time responses to environmental or chemical changes, a genetically encoded reporter system that allows for a time-resolved quantification of miRNA activity is required. Previously, several luciferase-based reporter systems were generated [34,39–41]. These reporters allowed for the quantification of miRNA levels without cell lysis but required the addition of luciferin or a luciferin synthesis plasmid. Luciferase-based reporters provide a snapshot of cellular miRNA levels, rather than the ability to continuously report miRNA activity over an uninterrupted time course.

To generate a reporter system that is independent of external supplementation with chemicals and allows for a time-resolved quantification of miRNA activity in the cell, we developed a GFP-based miRNA activity reporter. The 3′-UTR of KRas contains several let-7 miRNA binding sites (Figure 1a), and KRas mRNA stability is well known to be regulated by let-7 miRNA [16]. We showed that our reporter accurately reports cellular miRNA concentrations via an inverse response from MiRAR fluorescence. Co-transfected miRNA let-7 or anti-let-7 efficiently reduced or elevated GFP fluorescence in the anticipated inverse relationship. The let-7 reporter displayed high sensitivity, with changes in fluorescence corresponding to pM of miRNA transfected. Thus, we can utilize the reporter to observe small changes in miRNA activity resulting from increasing and decreasing let-7 levels.

Our MiRAR system can also monitor changes in cellular miRNA metabolism. A 2.4-fold decrease in fluorescence corresponds to a 2.7-fold increase in miRNA content upon depletion of the let-7 degradative enzyme Tut4. MiRAR may thus be a highly useful tool to screen small molecule inhibitor libraries for compounds active in altering miRNA metabolism. Previous studies relied on biochemical assays to identify inhibitors of let-7 degradative enzymes Lin28 [27,28] and Tut4 [29]. Utilizing a live-cell reporter to quantify the impact of small molecule inhibitors may lead to a more efficient screening in a physiologically relevant model system. Screening drugs with MiRAR can circumvent time and effort spent on in vitro hits that subsequently fail to permeate the cell or increase let-7 levels in the cell.

For the miRNA let-7, we observed a less pronounced change in GFP fluorescence when adding anti-miR RNA complementary to let-7. This decrease in sensitivity compared to supplementing additional let-7 is most likely due to reaching the level of endogenous let-7 levels in HEK 293 cells. Supplementation of the anti-miRNA will eventually bind and deactivate free cellular let-7, and if provided in excess will not further decrease GFP fluorescence. Thus, the reporter system can also be utilized as a tool to not only quantify changes in let-7 levels, but our data suggest that MiRAR may also be a valuable tool to quantify absolute let-7 levels by titrating increasing concentrations of anti-let-7 until no further increase in GFP fluorescence is observed.

To demonstrate the generality of our approach, we generated a MiRAR for another miRNA, miR-122. miR-122 is associated with both Hepatitis B and C infections [42–44] and hepatic cancer [45–49]. We used an miR-122-responsive CBEB 3′-UTR in generating this MiRAR contruct. The CBEB 3′-UTR contains only two miR-122 binding sites [38], compared to the five binding sites of let-7 in our KRas-derived MiRAR. Consistent with a reduced number of miRNA binding sites, we observed that a significant change in GFP fluorescence was only observed upon an increase to 100 nM of transfected miR-122 compared to 80 pM in the MiRAR-let-7. The MiRAR-miR-122 reporter is thus approximately 1000-fold less sensitive to changes in miRNA concentration. Thus, our data indicate that the sensitivity of the MiRAR can be tuned depending on the concentration range of miRNA of interest. In the future, we envision that changes in sensitivity or adaptation to other miRNA species can be achieved by either fusing the 3′-UTR of a miRNA-regulated gene to *gfp*, or by mutating the binding sites of e.g., miR-122 or let-7 in the existing constructs to another seed sequence.

In summary, we present a versatile reporter system that is adaptable to different miRNAs and is scalable in terms of sensitivity. Changes in miRNA metabolism in response to extracellular stimuli or over the lifetime of a cell can be monitored in a time resolved manner at the single-cell level without further interfering with the cellular environment. This tool will allow unprecedented insight into miRNA metabolism and biology. Future efforts will include the generation of stable cell lines containing MiRARs directed at certain miRNAs, and adaptation of the current constructs to other target miRNAs. Compared to luciferase-based reporters, our GFP-based reporter circumvents use of chemicals (luciferin) as an additional screening step and may provide a useful tool for high-throughput screening of small molecule chemotherapeutics that are effective in altering miRNA content or activity.

Supplementary Materials: The following are available online at http://www.mdpi.com/2073-4425/9/6/305/s1, File S1: Reporter gene constructs.

Author Contributions: Conceptualization, I.U.H.; Data curation, M.A.T., C.Z.C., E.M., A.E. and I.U.H.; Formal analysis, M.A.T., C.Z.C. and E.M.; Funding acquisition, I.U.H.; Investigation, M.A.T., C.Z.C., E.M., S.A.Z., A.E., Y.B. (Yasaman Badakhshi) and Y.B. (Yumin Bi); Methodology, C.Z.C., S.A.Z., A.E. and I.U.H.; Supervision, I.U.H.; Validation, M.A.T., C.Z.C. and E.M.; Visualization, M.A.T. and E.M.; Writing—original draft, M.A.T. and I.U.H.; Writing—review and editing, C.Z.C., E.M. and I.U.H.

Acknowledgments: We are thankful to Lauren Seidl and Jeremy Lant for advice and discussion and to Patrick O'Donoghue for careful reading of the manuscript. This research was funded by the Natural Sciences and Engineering Research Council of Canada (grant number RGPIN 04776-2014 to I.U.H.), the J. P. Bickell Foundation (to I.U.H.), an Ontario Graduate Scholarship (to C.Z.C.), an Alexander Graham Bell Canada Graduate Scholarship

(Doctoral) from the Natural Sciences and Engineering Research Council of Canada (to C.Z.C.), and a scholarship from the Saudi Arabian Cultural Bureau (to E.M).

Conflicts of Interest: The authors declare no conflict of interest.

References

1. Chung, C.Z.; Seidl, L.E.; Mann, M.R.; Heinemann, I.U. Tipping the balance of RNA stability by 3′ editing of the transcriptome. *Biochim Biophys Acta* **2017**, *1861*, 2971–2979. [CrossRef] [PubMed]
2. Lewis, B.P.; Shih, I.H.; Jones-Rhoades, M.W.; Bartel, D.P.; Burge, C.B. Prediction of mammalian miRNA targets. *Cell* **2003**, *115*, 787–798. [CrossRef]
3. Loganantharaj, R.; Randall, T.A. The limitations of existing approaches in improving miRNA target prediction accuracy. *Methods Mol. Biol.* **2017**, *1617*, 133–158. [PubMed]
4. Cammarata, G.; Augugliaro, L.; Salemi, D.; Agueli, C.; La Rosa, M.; Dagnino, L.; Civiletto, G.; Messana, F.; Marfia, A.; Bica, M.G.; et al. Differential expression of specific miRNA and their targets in acute myeloid leukemia. *Am. J. Hematol.* **2010**, *85*, 331–339. [PubMed]
5. Kobayashi, M.; Salomon, C.; Tapia, J.; Illanes, S.E.; Mitchell, M.D.; Rice, G.E. Ovarian cancer cell invasiveness is associated with discordant exosomal sequestration of let-7 mirna and mir-200. *J. Transl. Med.* **2014**, *12*, 4. [CrossRef] [PubMed]
6. Mulrane, L.; McGee, S.F.; Gallagher, W.M.; O'Connor, D.P. MiRNA dysregulation in breast cancer. *Cancer Res.* **2013**, *73*, 6554–6562. [CrossRef] [PubMed]
7. Zhu, X.M.; Wu, L.J.; Xu, J.; Yang, R.; Wu, F.S. Let-7c miRNA expression and clinical significance in hepatocellular carcinoma. *J. Int. Med. Res.* **2011**, *39*, 2323–2329. [CrossRef] [PubMed]
8. Provost, P. MicroRNAs as a molecular basis for mental retardation, Alzheimer's and prion diseases. *Brain Res.* **2010**, *1338*, 58–66. [CrossRef] [PubMed]
9. Hagiwara, S.; McClelland, A.; Kantharidis, P. miRNA in diabetic nephropathy: Renin angiotensin, age/rage, and oxidative stress pathway. *J. Diabetes Res.* **2013**, *2013*, 173783. [CrossRef] [PubMed]
10. Yin, J.; Lin, J.; Luo, X.; Chen, Y.; Li, Z.; Ma, G.; Li, K. mir-137: A new player in schizophrenia. *Int. J. Mol. Sci.* **2014**, *15*, 3262–3271. [CrossRef] [PubMed]
11. Roush, S.; Slack, F.J. The let-7 family of microRNAs. *Trends Cell Biol.* **2008**, *18*, 505–516. [CrossRef] [PubMed]
12. Wang, X.; Cao, L.; Wang, Y.; Wang, X.; Liu, N.; You, Y. Regulation of let-7 and its target oncogenes. *Oncol. Lett.* **2012**, *3*, 955–960. [CrossRef] [PubMed]
13. Boyerinas, B.; Park, S.M.; Hau, A.; Murmann, A.E.; Peter, M.E. The role of let-7 in cell differentiation and cancer. *Endocr.-Relat. Cancer* **2010**, *17*, F19–F36. [CrossRef] [PubMed]
14. Boyerinas, B.; Park, S.M.; Shomron, N.; Hedegaard, M.M.; Vinther, J.; Andersen, J.S.; Feig, C.; Xu, J.; Burge, C.B.; Peter, M.E. Identification of let-7-regulated oncofetal genes. *Cancer Res.* **2008**, *68*, 2587–2591. [CrossRef] [PubMed]
15. Johnson, C.D.; Esquela-Kerscher, A.; Stefani, G.; Byrom, M.; Kelnar, K.; Ovcharenko, D.; Wilson, M.; Wang, X.; Shelton, J.; Shingara, J.; et al. The let-7 miRNA represses cell proliferation pathways in human cells. *Cancer Res.* **2007**, *67*, 7713–7722. [CrossRef] [PubMed]
16. Johnson, S.M.; Grosshans, H.; Shingara, J.; Byrom, M.; Jarvis, R.; Cheng, A.; Labourier, E.; Reinert, K.L.; Brown, D.; Slack, F.J. Ras is regulated by the let-7 miRNA family. *Cell* **2005**, *120*, 635–647. [CrossRef] [PubMed]
17. Tsang, W.P.; Kwok, T.T. Let-7a miRNA suppresses therapeutics-induced cancer cell death by targeting caspase-3. *Apoptosis* **2008**, *13*, 1215–1222. [CrossRef] [PubMed]
18. Esquela-Kerscher, A.; Slack, F.J. Oncomirs—MicroRNA with a role in cancer. *Nat. Rev. Cancer* **2006**, *6*, 259–269. [CrossRef] [PubMed]
19. Kloosterman, W.P.; Plasterk, R.H. The diverse functions of miRNA in animal development and disease. *Dev. Cell* **2006**, *11*, 441–450. [CrossRef] [PubMed]
20. Akao, Y.; Nakagawa, Y.; Naoe, T. Let-7 miRNA functions as a potential growth suppressor in human colon cancer cells. *Biol. Pharm. Bull.* **2006**, *29*, 903–906. [CrossRef] [PubMed]
21. Sampson, V.B.; Rong, N.H.; Han, J.; Yang, Q.; Aris, V.; Soteropoulos, P.; Petrelli, N.J.; Dunn, S.P.; Krueger, L.J. miRNA let-7a down-regulates *MYC* and reverts *MYC*-induced growth in Burkitt lymphoma cells. *Cancer Res.* **2007**, *67*, 9762–9770. [CrossRef] [PubMed]

22. Zhang, H.H.; Wang, X.J.; Li, G.X.; Yang, E.; Yang, N.M. Detection of let-7a miRNA by real-time PCR in gastric carcinoma. *World J. Gastroenterol.* **2007**, *13*, 2883–2888. [CrossRef] [PubMed]

23. Takamizawa, J.; Konishi, H.; Yanagisawa, K.; Tomida, S.; Osada, H.; Endoh, H.; Harano, T.; Yatabe, Y.; Nagino, M.; Nimura, Y.; et al. Reduced expression of the let-7 miRNA in human lung cancers in association with shortened postoperative survival. *Cancer Res.* **2004**, *64*, 3753–3756. [CrossRef] [PubMed]

24. Mayr, C.; Hemann, M.T.; Bartel, D.P. Disrupting the pairing between let-7 and Hmga2 enhances oncogenic transformation. *Science* **2007**, *315*, 1576–1579. [CrossRef] [PubMed]

25. Barh, D.; Malhotra, R.; Ravi, B.; Sindhurani, P. MicroRNA let-7: An emerging next-generation cancer therapeutic. *Curr. Oncol.* **2010**, *17*, 70–80. [CrossRef] [PubMed]

26. Haussecker, D. Current issues of RNAi therapeutics delivery and development. *J. Control Release* **2014**, *195*, 49–54. [CrossRef] [PubMed]

27. Lightfoot, H.L.; Miska, E.A.; Balasubramanian, S. Identification of small molecule inhibitors of the lin28-mediated blockage of pre-let-7g processing. *Org. Biomol. Chem.* **2016**, *14*, 10208–10216. [CrossRef] [PubMed]

28. Qiu, Z.; Zhou, J.; Zhang, C.; Cheng, Y.; Hu, J.; Zheng, G. Antiproliferative effect of urolithin A, the ellagic acid-derived colonic metabolite, on hepatocellular carcinoma Hepg2.2.15 cells by targeting Lin28a/let-7a axis. *Braz. J. Med. Biol. Res.* **2018**, *51*, e7220. [CrossRef] [PubMed]

29. Lin, S.; Gregory, R.I. Identification of small molecule inhibitors of Zcchc11 TUTase activity. *RNA Biol.* **2015**, *12*, 792–800. [CrossRef] [PubMed]

30. Redshaw, N.; Wilkes, T.; Whale, A.; Cowen, S.; Huggett, J.; Foy, C.A. A comparison of miRNA isolation and RT-qPCR technologies and their effects on quantification accuracy and repeatability. *Biotechniques* **2013**, *54*, 155–164. [CrossRef] [PubMed]

31. Verma, R.; Sharma, P.C. Next generation sequencing-based emerging trends in molecular biology of gastric cancer. *Am. J. Cancer Res.* **2018**, *8*, 207–225. [PubMed]

32. Jones, M.R.; Blahna, M.T.; Kozlowski, E.; Matsuura, K.Y.; Ferrari, J.D.; Morris, S.A.; Powers, J.T.; Daley, G.Q.; Quinton, L.J.; Mizgerd, J.P. Zcchc11 uridylates mature miRNA to enhance neonatal IGF-1 expression, growth, and survival. *PLoS Genet.* **2012**, *8*, e1003105. [CrossRef] [PubMed]

33. Remuzgo-Martinez, S.; Aranzamendi-Zaldunbide, M.; Pilares-Ortega, L.; Icardo, J.M.; Acosta, F.; Martinez-Martinez, L.; Ramos-Vivas, J. Interaction of macrophages with a cytotoxic Serratia liquefaciens human isolate. *Microbes Infect.* **2013**, *15*, 480–490. [CrossRef] [PubMed]

34. Kumar, M.S.; Lu, J.; Mercer, K.L.; Golub, T.R.; Jacks, T. Impaired miRNA processing enhances cellular transformation and tumorigenesis. *Nat. Genet.* **2007**, *39*, 673–677. [CrossRef] [PubMed]

35. Varkonyi-Gasic, E.; Wu, R.; Wood, M.; Walton, E.F.; Hellens, R.P. Protocol: A highly sensitive RT-PCR method for detection and quantification of miRNA. *Plant Methods* **2007**, *3*, 12. [CrossRef] [PubMed]

36. Hagan, J.P.; Piskounova, E.; Gregory, R.I. Lin28 recruits the TUTase Zcchc11 to inhibit let-7 maturation in mouse embryonic stem cells. *Nat. Struct. Mol. Biol.* **2009**, *16*, 1021–1025. [CrossRef] [PubMed]

37. Thornton, J.E.; Chang, H.M.; Piskounova, E.; Gregory, R.I. Lin28-mediated control of let-7 miRNA expression by alternative TUTases Zcchc11 (TUT4) and Zcchc6 (TUT7). *RNA* **2012**, *18*, 1875–1885. [CrossRef] [PubMed]

38. Burns, D.M.; D'Ambrogio, A.; Nottrott, S.; Richter, J.D. CPEB and two poly(A) polymerases control miR-122 stability and p53 mRNA translation. *Nature* **2011**, *473*, 105–108. [CrossRef] [PubMed]

39. Brustikova, K.; Sedlak, D.; Kubikova, J.; Skuta, C.; Solcova, K.; Malik, R.; Bartunek, P.; Svoboda, P. Cell-based reporter system for high-throughput screening of miRNA pathway inhibitors and its limitations. *Front. Genet.* **2018**, *9*, 45. [CrossRef] [PubMed]

40. Guo, R.; Abdelmohsen, K.; Morin, P.J.; Gorospe, M. Novel miRNA reporter uncovers repression of let-7 by GSK-3β. *PLoS ONE* **2013**, *8*, e66330.

41. Sano, M.; Ohtaka, M.; Iijima, M.; Nakasu, A.; Kato, Y.; Nakanishi, M. Sensitive and long-term monitoring of intracellular miRNA using a non-integrating cytoplasmic RNA vector. *Sci. Rep.* **2017**, *7*, 12673. [CrossRef] [PubMed]

42. Flor, T.B.; Blom, B. Pathogens use and abuse miRNA to deceive the immune system. *Int. J. Mol. Sci.* **2016**, *17*, 538. [CrossRef] [PubMed]

43. Peng, F.; Xiao, X.; Jiang, Y.; Luo, K.; Tian, Y.; Peng, M.; Zhang, M.; Xu, Y.; Gong, G. HBx down-regulated Gld2 plays a critical role in HBV-related dysregulation of miR-122. *PLoS ONE* **2014**, *9*, e92998. [CrossRef] [PubMed]

44. Van der Ree, M.H.; van der Meer, A.J.; van Nuenen, A.C.; de Bruijne, J.; Ottosen, S.; Janssen, H.L.; Kootstra, N.A.; Reesink, H.W. Miravirsen dosing in chronic hepatitis C patients results in decreased microRNA-122 levels without affecting other microRNAs in plasma. *Aliment. Pharmacol. Ther.* **2016**, *43*, 102–113. [CrossRef] [PubMed]

45. Bai, S.; Nasser, M.W.; Wang, B.; Hsu, S.H.; Datta, J.; Kutay, H.; Yadav, A.; Nuovo, G.; Kumar, P.; Ghoshal, K. miRNA-122 inhibits tumorigenic properties of hepatocellular carcinoma cells and sensitizes these cells to sorafenib. *J. Biol. Chem.* **2009**, *284*, 32015–32027. [CrossRef] [PubMed]

46. Fornari, F.; Gramantieri, L.; Giovannini, C.; Veronese, A.; Ferracin, M.; Sabbioni, S.; Calin, G.A.; Grazi, G.L.; Croce, C.M.; Tavolari, S.; et al. mir-122/cyclin G1 interaction modulates p53 activity and affects doxorubicin sensitivity of human hepatocarcinoma cells. *Cancer Res.* **2009**, *69*, 5761–5767. [CrossRef] [PubMed]

47. Lian, J.H.; Wang, W.H.; Wang, J.Q.; Zhang, Y.H.; Li, Y. miRNA-122 promotes proliferation, invasion and migration of renal cell carcinoma cells through the PI3k/Akt signaling pathway. *Asian Pac. J. Cancer Prev.* **2013**, *14*, 5017–5021. [CrossRef] [PubMed]

48. Manfe, V.; Biskup, E.; Rosbjerg, A.; Kamstrup, M.; Skov, A.G.; Lerche, C.M.; Lauenborg, B.T.; Odum, N.; Gniadecki, R. mir-122 regulates p53/Akt signalling and the chemotherapy-induced apoptosis in cutaneous T-cell lymphoma. *PLoS ONE* **2012**, *7*, e29541. [CrossRef] [PubMed]

49. Wang, W.; Yang, J.; Yu, F.; Li, W.; Wang, L.; Zou, H.; Long, X. miRNA-122-3p inhibits tumor cell proliferation and induces apoptosis by targeting Forkhead box O in A549 cells. *Oncol. Lett.* **2018**, *15*, 2695–2699. [PubMed]

genes

MDPI

Article

Genome-Wide Quantification of the Effect of Gene Overexpression on *Escherichia coli* Growth

Hao Chen [1], Sumana Venkat [1], Jessica Wilson [2], Paige McGuire [3], Abigail L. Chang [4], Qinglei Gan [5] and Chenguang Fan [1,5,*]

[1] Cell and Molecular Biology Program, University of Arkansas, Fayetteville, AR 72701, USA; hc019@uark.edu (H.C.); sv009@uark.edu (S.V.)
[2] Department of Chemistry, University of Arkansas, Fort Smith, AR 72913, USA; jwilso18@g.uafs.edu
[3] Department of Biological Sciences, University of Arkansas, Fayetteville, AR 72701, USA; plmcguir@uark.edu
[4] Fayetteville High School, Fayetteville, AR 72701, USA; a.chang@g.fayar.net
[5] Department of Chemistry and Biochemistry, University of Arkansas, Fayetteville, AR 72701, USA; qingleig@uark.edu
* Correspondence: cf021@uark.edu; Tel.: +1-479-575-4653

Received: 19 July 2018; Accepted: 10 August 2018; Published: 16 August 2018

Abstract: Recombinant protein production plays an essential role in both biological studies and pharmaceutical production. *Escherichia coli* is one of the most favorable hosts for this purpose. Although a number of strategies for optimizing protein production have been developed, the effect of gene overexpression on host cell growth has been much less studied. Here, we performed high-throughput tests on the *E. coli* a complete set of *E. coli* K-12 ORF archive (ASKA) collection to quantify the effects of overexpressing individual *E. coli* genes on its growth. The results indicated that overexpressing membrane-associated proteins or proteins with high abundances of branched-chain amino acids tended to impair cell growth, the latter of which could be remedied by amino acid supplementation. Through this study, we expect to provide an index for a fast pre-study estimate of host cell growth in order to choose proper rescuing approaches when working with different proteins.

Keywords: *Escherichia coli*; recombinant protein production; gene overexpression; growth effect; ASKA collection; codon bias; branched-chain amino acids; gene ontology

1. Introduction

After the whole-genome sequences of thousands of organisms have been well documented, overexpressing genes to get highly pure proteins for further characterization and engineering becomes an indispensable part of biochemistry, molecular biology, cell biology, and synthetic biology. Moreover, among the 239 US-FDA (Food and Drug Administration) approved therapeutic peptides and proteins, as well as their 380 drug variants, the majority are manufactured by recombinant protein production [1].

In both basic research and drug production, *Escherichia coli* is one of the most widely-used hosts to express recombinant proteins due to a number of advantages. First, it grows quickly, with a doubling time of about 20 min in rich growth media [2], which means the total time of expressing target proteins, from inoculation to cell harvest, is only a few hours in most circumstances. Second, it readily reaches a high cell density for good protein yields. Commonly, 1 to 2 g dry cell weight or 10^{13} cells could be obtained from 1 L of liquid Lysogeny broth (LB) medium [2]. Third, it is cheap and easy to make growth media for *E. coli* such as the LB medium and the Terrific Broth (TB) medium. Fourth, the genetics of *E. coli* is well known, and it is convenient to remove certain genes from the genome for different purposes [3]. Fifth, it is easy to introduce heterologous genes into *E. coli* by plasmid transformation. Last but not least, a large number of vectors, fusion tags, and mutant strains have been developed for optimal expression of target proteins in *E. coli*. Several review articles have been published recently to cover these topics [4,5].

A commonly encountered problem for recombinant protein production is impeded cell growth or reduced biomass accumulation. There are two major reasons for this phenomenon. The first is the general metabolic burden, which could be explained as the competition between biomass accumulation and recombinant protein production for metabolic materials such as cellular energy, ATP, and substrates, amino acids, [6]. This competition leads to stress responses including the stringent response and RNA polymerase subunit S-mediated stress responses, which could further decrease or even inhibit cell growth [7]. This competition also causes increased protease activities for the overexpressed proteins [8]. The second reason is the specific protein toxicity, which is caused by the harmful functions of overexpressed proteins on normal proliferation and homeostasis of host cells [9]. For the metabolic burden, several improving approaches have been developed such as decreasing inducer concentrations, lowering plasmid copy numbers, and adding more nutrients in growth media [10,11]. Although the protein toxicity could depend on individual proteins case by case, we aimed to find general features to facilitate recombinant protein production.

For this purpose, we utilized the a complete set of *E. coli* K-12 ORF archive (ASKA) collection, which is a complete set of *E. coli* strains for overexpressing individual *E. coli* K-12 genes [12]. Although the authors who constructed the ASKA collection also tested the cell growth of individual strains in the library qualitatively by using the LB agar plate, the determination of growth effects was not clearly described, and the list of genes which impaired cell growth was not provided [12]. Thus, in this study, we quantified the effects of overexpressing individual *E. coli* genes on its own cell growth and combined the results with bioinformatical analyses to identify shared features of proteins, which could hamper cell growth.

2. Materials and Methods

2.1. Strain and Plasmid Construction

The ASKA (−) collection was obtained originally from the Coli Genetic Stock Center at Yale University. The no insert control of pCA24N was constructed by the Q5 Site-Directed Mutagenesis Kit (New England BioLabs, Ipswich, MA USA), with the F primer: 5′-taagggtcgacctgcagccaagc-3′ and the R primer: 5′-atccgtatggtgatggtgatggtgagatcc-3′. The plasmid pCA24N-*gfp* was constructed by the HiFi DNA Assembly Cloning Kit (New England BioLabs) with the F primer: 5′-gaattcattaaagaggagaaattaactatgagcaagggcgaagaactgtttacgg-3′ and the R primer: 5′-ctaattaagcttggctgcaggtcgacccttaatgatgatgatgatgatgtgagcctttatacag-3′. The gene of green fluorescent protein (GFP) was expressed under the control of the same promoter used for the ASKA strains. The *E. coli* AG1 strain, which is the host strain of the ASKA collection was purchased from Agilent Technologies (Wilmington, DE, USA).

2.2. Cell Growth Experiments

Individual plates of the ASKA collection were replicated by inoculating 3 μL stock culture into 150 μL fresh LB media with 50 μg/mL chloramphenicol in each well of 96-well plates, and incubated at 37 °C overnight. The absorbance at 600 nm of each well was then read by the microplate reader. The overnight culture in each well was diluted to $OD_{600nm} = 0.15$ with a total volume 150 μL of fresh LB media with 50 μg/mL chloramphenicol. Each plate had three biological replicates. The 96-well plates were sealed with oxygen-permeable membranes (Sigma-Aldrich, St. Louis, MO, USA). The cell growth was monitored by reading the absorbance at 600 nm with microplate readers at 37 °C continuously. The doubling time was calculated by the equation: Doubling time = lg2/lgX. X is the growth rate in the exponential phase, which was automatically provided by Gen5 software designed for the BioTek microplate reader (Winooski, VT, USA). The monitoring of GFP expression by fluorescence followed previous studies [13,14].

2.3. Bioinformatical Analyses

The software and online resources used for bioinformatical analyses were described in each subsection of Results and Discussion.

3. Results and Discussion

3.1. Growth Condition Selection

First, GFP was used as a reporter to determine the optimal concentration of the inducer isopropyl β-D-1-thiogalactopyranoside (IPTG) for high-throughput growth tests. We monitored both recombinant protein production by the fluorescence intensity (Figure 1a,b) and biomass accumulation by OD_{600nm} (Figure 1c). Interestingly, there was a high fluorescence reading, even without IPTG in the growth medium, indicating that the pCA24N vector is not tightly controlled. Lower concentrations of IPTG (0.05 to 0.2 mM) significantly increased the GFP expression ($p < 0.01$ by the *t*-test), while commonly used concentrations of IPTG (0.5 to 1 mM) decreased the GFP expression significantly (Figure 1a). This result was consistent with previous studies, which showed that decreasing concentrations of inducers could enhance recombinant protein production [6]. On the other hand, the concentration of IPTG also affected cell growth. Starting from 0.2 mM, higher concentrations of IPTG hindered normal cell growth (Figure 1b). Considering both recombinant protein production and biomass accumulation, 0.05 mM IPTG was chosen for high-throughput growth tests, since this concentration provided the best protein yield without negative effects on cell growth.

3.2. High-Throughput Growth Tests of the ASKA Collection

With 0.05 mM IPTG as the inducer, the doubling time of the strain containing the no insert control of the pCA24N vector was 42 min, which is longer than the previously reported 20 min [2]. This was because that 96-well plates used in this study have relatively smaller top space and lower oxygen supply than regular culture tubes. To better demonstrate the growth effect of overexpressing individual genes on cell growth, delay factor was defined as the ratio of the doubling time of individual strains in the ASKA collection over the doubling time of the strain harboring the no insert control of the pCA24N vector. The delay factors for all the 4071 *E. coli* genes tested in this study were summarized in Table 1 and listed in Table S1, respectively. Among them, 921 strains had no or moderate growth effects (delay factor < 2), 3049 strains had significant growth effects (delay factor between 2 and 7), and 101 strains had severe growth effects (delay factor > 7). More than 75% of strains had significant or severe growth effects, indicating that the metabolic burden could be a general issue in recombinant protein production. Only a small portion of strains severely impaired cell growth, possibly due to both the metabolic burden and specific protein toxicity.

3.3. Bioinformatical Analyses of Factors Affecting Cell Growth

3.3.1. The Effect of Protein Length on Cell Growth

We expected that a longer gene length or protein length needs more materials, thus affecting cell growth. To test this factor, the delay factor versus the protein length of each ASKA collection strains was plotted (Figure 2). The median *E. coli* protein length is 280 aa. The median delay factor for *E. coli* proteins less than 280 aa is 2.37, while that for *E. coli* proteins more than 280 aa is 2.44. The difference is not significant, consistent with the trend line which shows only a slight rise of the delay factor with increasing protein length (Figure 2).

(a)

(b)

(c)

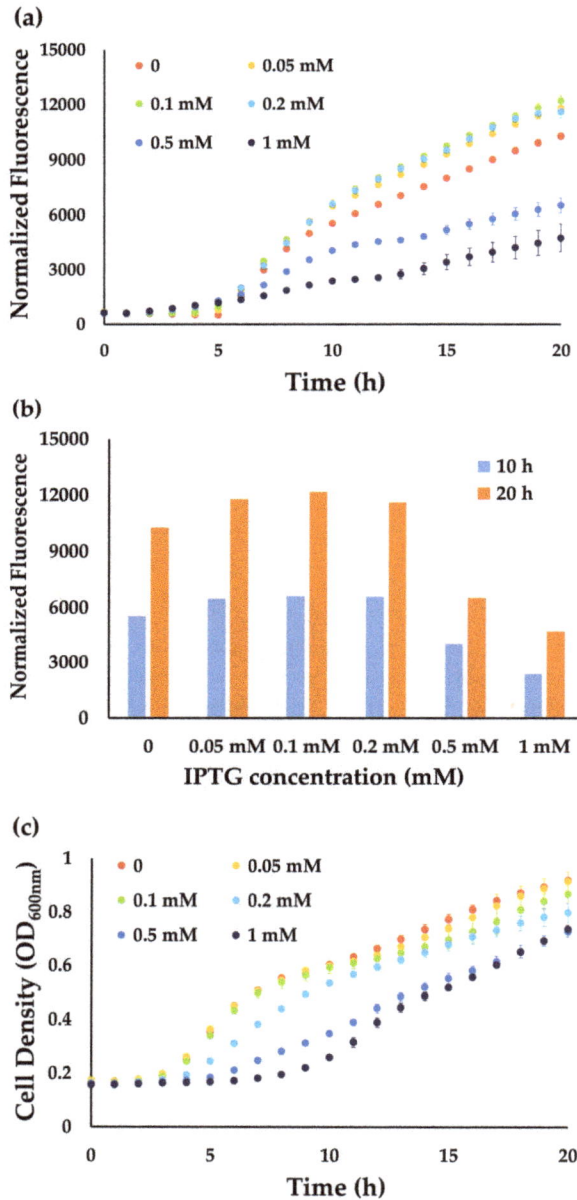

Figure 1. The effects of isopropyl β-D-1-thiogalactopyranoside (IPTG) concentrations on green fluorescent protein (GFP) expression and cell growth. (**a**) The normalized fluorescence intensity was used for determining protein production, which was the absolute fluorescent intensity subtracted with the fluorescence background of AG1 cells in the same growth condition over the cell density (OD_{600nm}); (**b**) The effects of IPTG concentrations on the normalized fluorescence intensity at 10 h and 20 h after inoculation; (**c**) The cell density was used for determining cell growth. The mean values and standard deviations were calculated from three biological replicates.

Table 1. The summary of delay factors of individual a complete set of *E. coli* K-12 ORF archive (ASKA) strains.

Growth Effects [1]	Delay Factors [2]	Numbers of Strains	Percentage
No or moderate	<2	921	22.6%
Significant	2 to 7	3049	74.9%
Severe	>7	101	2.5%

[1] The classification of growth effects was based on delay factors subjectively. [2] Delay factors were calculated by dividing the doubling time of individual strains in the ASKA collection by the doubling time of the strain harboring the no insert control of the pCA24N vector.

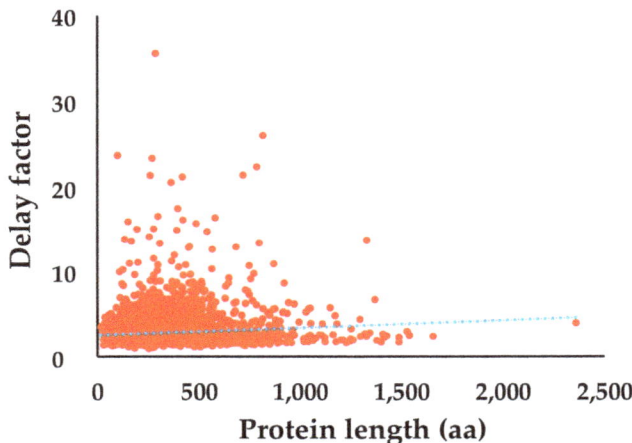

Figure 2. The effects of protein length on cell growth. The delay factor and protein length of individual ASKA collection strains were plotted with red color. The blue dot line is the trend line.

3.3.2. The Effect of Amino Acid Compositions on Cell Growth

Because the amount of individual free amino acids in cells are different, we assumed that the abundance of each amino acids in a target protein might affect cell growth when overexpressing it. Thus, the amino acid compositions of proteins overexpressed in the strains that had severe growth effects were calculated (Table S2). Comparing with the mean values of all the *E. coli* proteins, the abundances of isoleucine (Ile), leucine (Leu), and valine (Val) are significantly increased in proteins which had severe effects on cell growth (Figure 3a). Interestingly, these three amino acids all belong to branched-chain amino acids (BCAAs), which have been shown to be essential for bacterial growth [15,16].

To test if the severe growth effect was really caused by insufficient intracellular BCAAs, we randomly selected ten strains from the severe growth group, which overexpress proteins with high abundance of BCAAs, and tested their growth in LB media supplemented with 2 mM (each) of Ile, Leu, and Val. Most of the strains had improved growth, indicating that overexpressing proteins with high abundance of BCAAs could indeed impair cell growth, which could be then remedied by adding those BCAAs in growth media. (Figure 3b). We also tested the effect of BCAA supplementation on growth of strains expressing proteins with average abundance of BCAAs. The results showed that the improvement was not as significant as that for proteins with high abundance BCAAs (Figure S1).

(a)

(b)

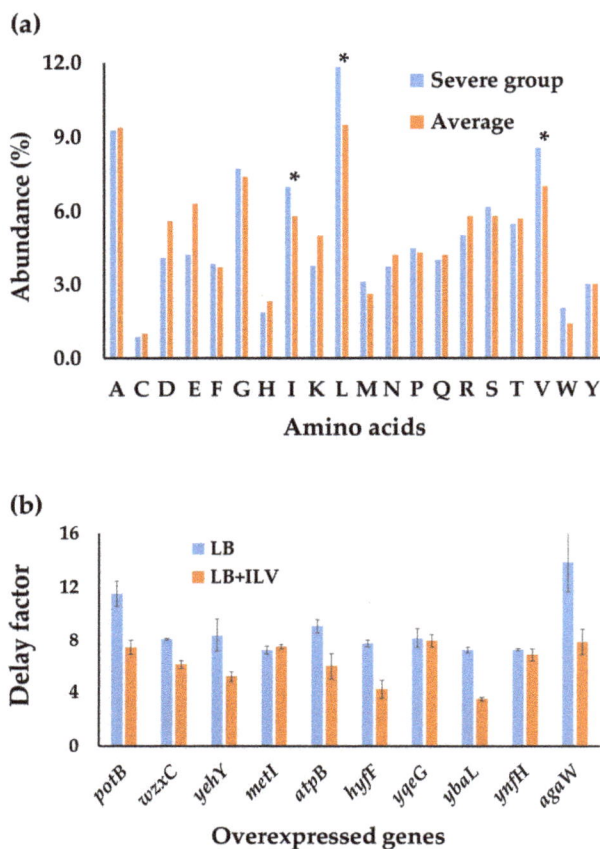

Figure 3. The effects of amino acid compositions on cell growth. (a) The abundance of each amino acids in proteins with severe growth effects. One-letter abbreviations of amino acids were used. The *t*-test was used and significant differences ($p < 0.05$) were marked with *. (b) The delay factors of selected strains grown in media with or without supplementary branched-chain amino acids (BCAAs). The abundances of ILV in the selected genes are *potB* (37.46%), *wzxC* (36.19%), *yehY* (35.58%), *metI* (35.02%), *atpB* (34.44%), *hyfF* (33.64%), *yqeG* (33.25%), *ybaL* (33.16%), *ynfH* (33.10%), and *agaW* (33.08%). The delay factor was defined as the ratio of the doubling time of individual strains in the ASKA collection over the doubling time of the strain harboring the no insert control of the pCA24N vector. ILV means isoleucine, leucine, and valine. The mean values and deviations were calculated from three biological replicates.

3.3.3. The Effect of Codon Bias on Cell Growth

Another common issue in recombinant protein production is codon bias, which means the occurrence of synonymous codons in target genes is largely different from that of host cells. The depletion of rare tRNAs by overexpressing recombinant proteins could cause early termination or mistranslation of recombinant proteins [17,18]. We expected that the shortage of rare tRNAs could also cause similar problems for native protein production, thus affecting cell growth. To test this assumption, the rare codon usage in all the *E. coli* K12 proteins and in the group with severe growth effects was compared (Figure 4, listed in Table S3). We focused on the seven rare codons in *E. coli* K12 cells, which are AGG, AGA, CGA, and CGG for arginine, AUA for isoleucine, CUA for leucine, and CCC for proline [19]. Unexpectedly, no significant differences were observed between the severe group and all the *E. coli* K12 proteins in rare codon usage.

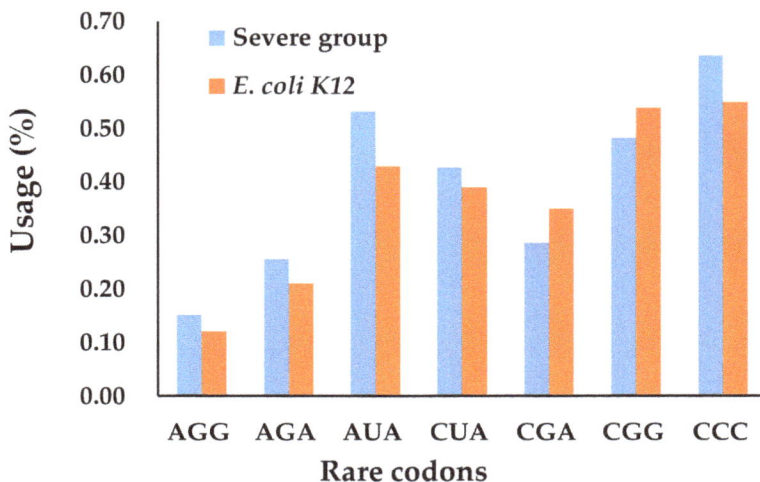

Figure 4. The effects of rare codon usage on cell growth. The codon usage of all the *E. coli* K12 genes was cited from Kazusa DNA Research Institute, Japan (http://www.kazusa.or.jp). The *t*-test was used.

The result is consistent with previous studies, which have shown that tRNA availability for rare codons is not the most important factor for protein production during gene overexpression [20,21]. Common strategies for dealing with codon bias include codon optimization and special strains harboring rare tRNAs in plasmids [22]. However, these approaches were reported to cause mRNA instability and protein aggregation [23,24], as rare codons could play important roles in forming specific RNA secondary structures for its stability and interaction with ribosomes [20,25,26], and in translational pausing which could help proper protein folding [27].

3.3.4. Gene Ontology Analyses

In the above three subsections, we focused on the metabolic burden resulting from the general properties of proteins rather than their functions. From this subsection, we started to consider the protein toxicity associated with their functions. We first categorized proteins which were overexpressed in strains with severe growth effects into different groups according to their annotated molecular functions, cellular components, and biological processes in UniProt-GOA database [28] (Figure 5).

For molecular functions, those proteins are distributed evenly in the three major categories: Enzymes, transporters, and binding proteins. Compared with the analysis of all the *E. coli* genes (Figure S2a), the fraction in transporters was significantly higher in the group with severe effects ($p < 0.01$ by the *t*-test). For cellular localization, most of them are associated with membranes, which is consistent with the previous analysis of amino acid compositions, since membrane proteins tend to have higher abundances of nonpolar amino acids, including BCAAs due to their interactions with membrane lipids. Compared with the analysis of all the *E. coli* genes (Figure S2b), the fraction in membranes was significantly higher in the group with severe effects ($p < 0.01$ by the *t*-test). For biological processes, they span on all essential cellular processes. Compared with the analysis of all the *E. coli* genes (Figure S2c), the fraction in metabolism was significantly lower in the group with severe effects ($p < 0.01$ by the *t*-test).

(a) Molecular Functions

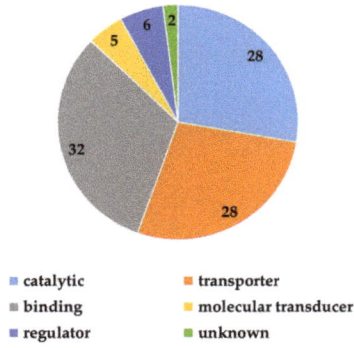

- catalytic
- transporter
- binding
- molecular transducer
- regulator
- unknown

(b) Cellular Components

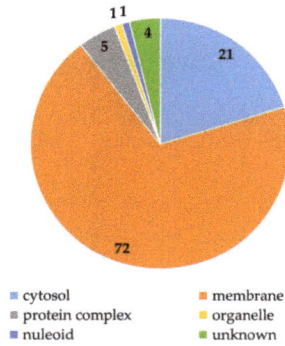

- cytosol
- membrane
- protein complex
- organelle
- nuleoid
- unknown

(c) Biological Processes

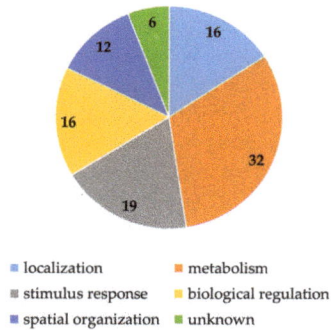

- localization
- metabolism
- stimulus response
- biological regulation
- spatial organization
- unknown

Figure 5. Gene ontology of proteins overexpressed in strains with severe growth effects. (**a**) Molecular functions; (**b**) Cellular components; and (**c**) Biological processes.

3.3.5. Protein Functional Interaction Network Analyses

Next, we analyzed the functional interaction network of proteins in the severe growth group by STRING database (http://string-db.org) [29] (Figure 6). Consistent with the gene ontology analyses, which demonstrated that target proteins are distributed evenly in different functional categories and biological processes, the interaction map only showed three clusters of proteins with five–six members.

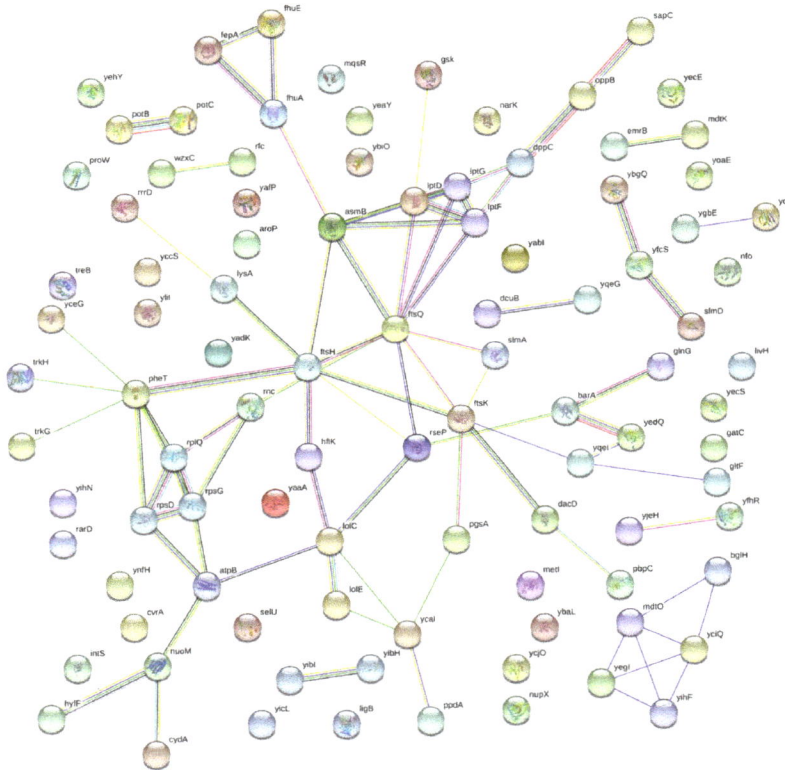

Figure 6. The functional interaction network analysis of proteins with severe growth effects by using STRING database. Network nodes represent proteins, and splice isoforms or post-translational modifications are collapsed. Edges represent protein-protein associations, and associations are meant to be specific and meaningful.

One cluster includes *asmB*, *lptD*, *lptG*, *dppC*, *lptF*, and *ftsQ*. LptD, LptF, and LptG are three essential proteins in the lipopolysaccharide transport system [29–31]. DppC is a membrane subunit for dipeptide ABC transporter [32]. FtsO is an essential cell division protein, which is required for localization of transporter proteins to the cell poles [33]. AsmB is also associated with cell division and involved in lipid A biosynthesis. Clearly, overexpression of these genes could interfere with normal cell division and membrane formation, which are essential for cell growth.

Another cluster contains *pheT*, *rplQ*, *rpsD*, *rpsG*, and *rnc*. RplO, RpsD, and RpsG are components of ribosomes [34–36]. RNase III (Rnc) is a key enzyme in rRNA processing [37]. PheT is one subunit of phenylalanyl-tRNA synthetase, which has the binding site with tRNA^{Phe} and editing activity [38]. Overexpressing these genes may affect the proper assembly of ribosomes and translation fidelity, thus impeding protein biosynthesis and cell growth. Actually, growth effects have also been observed in our studies with other aminoacyl-tRNA synthetases [39].

The last cluster includes *bglH*, *mdtO*, *yciQ*, *yegI*, and *yihF*. BglH is a carbohydrate-specific outer membrane porin [40]. MdtO is a component of a putative multidrug efflux pump [41]. YciQ is involved in membrane integration [42]. YihF and YegI have unknown functions, but they have high gene co-occurrences across genomes with both MdtO and YciO [43]. Again, overexpressing membrane-associated proteins tends to have growth effects, which is consistent with previous plate tests [12].

4. Conclusions

In summary, we quantified the effect of overexpressing individual *E. coli* genes on its cell growth. Overexpression of membrane-associated proteins, or proteins with high BCAA abundances, tended to hinder cell growth. For recombinant protein production, it is suggested that the first thing is to check BCAA abundances, and supplementing BCAAs in growth media could recover cell growth when overexpressing proteins with high BCAA abundances. For membrane-associated proteins or proteins related to protein biosynthesis, it is recommended to reduce the rate of protein production with lower inducer concentrations, weaker promoters, or lower copy numbers of vector to improve cell growth for an increased total protein yields.

Supplementary Materials: The following are available online at http://www.mdpi.com/2073-4425/9/8/414/s1, **Figure S1.** The delay factors of selected strains with average BCAA abundance. **Figure S2.** Gene ontology of all *E. coli* K12 genes. **Table S1.** The complete list of delay factors for the ASKA collection. **Table S2.** The list of amino acid compositions of proteins overexpressed in strains with severe growth effects. **Table S3.** The list of rare codon abundances in proteins overexpressed in strains with severe growth effects.

Author Contributions: Conceptualization, C.F.; Data curation, H.C., S.V., J.W., P.M., A.C., Q.G. and C.F.; Formal analysis, H.C., S.V., Q.G. and C.F.; Funding acquisition, C.F.; Investigation, H.C., J.W., P.M., A.C., Q.G. and C.F.; Methodology, C.F.; Project administration, C.F.; Resources, Q.G. and C.F.; Supervision, C.F.; Validation, Q.G. and C.F.; Visualization, C.F.; Writing—original draft, C.F.; Writing—review & editing, H.C., S.V., J.W., P.M., A.C., Q.G. and C.F.

Funding: This work was supported by Ralph E. Powe Junior Faculty Enhancement Awards to C.F. from Oak Ridge Associated Universities, the seed fund to C.F. from Arkansas Biosciences Institute, and the start-up fund to C.F. from University of Arkansas.

Conflicts of Interest: All authors declare no conflict of interest.

References

1. Usmani, S.S.; Bedi, G.; Samuel, J.S.; Singh, S.; Kalra, S.; Kumar, P.; Ahuja, A.A.; Sharma, M.; Gautam, A.; Raghava, G.P.S. THPdb: Database of FDA-approved peptide and protein therapeutics. *PLoS ONE* **2017**, *12*, e0181748. [CrossRef] [PubMed]

2. Sezonov, G.; Joseleau-Petit, D.; D'Ari, R. Escherichia coli physiology in Luria-Bertani broth. *J. Bacteriol.* **2007**, *189*, 8746–8749. [CrossRef] [PubMed]

3. Datsenko, K.A.; Wanner, B.L. One-step inactivation of chromosomal genes in Escherichia coli K-12 using PCR products. *Proc. Natl. Acad. Sci. USA* **2000**, *97*, 6640–6645. [CrossRef] [PubMed]

4. Jia, B.; Jeon, C.O. High-throughput recombinant protein expression in Escherichia coli: Current status and future perspectives. *Open Biol.* **2016**, *6*. [CrossRef] [PubMed]

5. Rosano, G.L.; Ceccarelli, E.A. Recombinant protein expression in Escherichia coli: Advances and challenges. *Front Microbiol.* **2014**, *5*, 172. [CrossRef] [PubMed]

6. Bentley, W.E.; Kompala, D.S. Optimal induction of protein synthesis in recombinant bacterial cultures. *Ann. N. Y. Acad. Sci.* **1990**, *589*, 121–138. [CrossRef] [PubMed]

7. Hoffmann, F.; Rinas, U. Stress induced by recombinant protein production in Escherichia coli. *Adv. Biochem. Eng. Biotechnol.* **2004**, *89*, 73–92. [PubMed]

8. Dong, H.; Nilsson, L.; Kurland, C.G. Gratuitous overexpression of genes in Escherichia coli leads to growth inhibition and ribosome destruction. *J. Bacteriol.* **1995**, *177*, 1497–1504. [CrossRef] [PubMed]

9. Dumon-Seignovert, L.; Cariot, G.; Vuillard, L. The toxicity of recombinant proteins in *Escherichia coli*: A comparison of overexpression in BL21(DE3), C41(DE3), and C43(DE3). *Protein Expr. Purif.* **2004**, *37*, 203–206. [CrossRef] [PubMed]

10. Trepod, C.M.; Mott, J.E. A spontaneous runaway vector for production-scale expression of bovine somatotropin from *Escherichia coli. Appl. Microbiol. Biotechnol.* **2002**, *58*, 84–88. [PubMed]

11. Chang, D.E.; Smalley, D.J.; Conway, T. Gene expression profiling of *Escherichia coli* growth transitions: An expanded stringent response model. *Mol. Microbiol.* **2002**, *45*, 289–306. [CrossRef] [PubMed]

12. Kitagawa, M.; Ara, T.; Arifuzzaman, M.; Ioka-Nakamichi, T.; Inamoto, E.; Toyonaga, H.; Mori, H. Complete set of ORF clones of *Escherichia coli* ASKA library (a complete set of *E. coli* K-12 ORF archive): Unique resources for biological research. *DNA Res.* **2005**, *12*, 291–299. [CrossRef] [PubMed]

13. Venkat, S.; Nannapaneni, D.T.; Gregory, C.; Gan, Q.; McIntosh, M.; Fan, C. Genetically encoding thioacetyl-lysine as a non-deacetylatable analog of lysine acetylation in *Escherichia coli. FEBS Open Biol.* **2017**, *7*, 1805–1814. [CrossRef] [PubMed]

14. Venkat, S.; Sturges, J.; Stahman, A.; Gregory, C.; Gan, Q.; Fan, C. Genetically incorporating two distinct post-translational modifications into one protein simultaneously. *ACS Synth. Biol.* **2018**, *7*, 689–695. [CrossRef] [PubMed]

15. Yamamoto, K.; Tsuchisaka, A.; Yukawa, H. Branched-chain amino acids. *Adv. Biochem. Eng. Biotechnol.* **2017**, *159*, 103–128. [PubMed]

16. Massey, L.K.; Sokatch, J.R.; Conrad, R.S. Branched-chain amino acid catabolism in bacteria. *Bacteriol. Rev.* **1976**, *40*, 42–54. [PubMed]

17. Kurland, C.; Gallant, J. Errors of heterologous protein expression. *Curr. Opin. Biotechnol.* **1996**, *7*, 489–493. [CrossRef]

18. Gustafsson, C.; Govindarajan, S.; Minshull, J. Codon bias and heterologous protein expression. *Trends Biotechnol.* **2004**, *22*, 346–353. [CrossRef] [PubMed]

19. Kane, J.F. Effects of rare codon clusters on high-level expression of heterologous proteins in *Escherichia coli. Curr. Opin. Biotechnol.* **1995**, *6*, 494–500. [CrossRef]

20. Li, G.W.; Oh, E.; Weissman, J.S. The anti-Shine-Dalgarno sequence drives translational pausing and codon choice in bacteria. *Nature* **2012**, *484*, 538–541. [CrossRef] [PubMed]

21. Pedersen, S. *Escherichia coli* ribosomes translate in vivo with variable rate. *EMBO J.* **1984**, *3*, 2895–2898. [PubMed]

22. Sorensen, H.P.; Mortensen, K.K. Advanced genetic strategies for recombinant protein expression in *Escherichia coli. J. Biotechnol.* **2005**, *115*, 113–128. [CrossRef] [PubMed]

23. Wu, X.; Jornvall, H.; Berndt, K.D.; Oppermann, U. Codon optimization reveals critical factors for high level expression of two rare codon genes in *Escherichia coli*: RNA stability and secondary structure but not tRNA abundance. *Biochem. Biophys. Res. Commun.* **2004**, *313*, 89–96. [CrossRef] [PubMed]

24. Rosano, G.L.; Ceccarelli, E.A. Rare codon content affects the solubility of recombinant proteins in a codon bias-adjusted *Escherichia coli* strain. *Microb. Cell Fact.* **2009**, *8*, 41. [CrossRef] [PubMed]

25. Chu, D.; Kazana, E.; Bellanger, N.; Singh, T.; Tuite, M.F.; von der Haar, T. Translation elongation can control translation initiation on eukaryotic mRNAs. *EMBO J.* **2014**, *33*, 21–34. [CrossRef] [PubMed]

26. Oresic, M.; Shalloway, D. Specific correlations between relative synonymous codon usage and protein secondary structure. *J. Mol. Biol.* **1998**, *281*, 31–48. [CrossRef] [PubMed]

27. Tsai, C.J.; Sauna, Z.E.; Kimchi-Sarfaty, C.; Ambudkar, S.V.; Gottesman, M.M.; Nussinov, R. Synonymous mutations and ribosome stalling can lead to altered folding pathways and distinct minima. *J. Mol. Biol.* **2008**, *383*, 281–291. [CrossRef] [PubMed]

28. Huntley, R.P.; Sawford, T.; Mutowo-Meullenet, P.; Shypitsyna, A.; Bonilla, C.; Martin, M.J.; O'Donovan, C. The GOA database: Gene ontology annotation updates for 2015. *Nucleic Acids Res.* **2015**, *43*, D1057–D1063. [CrossRef] [PubMed]

29. Wu, T.; McCandlish, A.C.; Gronenberg, L.S.; Chng, S.S.; Silhavy, T.J.; Kahne, D. Identification of a protein complex that assembles lipopolysaccharide in the outer membrane of *Escherichia coli. Proc. Natl. Acad. Sci. USA* **2006**, *103*, 11754–11759. [CrossRef] [PubMed]

30. Daley, D.O.; Rapp, M.; Granseth, E.; Melen, K.; Drew, D.; von Heijne, G. Global topology analysis of the *Escherichia coli* inner membrane proteome. *Science* **2005**, *308*, 1321–1323. [CrossRef] [PubMed]

31. Narita, S.; Tokuda, H. Biochemical characterization of an ABC transporter LptBFGC complex required for the outer membrane sorting of lipopolysaccharides. *FEBS Lett.* **2009**, *583*, 2160–2164. [CrossRef] [PubMed]
32. Abouhamad, W.N.; Manson, M.D. The dipeptide permease of *Escherichia coli* closely resembles other bacterial transport systems and shows growth-phase-dependent expression. *Mol. Microbiol.* **1994**, *14*, 1077–1092. [CrossRef] [PubMed]
33. Fixen, K.R.; Janakiraman, A.; Garrity, S.; Slade, D.J.; Gray, A.N.; Karahan, N.; Hochschild, A.; Goldberg, M.B. Genetic reporter system for positioning of proteins at the bacterial pole. *MBio* **2012**, *3*, e00251-11. [CrossRef] [PubMed]
34. Robert, F.; Gagnon, M.; Sans, D.; Michnick, S.; Brakier-Gingras, L. Mapping of the RNA recognition site of *Escherichia coli* ribosomal protein S7. *RNA* **2000**, *6*, 1649–1659. [CrossRef] [PubMed]
35. Ramaswamy, P.; Woodson, S.A. Global stabilization of rRNA structure by ribosomal proteins S4, S17, and S20. *J. Mol. Biol.* **2009**, *392*, 666–677. [CrossRef] [PubMed]
36. Metspalu, E.; Maimets, T.; Ustav, M.; Villems, R. A quaternary complex consisting of two molecules of tRNA and ribosomal proteins L2 and L17. *FEBS Lett.* **1981**, *132*, 105–108. [CrossRef]
37. Robertson, H.D.; Dunn, J.J. Ribonucleic acid processing activity of *Escherichia coli* ribonuclease III. *J. Biol. Chem.* **1975**, *250*, 3050–3056. [PubMed]
38. Ling, J.; Roy, H.; Ibba, M. Mechanism of tRNA-dependent editing in translational quality control. *Proc. Natl. Acad. Sci. USA* **2007**, *104*, 72–77. [CrossRef] [PubMed]
39. Venkat, S.; Gregory, C.; Gan, Q.; Fan, C. Biochemical characterization of the lysine acetylation of tyrosyl-tRNA synthetase in *Escherichia coli*. *Chembiochem* **2017**, *18*, 1928–1934. [CrossRef] [PubMed]
40. Andersen, C.; Rak, B.; Benz, R. The gene *bglH* present in the *bgl* operon of *Escherichia coli*, responsible for uptake and fermentation of β-glucosides encodes for a carbohydrate-specific outer membrane porin. *Mol. Microbiol.* **1999**, *31*, 499–510. [CrossRef] [PubMed]
41. Shimada, T.; Yamamoto, K.; Ishihama, A. Involvement of the leucine response transcription factor LeuO in regulation of the genes for sulfa drug efflux. *J. Bacteriol.* **2009**, *191*, 4562–4571. [CrossRef] [PubMed]
42. Skretas, G.; Georgiou, G. Simple genetic selection protocol for isolation of overexpressed genes that enhance accumulation of membrane-integrated human G protein-coupled receptors in *Escherichia coli*. *Appl. Environ. Microbiol.* **2010**, *76*, 5852–5859. [CrossRef] [PubMed]
43. Kim, P.J.; Price, N.D. Genetic co-occurrence network across sequenced microbes. *PLoS Comput. Biol.* **2011**, *7*, e1002340. [CrossRef] [PubMed]

genes

MDPI

Article

Phosphorylation-Dependent Inhibition of Akt1

Nileeka Balasuriya [1], McShane McKenna [1], Xuguang Liu [1], Shawn S. C. Li [1] and Patrick O'Donoghue [1,2,]*

[1] Department of Biochemistry, Schulich School of Medicine and Dentistry, The University of Western Ontario, London, ON N6A 5C1, Canada; bbalasur@uwo.ca (N.B.); mmcken22@uwo.ca (M.M.); xliu329@uwo.ca (X.L.); sli@uwo.ca (S.S.C.L.)
[2] Department of Chemistry, Faculty of Science, The University of Western Ontario, London, ON N6A 5C1, Canada
* Correspondence: patrick.odonoghue@uwo.ca; Tel.: +1-519-850-2373

Received: 20 July 2018; Accepted: 10 August 2018; Published: 7 September 2018

Abstract: Protein kinase B (Akt1) is a proto-oncogene that is overactive in most cancers. Akt1 activation requires phosphorylation at Thr308; phosphorylation at Ser473 further enhances catalytic activity. Akt1 activity is also regulated via interactions between the kinase domain and the N-terminal auto-inhibitory pleckstrin homology (PH) domain. As it was previously difficult to produce Akt1 in site-specific phosphorylated forms, the contribution of each activating phosphorylation site to auto-inhibition was unknown. Using a combination of genetic code expansion and in vivo enzymatic phosphorylation, we produced Akt1 variants containing programmed phosphorylation to probe the interplay between Akt1 phosphorylation status and the auto-inhibitory function of the PH domain. Deletion of the PH domain increased the enzyme activity for all three phosphorylated Akt1 variants. For the doubly phosphorylated enzyme, deletion of the PH domain relieved auto-inhibition by 295-fold. We next found that phosphorylation at Ser473 provided resistance to chemical inhibition by Akti-1/2 inhibitor VIII. The Akti-1/2 inhibitor was most effective against pAkt1^{T308} and showed four-fold decreased potency with Akt1 variants phosphorylated at Ser473. The data highlight the need to design more potent Akt1 inhibitors that are effective against the doubly phosphorylated and most pathogenic form of Akt1.

Keywords: genetic code expansion; protein kinase B; phosphoinositide dependent kinase 1; phosphoseryl-tRNA synthetase; tRNA^{Sep}

1. Introduction

Protein kinase B (Akt) is a human serine–threonine kinase and a member of the AGC family of protein kinases [1,2]. The pathway regulated by Akt is the most commonly activated signaling pathway in human cancers [3]. Given that more than 50% of human tumors contain hyperactivated Akt [2], effective inhibition of active Akt has the potential to treat several distinct cancers. There are three *AKT* genes in humans, encoding the isozymes Akt1, Akt2, and Akt3. The Akt1 isozyme has well-established roles in many human cancers. Overactive Akt1 is a hallmark of diverse human malignancies [3,4] and linked to reduced survival outcomes [5,6]. Indeed, Akt1 is as a leading drug target in cancer [7,8]. Over 300 clinical trials have been completed or are under way that involve targeting the Akt1 signaling pathway [9,10].

Akt1 is a key regulator of the phosphoinositide 3 kinase (PI3K)/Akt1 signaling cascade that controls cell growth and survival [1]. In human cells, the activation of Akt1 occurs in response to growth factor stimulation. Following activation by a receptor tyrosine kinase at the plasma membrane, PI3K phosphorylates its immediate downstream target, a lipid second messenger called phosphatidylinositol-4,5-bisphosphate (PIP$_2$), converting PIP$_2$ into phosphatidylinositol-3,4,5-triphosphate (PIP$_3$) (Figure 2) [2]. Membrane-anchored PIP$_3$ is a binding site for pleckstrin homology (PH) domain–containing proteins such as Akt1 and one of the upstream kinases that activates Akt1, phosphoinositide-dependent kinase 1 (PDK1) [11]. Co-localization of Akt1 with PDK1 leads to partial activation of Akt1 by PDK1-mediated phosphorylation of Thr308 in the kinase domain of Akt1. Mechanistic target of rapamycin complex 2 (mTORC2) is responsible for phosphorylating Akt1 at Ser473 in the C-terminal hydrophobic motif of Akt1 (Figure 2). The phosphorylation of Ser473 further increases the kinase activity of Akt1. Although the cellular role of Ser473 phosphorylation is not well defined, several studies point to the idea that pSer473 may impact Akt1 substrate selectivity (reviewed in [2]).

Leveraging our ability to produce Akt1 protein in specifically phosphorylated forms (Figure 1), we recently quantified the precise contribution of pThr308 and pSer473 to Akt1 activity in vitro and in mammalian cells [12]. In studies with purified full-length Akt1, we found that pAkt1^{T308} achieves 30% of the activity of the doubly phosphorylated kinase. We also observed in COS-7 cells that pSer473 is dispensable for Akt1 signaling [12]. In close agreement with our findings, studies in adipocytes found that half-maximal Akt1 cellular signaling activity is achieved when the kinase is 5–22% phosphorylated [13]. In studies of human tumors [12–15] and in clinical diagnostic settings [14,15], the phosphorylation status of Akt1 at Ser473 only is often used as a marker or proxy for Akt1 activity. Our work [12] and work by others [16] found that phosphorylation of Thr308 alone was sufficient for maximal Akt1 signal prorogation in cells. These results indicated that compared to Ser473, Thr308 phosphorylation status is a superior biomarker for Akt1 activity.

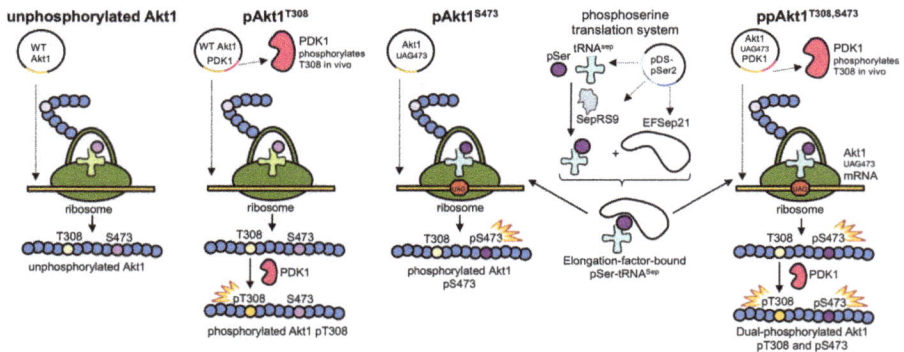

Figure 1. Production of Akt1 variants with programmed phosphorylation. To produce pAkt1^{T308}, PDK1 (Akt1's natural upstream kinase) was co-expressed along with Akt1. To produce pAkt1^{S473}, the phosphoserine orthogonal translation system was used to genetically incorporate phosphoserine at position 473 in response to an amber (UAG) codon. The ppAkt1^{T308,S473} variant was produced by combining both methods. WT: wild type.

The N-terminal PH domain is auto-inhibitory to Akt1 activity. Due to previous roadblocks in preparing Akt1 with programmed phosphorylation(s), there are no reports that measure the contribution of a phosphate at each key regulatory site to the auto-inhibition of Akt1 by the PH domain. In the current model, Akt1 exists in an auto-inhibited (PH-in) and activated (PH-out) conformation. In its auto-inhibited conformation (PH-in), the PH domain binds between the N- and C-terminal lobes of the unphosphorylated Akt1 kinase domain. During growth factor–mediated activation of Akt1, the PH domain forms a new interaction with PIP$_3$, causing it to move outward, away from the now more accessible kinase domain. This PH-out conformation is readily activated via phosphorylation at Thr308 by PDK1 and Ser473 by mTORC2. Disruption of the PH and kinase domain interaction was identified as a plausible cause of increased Akt1 phosphorylation and subsequent activity in cancer [18].

Here we quantified the ability of the PH domain to auto-inhibit Akt1 variants phosphorylated at either or both key regulatory sites Thr308 and Ser473. We produced novel recombinant Akt1 variants lacking the PH domain (ΔPH-Akt1) that also contained programmed phosphorylation at each site separately or with both sites phosphorylated. In the context of full-length phosphorylated Akt1 variants, we quantified the contribution of both regulatory phosphorylation sites to chemical inhibition with the PH domain–dependent allosteric inhibitor Akti-1/2.

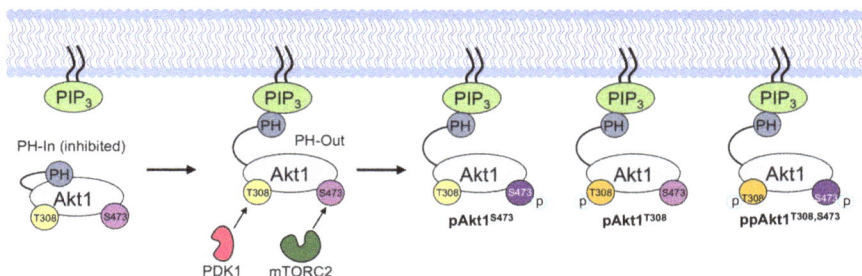

Figure 2. Simplified schematic of protein kinase B (Akt1) activation via phosphorylation of sites Thr308 and Ser473. The transition from Akt1's inactive state (PH-in) to its fully active state (ppAkt1$^{T308/S473}$) requires the release of pleckstrin homology (PH) domain–mediated auto-inhibition. This release occurs when Akt1's PH domain interacts with PIP$_3$ (PH-out). In the PH-out conformation, Akt1 is more susceptible to phosphorylation at Thr308 and Ser473 by phosphoinositide dependent kinase 1 (PDK1) and mechanistic target of rapamycin complex 2 (mTORC2), respectively. Upon release from PIP$_3$, Akt1 distributes rapidly in the cytosol and translocates to the nucleus to phosphorylate >100 cellular proteins [2,17].

2. Materials and Methods

2.1. Bacterial Strains and Plasmids

We designed a codon-optimized PH domain–deficient (ΔPH) human *AKT1* gene (residues 109–480), which was synthesized by ATUM (Newark, CA, USA). The ΔPH-*AKT1* gene was subcloned (*NcoI/NotI*) into an isopropyl β-D-1-thiogalactopyranoside (IPTG)–inducible T7*lac* promoter–driven pCDF-Duet1 vector with CloDF13-derived CDF replicon and streptomycin/spectinomycin resistance (pCDF-Duet1-ΔPHAkt1). The *PDPK1* gene was purchased from the Harvard PlasmidID repository service (plasmid ID: HsCD00001584; Boston, MA, USA) and subcloned (*KpnI/NdeI*) into the second multicloning site (MCS) of pCDF-Duet-1. Full-length pAkt1 variants were produced from pCDF-Duet1 plasmids as described previously [12]. The genetic code expansion system for phosphoserine (pSer) is encoded on the pDS-pSer2 plasmid [12,19,20], which contains 5 copies of tRNA^{Sep} [21], phosphoseryl-tRNA synthetase (SepRS9), and elongation factor Tu mutant (EFSep21) [22]. Incorporation of pSer also required site-directed mutagenesis of the Ser473 codon to TAG in the ΔPH-*akt1* constructs. Successful cloning was verified by DNA sequencing at the London Regional Genomics Centre (London, ON, Canada) and Genewiz (Cambridge, MA, USA).

2.2. Protein and Phosphoprotein Production

Recombinant Akt1 protein variants were expressed in BL21(DE3) (ThermoFisher Scientific, Waltham, MA, USA) (Figure 1). The pDS-pSer2 plasmid [20] was used as before [12] to incorporate pSer in response to a UAG codon at position 473 in ΔPHAkt1 and full-length Akt1 variants. To produce both full-length and PH domain–deficient Akt1 variants containing pSer473 (Supplementary Table S1), the pCDFDuet-1 Akt1–bearing plasmid was co-transformed with pDS-pSer2 into *Escherichia coli* Bl21(DE3) and plated on Luria broth (LB) agar plates with 25 μg/mL kanamycin and 50 μg/mL streptomycin. To produce pAkt1^{T308} variants, a pCDF-Duet1 plasmid containing both the Akt1 variant

(MSC 2) and PDK1 (MSC 1) was transformed into *E. coli* BL21(DE3) and plated on LB agar plates with 50 µg/mL streptomycin. To produce ppAkt1^{T308,S473} variants, a pCDF-Duet1 plasmid containing both the Akt1 variant, MSC 2, with a TAG codon at position 473 and PDK1 variant, MSC 1 [12] was co-transformed with pDS-pSer2 into *E. coli* Bl21(DE3) and plated on LB agar plates with 25 µg/mL kanamycin and 50 µg/mL streptomycin.

In all cases, a single colony was used to inoculate 70 mL of LB (with 50 µg/mL streptomycin and, if needed, 25 µg/mL kanamycin), which was grown, shaking, overnight at 37 °C. From this starter culture, a 10 mL inoculum was added to 1 L of LB with antibiotics (as above) and, for pSer473-containing variants only, *O*-phospho-L-serine (Sigma Aldrich, Oakville, ON, Canada) was added to a final concentration of 2.5 mM. The cultures were grown at 37 °C until OD$_{600}$ = 0.6, at which point, for pSer473-containing variants only, 2.5 mM of additional pSer was added to the culture. Protein expression was induced by adding 300 µM of IPTG at OD$_{600}$ = 0.8. Cultures were then incubated at 16 °C for 18 h. Cells were grown and pelleted at 5000× *g* and stored at −80 °C until further analysis. Akt1 protein variants were purified from the cell pellets using Ni-nitrilotriacetic acid affinity column chromatography (see Supplementary Material, affinity column chromatography description).

2.3. Parallel-Reaction Monitoring Mass Spectrometry of ppAkt1

The ppAkt1 protein produced as described above and the commercially available active Akt1 (Abcam, lot 1, Cambridge, MA, USA) were precipitated in ice-cold acetone/ethanol/acetic acid (50/50/0.1, vol/vol/vol). The protein precipitate was resuspended in 8 M urea, then reduced in 5 mM dithiothreitol (DTT) at 37 °C for 1 h and alkylated in 14 mM iodoacetamide (IAA) in darkness at room temperature for 1 h. Unreacted IAA was neutralized by adding 5 mM DTT. The final protein concentration was determined by Bradford assay. Glu-C digestion was performed at 37 °C overnight with a Glu-C–to–protein ratio of 1:20 (*w/w*). The digest was desalted in C18 column (Phenomenex, Torance, CA, USA) according to the manufacturer's protocol and resuspended in mass spectrometry (MS)-grade water. A Q Exactive Hybrid Quadrupole Orbitrap MS (Thermo Fisher Scientific, Waltham, MA, USA) was used to analyze the peptides. Data were analyzed using Skyline software [23].

2.4. MALDI-TOF/TOF Mass Spectrometry Analysis

In-gel digestion was performed using MassPREP automated digester station (PerkinElmer, Downers Grove, IL, USA). Gel pieces were de-stained using 50 mM ammonium bicarbonate and 50% acetonitrile, which was followed by protein reduction using 10 mM DTT, alkylation using 55 mM IAA, and tryptic digestion in 50 mM ammonium bicarbonate, pH 8. Peptides were extracted using a solution of 1% formic acid and 2% acetonitrile and lyophilized. Prior to mass spectrometry analysis, dried peptide samples were re-dissolved in a 10% acetonitrile and 0.1% trifluoroacetic acid solution. MALDI matrix, a–cyano–4–hydroxycinnamic acid, was prepared as 5 mg/mL in 6 mM ammonium phosphate monobasic, 50% acetonitrile, 0.1% trifluoroacetic acid and mixed with the sample at a 1:1 ratio (*v/v*). Mass spectrometry data (Figure S1) were obtained using an AB Sciex 5800 MALDI TOF/TOF system (Framingham, MA, USA). Data acquisition and data processing were done using a TOF/TOF Series Explorer a nd Data Explorer (both from AB Sciex, Boston, MA, USA), respectively. The instrument was equipped with a 349 nm Nd:YLF OptiBeam On-Axis laser. The laser pulse rate was 400 Hz. Reflectron positive mode was used. Reflectron mode was externally calibrated at 50 ppm mass tolerance and internally at 10 ppm. Each mass spectrum was collected as a sum of 500 shots.

2.5. Akt1 Kinase Activity Assay

The activity of each Akt1 variant was characterized by performing kinase assays in the presence of 200 µM substrate peptide CKRPRAASFAE (SignalChem, Vancouver, BC, Canada) derived from the natural Akt1 substrate, glycogen synthase kinase (GSK-3β). Assays were performed in 3-(*N*-morpholino) propanesulfonic acid (MOPS, 25 mM, pH 7.0), β-glycerolphosphate (12.5 mM), MgCl$_2$ (25 mM), ethylene glycol-bis(β-aminoethyl ether)-*N*,*N*,*N'*,*N'*-tetraacetic acid (EGTA, 5 mM, pH 8.0), ethylenediaminetetraacetic acid (EDTA) (2 mM), ATP (0.02 mM), and 0.4 µCi (0.033 µM)

γ-[32P]-ATP in a 30 µL reaction volume. Reactions were incubated at 37 °C and time points were taken over 30 min time courses. As previously [12], reactions were initiated by the addition of 18 pmol of the indicated Akt1 variant to yield a final enzyme concentration of 600 nM and quenched by spotting on P81 paper [24]. For highly active Akt1 variants, the level of Akt1 was titrated to identify a linear range to accurately determine initial velocity (v_o) (Figure S2). For this reason, Akt1 activity was compared based on the apparent catalytic rate (k_{app}) = v_o/[Akt1]. Samples from each reaction (5 µL) were spotted on P81 paper at specified time points. Following washes with 1% phosphoric acid (3 × 10 min) and 95% ethanol (1 × 5 min), the P81 paper was air-dried. Incorporation of ^{32}P into the substrate peptide was detected by exposing the P81 paper to a phosphor-imaging screen. The ^{32}P-peptide products were imaged and quantitated using a Storm 860 Molecular Imager and ImageQuant TL software (GE Healthcare, Mississauga, ON, Canada).

2.6. Kinase Inhibition Assay

A concentration gradient (0.001, 0.05, 0.5, 1, 5, 10 µM) of Akti-1/2 inhibitor VIII (Sellekchem, Houston, TX, USA) [25] was selected to determine the concentration dependence of Akt1 inhibition. Kinase inhibition assays were performed exactly as described above (Section 2.5) and with the addition of dimethyl sulfoxide at 10% (vol/vol) (control) or the indicated concentration of Akti-1/2 inhibitor. For each condition, initial velocity was determined over a time course of 10 min with time points at 0, 2, 5, and 10 min (Figure S3). In the inhibition assays, following established protocols [18], the inhibitor was preincubated with the enzyme for 5 min at 37 °C before addition of ATP and Akt1 substrate peptide to start the reaction. The time courses were used to determine the initial velocity of each reaction. The fraction of enzyme inhibition was measured based on initial velocities of uninhibited and inhibited reactions. From these data, half maximal inhibitory concentration (IC$_{50}$) values were calculated using Sigma Plot (Systat Software, Inc., San Jose, CA, USA).

3. Results

3.1. Production of Recombinant Akt1 Variants

We recently established a method to produce full-length Akt1 variants with programmed phosphorylations [12]. Here we applied this method to produced site-specifically phosphorylated Akt1 variants lacking the auto-inhibitory PH domain. The approach combines in vivo enzymatic phosphorylation with genetic code expansion to produce Akt1 variants containing either or both pThr308 and pSer473 (Figure 1).

Using genetic code expansion [12,20], we incorporated pSer in response to a UAG stop codon at position 473 in the relevant Akt1 constructs. Thr308 was site-specifically phosphorylated by co-expression of the upstream kinase PDK1 in *E. coli*. We [12] and others [26] have demonstrated that PDK1 is strictly specific in phosphorylating only the 308 site in Akt1. We used these methods in isolation or in combination to produce the three physiologically relevant Akt1 variants pAkt1^{S473}, pAkt1^{T308}, and ppAkt1^{T308,S473} (Figure 1). We observed that the ΔPH-Akt1 variants were all more soluble and were produced at 1.5- to 7-fold greater yield per liter of *E. coli* culture than the corresponding full-length Akt1 variants (Table 1). As previously reported [12], we confirmed phosphorylation of the Thr308 site (Figure 3C) and the Ser473 site (Figure S1) by mass spectrometry.

Table 1. Protein yields for Akt1 variants.

Akt1 Variant	Protein Yields (µg/L *Escherichia coli* Culture)	
	Full Length	ΔPH Akt1
Akt1 (unphosphorylated)	46	330
pAkt1^{S473}	100	150
pAkt1^{T308}	37	235
ppAkt1^{T308,S473}	45	224

3.2. Recombinant Akt1 Produced in E. coli Versus Sf9 Cells

Earlier work established production of partially active and truncated Akt1 in *E. coli*, which was unsuccessful at producing a sufficient amount of full-length Akt1 to determine activity [27]. Although we recently overcame these difficulties and developed a robust protocol to produce full-length Akt1 from *E. coli* with programmed phosphorylation, studies in the intervening period relied on insect Sf9 cell culture to produce recombinant Akt1 [28]. The ability to generate ppAkt1 from insect cells, however, requires a complex and low-yield in vitro procedure to phosphorylate Akt with two additional purified upstream kinases in the presence of lipid vesicles [28]. Protein production in Sf9 cells failed to yeild Akt1 with site-specific or programmed phosphorylation. The resulting protein was a mixture of singly and doubly phosphorylated species [29].

In order to benchmark the activity of the recombinant phosphorylated Akt1 protein we produced in *E. coli*, we compared the activity of full-length ppAkt1 to active Akt1 purchased from Abcam. The commercially available full-length Akt1 is made by a protocol similar to that established previously [28], in which Akt1 protein production in Sf9 cells is followed by in vitro phosphorylation of the purified Akt1 with purified PDK1. In testing an initial lot of active Akt1 (lot 1), we found that the commercially available Akt1 enzyme was catalytically deficient by eight-fold (Table 2) compared to the full-length ppAkt1 we produced (Figure 3A).

Table 2. Activity of Akt1 variants.

Akt1 Variant	Akt1 Amount (pmol)	Initial Velocity v_o (fmol/min)	Apparent Catalytic Rate k_{app} (fmol/min/pmol Akt1)	Activation (Fold Increase)	Reference
Akt1 (unphosphorylated)	18	0.6 ± 0.2	0.03 ± 0.01	1.0 ± 0.3	[12]
pAkt1^{S473}	18	46 ± 5	2.6 ± 0.3	85 ± 9	[12]
pAkt1^{T308}	1.8	22 ± 4	12 ± 2	400 ± 70	[12]
ppAkt1^{T308,S473}	1.8	79 ± 11	44 ± 6	1500 ± 200	[12]
ΔPHAkt1 (unphosphorylated)	18	1.4 ± 0.1	0.079 ± 0.006	2.7 ± 0.2	this study
ΔPH pAkt1^{S473}	18	210 ± 20	12 ± 1	390 ± 40	this study
ΔPH pAkt1^{T308}	0.18	370 ± 100	2100 ± 600	6900 ± 1800	this study
ΔPH ppAkt1^{T308,S473}	0.18	2300 ± 600	$(13 \pm 3) \times 10^3$	$(4 \pm 1) \times 10^5$	this study
Commercial Akt1 (Abcam)					
Lot 1	18	100 ± 20	6 ± 1	200 ± 30	this study
Lot 2	18	2200 ± 700	120 ± 40	4000 ± 1000	this study

Parallel-reaction monitoring mass spectrometry (PRM-MS/MS) was used to determine the identity of the residue at position 308 and the level of phosphorylation in ppAkt1 we produced and in the purchased protein. Mass spectrometry revealed that the low activity of lot 1 enzyme was attributable to the low level of phosphorylation at position 308 (peak intensity 1.3×10^1) and a relatively high level of unphosphorylated Thr at position 308 (Figure 3B). In contrast, ppAkt1^{T308,S473} that we produced showed a high level (peak intensity 1.6×10^6) of Thr308 phosphorylation and unphosphorylated Thr308 was not detected (Figure 3C). We then purchased a second lot of Akt1 enzyme (lot 2), and this enzyme displayed significantly more but highly variable activity compared to lot 1 (Table 2). Presumably, the second lot was quantitatively phosphorylated during production. In contrast to the method we developed for pAkt1 production in *E. coli* (Figure 1), the protocol to generate active Akt1 relying on Sf9 cells and subsequent in vitro phosphorylation appears to lead to a variable level of phosphorylation and activity in the resulting Akt1 preparations.

3.3. Impact of PH Domain Deletion on Differentially Phosphorylated Akt1

In order to determine the impact of the PH domain on each phospho-form of Akt1, we next assayed the activity of specifically phosphorylated ΔPH-Akt1 variants using γ-[32P]-ATP and a substrate peptide for Akt1 (CKRPRAA**S**FAE) that was derived from GSK-3β, a well-established Akt1 substrate [12,30]. In kinase assay conditions that we previously optimized [12] to measure a wide range of Akt1 activity, we determined the apparent catalytic rate of each ΔPH-Akt1 variant (unphosphorylated, pAkt1^{S473}, pAkt1^{T308}, ppAkt1^{T308,S473}). We then compared the apparent catalytic rates to the measurements we made previously with full-length Akt1 and pAkt1 variants to ascertain the relative impact of each phosphorylation state on auto-inhibition by the PH domain.

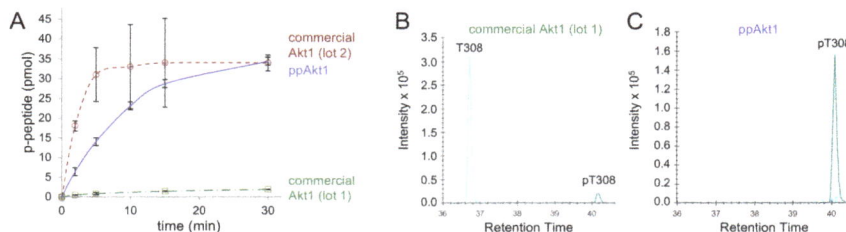

Figure 3. Activity of full-length ppAkt1 and commercially available active Akt1. (**A**) Kinase activity assays over a 30 min time course show substantially reduced activity of commercial Akt1 (lot 1, green squares) compared to full-length ppAkt1 (blue diamonds) produced in *E. coli*. Commercial Akt1 lot 2 (red circles) showed highly variable but similar activity to full-length ppAkt1. (**B**) Tryptic peptides from commercial Akt1 and ppAkt1 were analyzed by parallel-reaction monitoring mass spectrometry (PRM-MS). The purchased active Akt1 (lot 1) showed a low-intensity peak for phosphorylation at Thr308 (green peak at a retention time of ~40 min) and high-intensity peak for non-phosphorylated Thr308 (cyan peak, retention time of ~37 min). (**C**) PRM-MS analysis of ppAkt1^{T308,S473} showed a high-intensity peak for phosphorylation at Thr308 (green peak at a retention time of 40 min) and the non-phosphorylated Thr308 was undetectable.

As anticipated, all Akt1 variants lacking the PH domain were significantly more active than their full-length counterparts (Table 2). Doubly phosphorylated ΔPH-Akt1 (ppAkt^{T308,S473}) showed the highest activity among all three variants (Figure 4). The unphosphorylated ΔPH-Akt1 showed a basal level of activity that was significantly higher (2.7 ± 0.3-fold) than the background activity we recorded for full-length unphosphorylated Akt1 (Table 3). Although both unphosphorylated Akt1s would not have sufficient activity to induce Akt1-dependent signaling in cells [2,12], it is interesting to note that the auto-inhibitory effect of the PH domain is indeed measurable in the increased minimal activity of unphosphorylated ΔPH-Akt1 (Figure 4B).

Compared to unphosphorylated ΔPH-Akt1, the most active doubly phosphorylated Akt1 variant (ΔPH-ppAkt^{T308,S473}) showed a 150,000-fold increase in apparent catalytic rate (Table 2, Figure 4). Since rapid enzyme kinetics were observed with 18 pmol of enzyme with both of these variants containing Thr308 phosphorylation, the enzyme concentrations were subsequently reduced by 10-fold and 100-fold to obtain a highly accurate initial velocity with which to determine the apparent rate ($k_{app} = v_o/[enzyme]$) (Figure S2).

Table 3. Relative activity of full-length Akt1 versus variants lacking the PH domain (ΔPH Akt1) *.

Akt1 Variant	Full-Length Akt1	ΔPH-Akt1
Akt1 (unphosphorylated)	1.0 ± 0.3	2.7 ± 0.2
pAkt1^{S473}	1.0 ± 0.1	4.6 ± 0.4
pAkt1^{T308}	1.0 ± 0.2	175 ± 50
ppAkt1^{T308,S473}	1.0 ± 0.1	295 ± 68

* The fold increase in k_{app} for each ΔPH-Akt1 enzyme variant was calculated by normalizing to the corresponding full-length Akt1 variant k_{app}.

Interestingly, dual phosphorylation of ΔPH-Akt1 led to a 100-fold greater increase in activity than that observed upon dual phosphorylation of the full-length enzyme, which was 1500-fold more active than the unphosphorylated full-length enzyme. The data indicate that the PH domain significantly dampens the catalytic activity endowed by phosphorylation at Thr308 and Ser473 (Figure 5). The ΔPH-Akt1 variant with a single phosphorylation at the Thr308 site was robustly active,

2500-fold above the unphosphorylated ΔPH-Akt1, yet with 60-fold reduced activity compared to the doubly phosphorylated enzyme ΔPH-ppAkt$^{T308/S473}$ (Table 2). Phosphorylation at the C-terminal site Ser473 activated the enzyme 140-fold above the unphosphorylated control, but to a lesser extent (18-fold) than ΔPH-pAkt1^{T308}. We previously observed a quantitatively similar pattern of activity with the full-length Akt1 variants (Table 2, [12]).

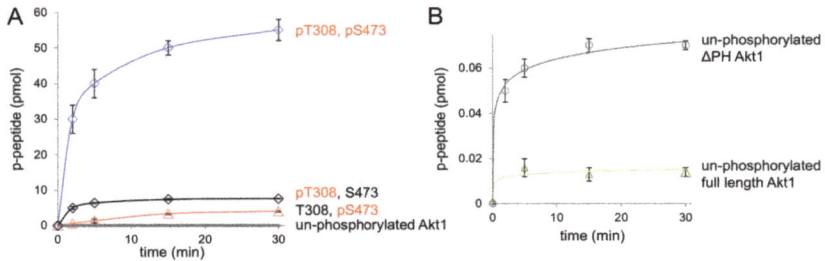

Figure 4. Enzyme activity of ΔPH-Akt1 variants. (**A**) The activity of differentially phosphorylated ΔPH Akt1 variants with the GSK-3β substrate peptide was measured over a 30 min time course. Akt1 phosphorylated at 308 and 473 (ΔPH-ppAkt^{S473,T308}, blue diamonds) showed maximal activity compared to unphosphorylated ΔPH-Akt1 (gray circles) and singly phosphorylated Akt1 variants ΔPH-pAkt1^{T308} (black diamonds) and ΔPH-pAkt1^{S473} (pink triangles). (**B**) The basal activity of unphosphorylated ΔPH-Akt1 (gray circles) was compared to full-length unphosphorylated Akt1 (green triangles). All reported values represent the mean of triplicate experiments, with error bars indicating one standard deviation.

We found that the PH domain exerts an auto-inhibitory effect, the strength of which depends on the phosphorylation status of the Akt1 enzyme. We compared apparent catalytic rates, k_{app}, (Figure 5A) and normalized relative catalytic rates (Figure 5B) between the full-length [12] and PH domain–deficient Akt1 variants. In the context of a single phosphorylation at Ser473, the PH domain is associated with an approximately five-fold reduction in activity (Table 3, Figure 5). In the singly phosphorylated pAkt1^{T308} enzyme, the auto-inhibition was far stronger (175-fold). The two phosphorylations together led to a super-additive inhibitory effect (295-fold) in the doubly phosphorylated enzyme (Figure 5).

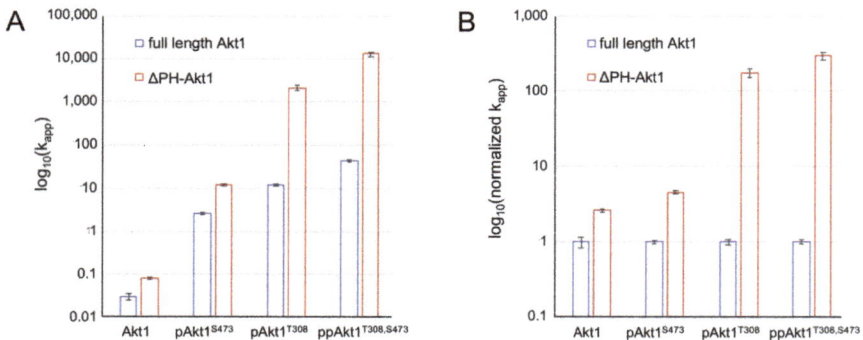

Figure 5. Impact of PH domain on the activity of differentially phosphorylated Akt1. (**A**) Apparent catalytic rates (k_{app}) and (**B**) normalized k_{app} values of full-length Akt1 variants (blue) and ΔPH-Akt1 variants (red) are shown. Error bars represent one standard deviation of triplicate measurements.

3.4. Chemical Inhibition of Phosphorylated Akt1 Variants

Given the differential impact of Akt1 phosphorylation on auto-inhibition, we next identified phosphorylation dependence in the interaction between Akt1 and a clinically relevant drug scaffold, Akti-1/2 inhibitor VIII. Several classes of chemical inhibitors were developed to repress aberrant Akt1 activity in cancer cells [10]. Early Akt1 inhibitors focused on ATP competitive compounds [31], yet these molecules suffered from significant cross-reactivity with other AGC family kinases [32]. For this reason, subsequent efforts focused on allosteric Akt1 inhibitors. Akti-1/2 is an allosteric inhibitor that binds Akt1 in a cleft between the kinase and PH domains, locking the enzyme in a noncatalytic conformation (Figure 6). Biochemical data with different Akt1 preparations has provided estimates of IC_{50} of Akt1 for the Akti-1/2 inhibitor ranging from 58 nM to 2.7 µM [24,33,34] (Table 4). We suspected that this range of values resulted from preparations of Akt1 with various levels of phosphorylation, with less active Akt1 preparations leading to an underestimate of IC_{50}. In addition, previous studies were unable to isolate the contribution of each regulatory phosphorylation site to inhibition by Akti-1/2.

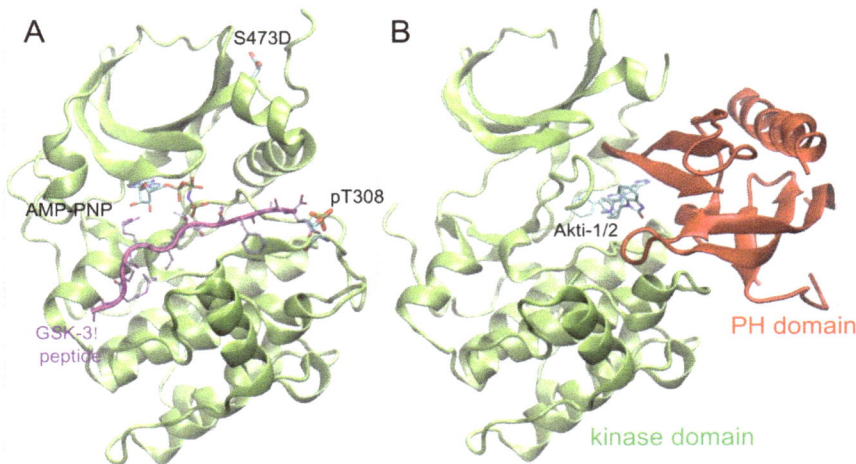

Figure 6. Structure of Akt1 in active and inhibitor bound forms. (**A**) Structure of ΔPH-pAkt1^{Thr308} Ser473Asp (PDB 1O6K [35]) is shown in complex with ATP analog (ANP-PNP) and substrate peptide (purple). (**B**) Structure of the full-length Akt1 (unphosphorylated) is shown in complex with the Akti-1/2 inhibitor VIII (PDB 1O96 [36]) binding in the cleft between the kinase domain (green) and the N-terminal PH domain (red).

We conducted Akt1 inhibition assays using Akti-1/2 concentrations varying from 0.01 to 10 µM. All three full-length phosphorylated Akt1 variants showed concentration-responsive enzyme inhibition (Figure 7, Figures S3–S6). Akti-1/2 was most potent (IC_{50} = 300 nM) when only the Thr308 site was phosphorylated (Table 4). Interestingly, at the lower concentrations of Akti-1/2 (0.001, 0.05, 0.5 µM), the compound was not effective at inhibiting the activity of pAkt1^{S473}. Akt1 phosphorylated at the C-terminal Ser473 site was overall more resistant to inhibitor VIII inhibition, resulting in an IC_{50} value of 1.3 µM. In a surprising finding, the doubly phosphorylated enzyme had an indistinguishable IC_{50} from the pAkt1^{S473} enzyme. Together the data indicate that Ser473 phosphorylation provides a four-fold increase in resistance to the Akti-1/2 inhibitor even in the presence of phosphorylation at Thr308. This is the first report demonstrating phosphorylation dependence in the interaction between Akt1 and an inhibitor.

Table 4. Akti-1/2 inhibitor half maximal inhibitory concentration (IC_{50}) for pAkt1 variants.

Akt1 Variant	Inhibitor Type	IC_{50} (μM)	Expression System	Reference
ppAkt1^{T308,S473}	Akti-1/2 inhibitor VIII	1.2 ± 0.3	*E. coli*	this study
pAkt1^{T308}	Akti-1/2 inhibitor VIII	0.3 ± 0.1	*E. coli*	this study
pAkt1^{S473}	Akti-1/2 inhibitor VIII	1.3 ± 0.1	*E. coli*	this study
active	Akti-1/2 inhibitor VIII	0.058	*Drosophila* S2 cells	[25]
active	Akti-1/2 inhibitor VIII	0.1	HEK 293 cells	[33]
active	Akti-1/2 inhibitor	2.7	*Drosophila* S2 cells	[34]
active	Akti-1 inhibitor	4.6	*Drosophila* S2 cells	[34]

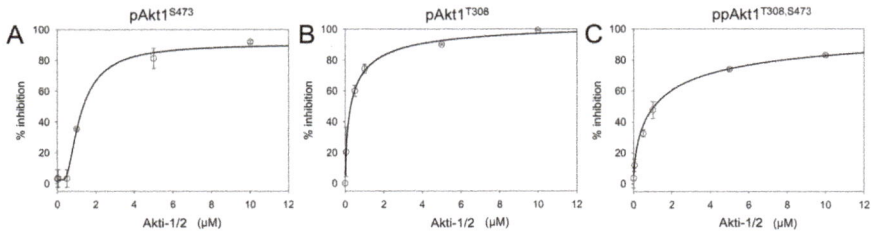

Figure 7. Chemical inhibition of full-length phosphorylated Akt1 variants. Inhibition of (**A**) pAkt1^{S473}, (**B**) pAkt1^{T308}, and (**C**) ppAkt1^{T308,S473} with varying concentrations (0.001, 0.05, 0.5, 1, 5, 10 μM) of Akti-1/2 inhibitor VIII. The resulting IC_{50} values are in Table 4.

4. Discussion

Akt1 is a prime target for therapeutic intervention due to its involvement in regulating a multitude of cellular pathways [37,38]. Aberrant Akt function has been linked to cancers and a variety of human diseases related to metabolic regulation, immune function, and neurological development [2]. The broad range of cellular processes governed by Akt, therefore, presents an opportunity for single-target therapeutic intervention in a variety of conditions. To capitalize on this opportunity, compounds that selectively attenuate Akt1 activity rather than global kinase inactivation are required. This need is reflected in clinical trials showing that nonspecific Akt inhibition in cancer therapy, due to inadvertent off-targeting of closely related kinases or multiple Akt isozymes, can result in detrimental side effects including liver damage and metabolic disorders [39–42]. Accordingly, the development of small molecule inhibitors for Akt reflects this sentiment of engineering greater target specificity [10]. From the earliest ATP-competitive inhibitors that unintentionally inhibited related AGC kinases, to more recent allosteric inhibitors that preferentially target Akt or even a subset of Akt isozymes, the continued pursuit for Akt1 inhibitor specificity requires an expanded understanding of the molecular basis of Akt1 activation and inhibition.

Previous work was unable to uncover the role of each regulatory site (Thr308 and Ser473) in the inhibition kinetics of Akt1. Given this critical knowledge gap, we investigated the role of Akt1 phosphorylation status on the auto-inhibitory effect of the PH domain and on allosteric inhibition by Akti-1/2.

4.1. Phosphorylation-Dependent Auto-Inhibition of Akt1

The N-terminal PH domain is shared among all three Akt isozymes and its direct function is to mediate protein–lipid interactions. It is well documented that PH domain binding to a lipid second messenger PIP_3 in cell membranes releases the inhibitory effect of the PH domain [2]. Here, we produced differentially phosphorylated Akt1 variants with and without the PH domain in *E. coli* using genetic code expansion to incorporate pSer473, and the upstream Thr308 kinase PDK1 was co-expressed in *E. coli*, leading to quantitative phosphorylation of the recombinant Akt1 protein. Compared to unphosphorylated Akt1, all phosphorylated variants showed increased enzyme activity.

Previously, live cell imaging studies revealed that, once phosphorylated, Akt1 favors a PH-out conformation that increases Akt1 activity in the downstream phosphorylation of substrate proteins in both the cytoplasm and the nucleus. In addition, binding of Akt1 to PIP₃ in the plasma membrane further releases the auto-inhibitory effect of the PH domain [43,44]. Although phosphorylation of the kinase domain at Thr308 is known to reduce the PH domain's affinity for the kinase domain [42], the contribution of Thr308 and Ser473 phosphorylation to auto-inhibition by the PH domain was unknown.

We quantitated the release of PH domain–mediated inhibition corresponding to each phosphorylation site and both sites in combination. Our data revealed that the magnitude of the auto-inhibitory effect of the PH domain was sensitively dependent on the phosphorylation status of Akt1. In comparison to the activity of full-length Akt1 phospho-variants, deletion of the PH domain significantly increased activity. In comparison to singly phosphorylated Akt1 variants, the auto-inhibitory effect of the PH domain was strongest (295-fold) in the most active and doubly phosphorylated Akt1 variant. Interestingly, the doubly phosphorylated Akt1 displayed increased activity upon PH domain deletion that was greater than the sum of the effects observed for the singly phosphorylated Akt1 variants. The data suggest a novel finding that the PH domain dampens the activity of Akt1 variants in a phosphorylation-dependent manner.

The combination of phosphorylation and PH domain release regulates the kinase activity of Akt1 over a 400,000-fold range. The data indicate the degree to which the unphosphorylated PH-in conformation of Akt1 is suppressed in comparison to the PH-out doubly phosphorylated version of the enzyme that exists at the plasma membrane in living cells. Recent in vitro biochemical work suggested that association with PIP₃-containing lipid vesicles stimulated the activity of a full-length Akt1 protein by seven-fold [44]. The authors suggested that this may be an underestimate due to low levels of phosphorylation (10% pThr308, 1% pSer473) in their Akt1 prepared from Sf9 cells. Our data suggest that release of the PH domain in the context of doubly phosphorylated Akt1 would lead to a maximal activity increase of 300-fold; a PH domain–dependent reduction lower than this value would indicate that there may be significant interaction between the Akt1 kinase and PH domains even in the PIP₃-bound state at the plasma membrane.

Indeed, a recent debate in the literature [45] is under way to explain how Akt1 activity can be maintained far from the membrane. Recent findings suggest that Akt1 is dephosphorylated on a time scale of ~10 min following dissociation from PIP₃ on the plasma membrane [46]. Multiple nuclear substrates of Akt1 are, however, known to be phosphorylated much more rapidly, on a time scale of seconds [47]. Although the authors suggested that all or perhaps most of the active Akt1 is membrane-associated [46], this finding disputes several independent studies identifying many cytosolic and nuclear targets of Akt1-dependent phosphorylation. Multiple studies clearly identified both phospho-Akt1 (e.g., [47–49]) and directly measured Akt1 activity in the nucleus [17]. Using a fluorescent reporter that quantifies Akt1 activity in live cells in real time, Kunkel and Newton observed Akt1 activity distributed in both the cytoplasm and nucleus on a time scale of seconds to minutes following growth factor stimulation. Their data further show similar levels of Akt1 activity in the nucleus and cytoplasm, with a delay in the peak activity in the nucleus of ~2 min. Additional experiments confirmed that this lag is related to the time it takes for active Akt1 to translocate from the plasma membrane to the nucleus [17]. In light of our data and these findings, membrane-associated Akt1 is likely significantly more active than doubly phosphorylated Akt1 found in the nucleus and cytosol. Phosphorylated Akt1, nevertheless, does persist away from the membrane, retaining significant activity in comparison to unphosphorylated Akt1.

4.2. Phosphorylation-Dependent Chemical Inhibition of Akt1

Posttranslational modifications, of which phosphorylation is the most common, are ubiquitous mechanisms that cells use to tune the activity of individual proteins and increase the functional diversity of the proteome. Phosphorylation can have a significant impact on the activity of proteins (kinases are a prime example), and the phosphorylation status of a protein target can accordingly impact the ability of small molecules to inhibit enzyme activity. In the experiments presented here, we determined precisely how the phosphorylation status of Akt1 influences the repression of kinase activity by chemical inhibition. We investigated the Akt-specific allosteric inhibitor Akti-1/2. Based on the results of our assays, the phosphorylation status of Akt1 can indeed influence the degree to which Akti-1/2 is able to act on Akt1. Akti-1/2 was most effective at inhibiting pAkt1^{T308}, whereas phosphorylation at position Ser473 in either pAkt^{S473} or ppAkt$^{T308/S473}$ reduced the relative effectiveness of the inhibitor by four-fold.

Ser473 phosphorylation appears to act as a built-in mechanism of drug resistance for Akt1. It is well established that hyperphosphorylation of Akt1 at both regulatory sites is linked to poor prognosis and therapeutic resistance [3–6,50]. The accepted model for Akti-1/2 activity suggests that the compound locks Akt1 into the auto-inhibited or PH-in conformation and prevents Akt1 from binding PIP$_3$ in the plasma membrane, inhibiting phosphorylation at Thr308 and Ser473 [51]. Our data also suggest that, although limited by Ser473 phosphorylation, Akti-1/2 has significant activity in inhibiting the pre-phosphorylated Akt1 variants that the inhibitor initially encounters in the cell. Studies in cell culture and mouse models show that the short-term impact of Akti-1/2 treatment is that pThr308 and pSer473 levels are depleted within hours. A related allosteric Akt inhibitor, and the most clinically promising [52] compound, MK-2206 significantly reduces Ser473 phosphorylation after 10 h in mouse models with patient-derived xenografts of endometrial tumors [53].

Surprisingly, breast cancer cells (BT474) treated with Akti-1/2 (1 μM) showed a rebound effect in Akt1 phosphorylation status following Akti-1/2 treatment. These cells have amplified *HER2* genes and constitutively activate the PI3K/Akt signaling cascade. In short-term experiments (minutes to hours), Akt1 phosphorylation reduced in response to inhibitor treatment, but at longer times (2–3 days) Akt1 phosphorylation status returned to stimulated levels in these cancer cells. In this longer time frame, phosphorylation of the downstream Akt1 target S6K was not restored, but PRAS40 phosphorylation was partially restored concomitant with increased Akt1 phospho-status. The study provided evidence that Akti-1/2 ultimately induces the expression and activation of receptor tyrosine kinases EGFR, HER3, and HER4, which in turn reactivate Akt1 [54]. Interestingly, in mouse models a similar rebound in Ser473 phosphorylation following treatment with MK-2206 was observed on time courses of 5 to 20 days [55]. In these cases, Ser473 phosphorylation may provide the tumor cell with a means to reduce the effectiveness of Akt1 inhibitors. The data suggest a need to develop an inhibitor that is more effective against the most active and doubly phosphorylated form of Akt1.

4.3. Synthetic Biology Approach to Generate Active Akt1

In our current study and previous work [12,19–21], we demonstrated that genetic code expansion with pSer in *E. coli* provides a simple route to produce designer phosphoproteins. Here, we produced differentially phosphorylated Akt1 variants with and without the PH domain in *E. coli* using genetic code expansion and enzymatic phosphorylation in *E. coli*. We found that the variants lacking the PH domain were produced at significantly higher yield compared to the full-length Akt1 protein. The ability to produce recombinant Akt1 protein with programmed phosphorylation(s) was essential for our investigation into the role of each phospho-site in Akt1 inhibition.

In agreement with our previous work [12], we found that the system we developed to produce Akt1 with either or both regulatory phosphorylation sites (Figure 1) leads to a consistently active Akt1 variant with indistinguishable batch-to-batch variability. This is in contrast to the commercially available Akt1 produced using the Sf9 cell system followed by in vitro PDK1 phosphorylation of Thr308 [28]. We observed vast batch-to-batch variability of enzyme activity in the commercially available Akt1, which warrants cautious attention in its experimental use. Previous work also established that Akt1 produced in Sf9 cells leads to a mixture of singly and doubly phosphorylated Akt1 variants [29], at times leading to low stoichiometry of phosphorylation [44] (Figure 3B). Our approach consistently results in site-specifically phosphorylated and active Akt1 variants that provide a reliable and consistent source of protein for biochemical studies and applications in inhibitor screening.

Supplementary Materials: The following are available online at http://www.mdpi.com/2073-4425/9/9/450/s1, Supplementary Materials and Methods, Table S1: Plasmids used in this study, Figure S1: Mass spectra confirming genetically encoded phosphoserine in ΔPH-ppAkt1, Figure S2: Initial velocity measurements for highly active ΔPH-Akt1 variants, Figure S3: Inhibition of full-length Akt1 variants incubated with Akti-1/2 inhibitor VIII, Figure S4: Autoradiographs of Akt1 inhibitor assays with ppAkt1T308, S473, Figure S5: Autoradiographs of Akt1 inhibitor assays with pAkt1T308, Figure S6: Autoradiographs of Akt1 inhibitor assays with pAkt1S473.

Author Contributions: Conceptualization, P.O., S.S.C.L.; methodology, N.B., X.L.; validation, N.B., M.M., X.L.; formal analysis, N.B.; investigation, N.B., M.M., X.L., P.O.; resources, P.O., S.S.C.L.; data curation, N.B., M.M.; writing—original draft preparation, N.B., M.M., P.O.; writing—review and editing, N.B., M.M., P.O.; visualization, N.B., M.M., X.L., P.O.; supervision, P.O., S.S.C.L.; project administration, P.O.; funding acquisition, P.O., S.S.C.L.

Funding: This work was supported by the Natural Sciences and Engineering Research Council of Canada (RGPIN 04282-2014 to P.O.), a Canadian Cancer Society Research Institute innovation grant (704324 to P.O. and S.S.C.L.), the Canada Foundation for Innovation (229917 to P.O.), the Ontario Research Fund (229917 to P.O.), Canada Research Chairs (950-229917 to P.O.), and the Canadian Breast Cancer Foundation (to S.S.C.L.).

Acknowledgments: We are grateful to Ilka Heinemann and Murray Junop for critical discussions and suggestions. We also thank Kristina Jurcic and Chaochao Chen at the MALDI Mass Spectrometry Facility at the University of Western Ontario for assistance with mass spectrometry analysis.

Conflicts of Interest: The authors declare no conflict of interest.

References

1. Manning, B.D.; Cantley, L.C. AKT/PKB signaling: Navigating downstream. *Cell* **2007**, *129*, 1261–1274. [CrossRef] [PubMed]

2. Manning, B.D.; Toker, A. AKT/PKB Signaling: Navigating the Network. *Cell* **2017**, *169*, 381–405. [CrossRef] [PubMed]

3. Agarwal, E.; Brattain, M.G.; Chowdhury, S. Cell survival and metastasis regulation by Akt signaling in colorectal cancer. *Cell. Signal* **2013**, *25*, 1711–1719. [CrossRef] [PubMed]

4. Spencer, A.; Yoon, S.S.; Harrison, S.J.; Morris, S.R.; Smith, D.A.; Brigandi, R.A.; Gauvin, J.; Kumar, R.; Opalinska, J.B.; Chen, C. The novel AKT inhibitor afuresertib shows favorable safety, pharmacokinetics, and clinical activity in multiple myeloma. *Blood* **2014**, *124*, 2190–2195. [CrossRef] [PubMed]

5. Antonelli, M.; Massimino, M.; Morra, I.; Garre, M.L.; Gardiman, M.P.; Buttarelli, F.R.; Arcella, A.; Giangaspero, F. Expression of pERK and pAKT in pediatric high grade astrocytomas: Correlation with YKL40 and prognostic significance. *Neuropathology* **2012**, *32*, 133–138. [CrossRef] [PubMed]

6. Suzuki, Y.; Shirai, K.; Oka, K.; Mobaraki, A.; Yoshida, Y.; Noda, S.E.; Okamoto, M.; Suzuki, Y.; Itoh, J.; Itoh, H.; et al. Higher pAkt expression predicts a significant worse prognosis in glioblastomas. *J. Radiat. Res.* **2010**, *51*, 343–348. [CrossRef] [PubMed]

7. Blachly, J.S.; Baiocchi, R.A. Targeting PI3-kinase (PI3K), AKT and mTOR axis in lymphoma. *Br. J. Haematol.* **2014**, *167*, 19–32. [CrossRef] [PubMed]

8. Westin, J.R. Status of PI3K/Akt/mTOR pathway inhibitors in lymphoma. *Clin. Lymphoma Myeloma Leuk* **2014**, *14*, 335–342. [CrossRef] [PubMed]

9. Jung-Testas, I.; Hu, Z.Y.; Baulieu, E.E.; Robel, P. Neurosteroids: Biosynthesis of pregnenolone and progesterone in primary cultures of rat glial cells. *Endocrinology* **1989**, *125*, 2083–2091. [CrossRef] [PubMed]

10. Nitulescu, G.M.; Margina, D.; Juzenas, P.; Peng, Q.; Olaru, O.T.; Saloustros, E.; Fenga, C.; Spandidos, D.; Libra, M.; Tsatsakis, A.M. Akt inhibitors in cancer treatment: The long journey from drug discovery to clinical use (Review). *Int. J. Oncol.* **2016**, *48*, 869–885. [CrossRef] [PubMed]

11. Altomare, D.A.; Testa, J.R. Perturbations of the AKT signaling pathway in human cancer. *Oncogene* **2005**, *24*, 7455–7464. [CrossRef] [PubMed]

12. Balasuriya, N.; Kunkel, M.T.; Liu, X.; Biggar, K.K.; Li, S.S.; Newton, A.C.; O'Donoghue, P. Genetic code expansion and live cell imaging reveal that Thr308 phosphorylation is irreplaceable and sufficient for Akt1 activity. *J. Biol. Chem.* **2018**, *293*, 10744–10756. [CrossRef] [PubMed]

13. Tan, S.X.; Ng, Y.; Meoli, C.C.; Kumar, A.; Khoo, P.S.; Fazakerley, D.J.; Junutula, J.R.; Vali, S.; James, D.E.; Stockli, J. Amplification and demultiplexing in insulin-regulated Akt protein kinase pathway in adipocytes. *J. Biol. Chem.* **2012**, *287*, 6128–6138. [CrossRef] [PubMed]

14. Parker, L.; Levinger, I.; Mousa, A.; Howlett, K.; de Courten, B. Plasma 25-hydroxyvitamin D is related to protein signaling involved in glucose homeostasis in a tissue-specific manner. *Nutrients* **2016**, *8*, 631. [CrossRef] [PubMed]

15. Tang, H.; Wu, Y.; Liu, M.; Qin, Y.; Wang, H.; Wang, L.; Li, S.; Zhu, H.; He, Z.; Luo, J.; et al. SEMA3B improves the survival of patients with esophageal squamous cell carcinoma by upregulating p53 and p21. *Oncol. Rep.* **2016**, *36*, 900–908. [CrossRef] [PubMed]

16. Hart, J.R.; Vogt, P.K. Phosphorylation of AKT: A mutational analysis. *Oncotarget* **2011**, *2*, 467–476. [CrossRef] [PubMed]

17. Kunkel, M.T.; Ni, Q.; Tsien, R.Y.; Zhang, J.; Newton, A.C. Spatio-temporal dynamics of protein kinase B/Akt signaling revealed by a genetically encoded fluorescent reporter. *J. Biol. Chem.* **2005**, *280*, 5581–5587. [CrossRef] [PubMed]

18. Parikh, C.; Janakiraman, V.; Wu, W.I.; Foo, C.K.; Kljavin, N.M.; Chaudhuri, S.; Stawiski, E.; Lee, B.; Lin, J.; Li, H.; et al. Disruption of PH-kinase domain interactions leads to oncogenic activation of AKT in human cancers. *Proc. Natl. Acad. Sci. USA* **2012**, *109*, 19368–19373. [CrossRef] [PubMed]

19. George, S.; Wang, S.M.; Bi, Y.; Treidlinger, M.; Barber, K.R.; Shaw, G.S.; O'Donoghue, P. Ubiquitin phosphorylated at Ser57 hyper-activates parkin. *Biochim. Biophys. Acta* **2017**, *1861*, 3038–3046. [CrossRef] [PubMed]

20. George, S.; Aguirre, J.D.; Spratt, D.E.; Bi, Y.; Jeffery, M.; Shaw, G.S.; O'Donoghue, P. Generation of phospho-ubiquitin variants by orthogonal translation reveals codon skipping. *FEBS Lett.* **2016**, *590*, 1530–1542. [CrossRef] [PubMed]

21. Aerni, H.R.; Shifman, M.A.; Rogulina, S.; O'Donoghue, P.; Rinehart, J. Revealing the amino acid composition of proteins within an expanded genetic code. *Nucleic Acids Res.* **2015**, *43*, e8. [CrossRef] [PubMed]

22. Lee, S.; Oh, S.; Yang, A.; Kim, J.; Soll, D.; Lee, D.; Park, H.S. A facile strategy for selective incorporation of phosphoserine into histones. *Angew. Chem. Int. Ed. Engl.* **2013**, *52*, 5771–5775. [CrossRef] [PubMed]

23. MacLean, B.; Tomazela, D.M.; Shulman, N.; Chambers, M.; Finney, G.L.; Frewen, B.; Kern, R.; Tabb, D.L.; Liebler, D.C.; MacCoss, M.J. Skyline: An open source document editor for creating and analyzing targeted proteomics experiments. *Bioinformatics* **2010**, *26*, 966–968. [CrossRef] [PubMed]

24. Turowec, J.P.; Duncan, J.S.; French, A.C.; Gyenis, L.; St Denis, N.A.; Vilk, G.; Litchfield, D.W. Protein kinase CK2 is a constitutively active enzyme that promotes cell survival: Strategies to identify CK2 substrates and manipulate its activity in mammalian cells. *Method Enzymol.* **2010**, *484*, 471–493.

25. Lindsley, C.W.; Zhao, Z.; Leister, W.H.; Robinson, R.G.; Barnett, S.F.; Defeo-Jones, D.; Jones, R.E.; Hartman, G.D.; Huff, J.R.; Huber, H.E.; et al. Allosteric Akt (PKB) inhibitors: Discovery and SAR of isozyme selective inhibitors. *Bioorg. Med. Chem. Lett.* **2005**, *15*, 761–764. [CrossRef] [PubMed]

26. Alessi, D.R.; James, S.R.; Downes, C.P.; Holmes, A.B.; Gaffney, P.R.; Reese, C.B.; Cohen, P. Characterization of a 3-phosphoinositide-dependent protein kinase which phosphorylates and activates protein kinase Bα. *Curr. Biol.* **1997**, *7*, 261–269. [CrossRef]

27. Klein, S.; Geiger, T.; Linchevski, I.; Lebendiker, M.; Itkin, A.; Assayag, K.; Levitzki, A. Expression and purification of active PKB kinase from *Escherichia coli*. *Protein Expr. Purif.* **2005**, *41*, 162–169. [CrossRef] [PubMed]

28. Zhang, X.; Zhang, S.; Yamane, H.; Wahl, R.; Ali, A.; Lofgren, J.A.; Kendall, R.L. Kinetic mechanism of AKT/PKB enzyme family. *J. Biol. Chem.* **2006**, *281*, 13949–13956. [CrossRef] [PubMed]

29. Fabbro, D.; Batt, D.; Rose, P.; Schacher, B.; Roberts, T.M.; Ferrari, S. Homogeneous purification of human recombinant GST-Akt/PKB from Sf9 cells. *Protein Expr. Purif.* **1999**, *17*, 83–88. [CrossRef] [PubMed]

30. Alessi, D.R.; Caudwell, F.B.; Andjelkovic, M.; Hemmings, B.A.; Cohen, P. Molecular basis for the substrate specificity of protein kinase B; comparison with MAPKAP kinase-1 and p70 S6 kinase. *FEBS Lett.* **1996**, *399*, 333–338. [CrossRef]

31. Reuveni, H.; Livnah, N.; Geiger, T.; Klein, S.; Ohne, O.; Cohen, I.; Benhar, M.; Gellerman, G.; Levitzki, A. Toward a PKB inhibitor: Modification of a selective PKA inhibitor by rational design. *Biochemistry* **2002**, *41*, 10304–10314. [CrossRef] [PubMed]

32. Yap, T.A.; Walton, M.I.; Grimshaw, K.M.; Te Poele, R.H.; Eve, P.D.; Valenti, M.R.; de Haven Brandon, A.K.; Martins, V.; Zetterlund, A.; Heaton, S.P.; et al. AT13148 is a novel, oral multi-AGC kinase inhibitor with potent pharmacodynamic and antitumor activity. *Clin. Cancer Res.* **2012**, *18*, 3912–3923. [CrossRef] [PubMed]

33. Green, C.J.; Goransson, O.; Kular, G.S.; Leslie, N.R.; Gray, A.; Alessi, D.R.; Sakamoto, K.; Hundal, H.S. Use of Akt inhibitor and a drug-resistant mutant validates a critical role for protein kinase B/Akt in the insulin-dependent regulation of glucose and system A amino acid uptake. *J. Biol. Chem.* **2008**, *283*, 27653–27667. [CrossRef] [PubMed]

34. Barnett, S.F.; Defeo-Jones, D.; Fu, S.; Hancock, P.J.; Haskell, K.M.; Jones, R.E.; Kahana, J.A.; Kral, A.M.; Leander, K.; Lee, L.L.; et al. Identification and characterization of pleckstrin-homology-domain-dependent and isoenzyme-specific Akt inhibitors. *Biochem. J.* **2005**, *385*, 399–408. [CrossRef] [PubMed]

35. Yang, J.; Cron, P.; Good, V.M.; Thompson, V.; Hemmings, B.A.; Barford, D. Crystal structure of an activated Akt/protein kinase B ternary complex with GSK3-peptide and AMP-PNP. *Nat. Struct. Biol.* **2002**, *9*, 940–944. [CrossRef] [PubMed]

36. Wu, W.I.; Voegtli, W.C.; Sturgis, H.L.; Dizon, F.P.; Vigers, G.P.; Brandhuber, B.J. Crystal structure of human AKT1 with an allosteric inhibitor reveals a new mode of kinase inhibition. *PLoS ONE* **2010**, *5*, e12913. [CrossRef] [PubMed]

37. Massihnia, D.; Avan, A.; Funel, N.; Maftouh, M.; van Krieken, A.; Granchi, C.; Raktoe, R.; Boggi, U.; Aicher, B.; Minutolo, F.; et al. Phospho-Akt overexpression is prognostic and can be used to tailor the synergistic interaction of Akt inhibitors with gemcitabine in pancreatic cancer. *J. Hematol. Oncol.* **2017**, *10*, 9. [CrossRef] [PubMed]

38. Slipicevic, A.; Holm, R.; Nguyen, M.T.; Bohler, P.J.; Davidson, B.; Florenes, V.A. Expression of activated Akt and PTEN in malignant melanomas: Relationship with clinical outcome. *Am. J. Clin. Pathol.* **2005**, *124*, 528–536. [CrossRef] [PubMed]

39. Wang, Q.; Chen, X.; Hay, N. Akt as a target for cancer therapy: More is not always better (lessons from studies in mice). *Br. J. Cancer* **2017**, *117*, 159–163. [CrossRef] [PubMed]

40. Lu, M.; Wan, M.; Leavens, K.F.; Chu, Q.; Monks, B.R.; Fernandez, S.; Ahima, R.S.; Ueki, K.; Kahn, C.R.; Birnbaum, M.J. Insulin regulates liver metabolism in vivo in the absence of hepatic Akt and Foxo1. *Nat. Med.* **2012**, *18*, 388–395. [CrossRef] [PubMed]

41. Crouthamel, M.C.; Kahana, J.A.; Korenchuk, S.; Zhang, S.Y.; Sundaresan, G.; Eberwein, D.J.; Brown, K.K.; Kumar, R. Mechanism and management of AKT inhibitor-induced hyperglycemia. *Clin. Cancer Res.* **2009**, *15*, 217–225. [CrossRef] [PubMed]

42. Cho, H.; Mu, J.; Kim, J.K.; Thorvaldsen, J.L.; Chu, Q.; Crenshaw, E.B.; Kaestner, K.H.; Bartolomei, M.S.; Shulman, G.I.; Birnbaum, M.J. Insulin resistance and a diabetes mellitus-like syndrome in mice lacking the protein kinase Akt2 (PKBβ). *Science* **2001**, *292*, 1728–1731. [CrossRef] [PubMed]

43. Calleja, V.; Alcor, D.; Laguerre, M.; Park, J.; Vojnovic, B.; Hemmings, B.A.; Downward, J.; Parker, P.J.; Larijani, B. Intramolecular and intermolecular interactions of protein kinase B define its activation in vivo. *PLoS Biol.* **2007**, *5*, e95. [CrossRef] [PubMed]

44. Lucic, I.; Rathinaswamy, M.K.; Truebestein, L.; Hamelin, D.J.; Burke, J.E.; Leonard, T.A. Conformational sampling of membranes by Akt controls its activation and inactivation. *Proc. Natl. Acad. Sci. USA* **2018**, *115*, E3940–E3949. [CrossRef] [PubMed]

45. Agarwal, A.K. How to explain the AKT phosphorylation of downstream targets in the wake of recent findings. *Proc. Natl. Acad. Sci. USA* **2018**, *115*, E6099–E6100. [CrossRef] [PubMed]

46. Ebner, M.; Lucic, I.; Leonard, T.A.; Yudushkin, I. PI(3,4,5)P3 engagement restricts akt activity to cellular membranes. *Mol. Cell* **2017**, *65*, 416–431. [CrossRef] [PubMed]

47. Humphrey, S.J.; Azimifar, S.B.; Mann, M. High-throughput phosphoproteomics reveals in vivo insulin signaling dynamics. *Nat. Biotechnol.* **2015**, *33*, 990–995. [CrossRef] [PubMed]

48. Zhu, L.; Hu, C.; Li, J.; Xue, P.; He, X.; Ge, C.; Qin, W.; Yao, G.; Gu, J. Real-time imaging nuclear translocation of Akt1 in HCC cells. *Biochem. Biophys. Res. Commun.* **2007**, *356*, 1038–1043. [CrossRef] [PubMed]

49. Hixon, M.L.; Boekelheide, K. Expression and localization of total Akt1 and phosphorylated Akt1 in the rat seminiferous epithelium. *J. Androl.* **2003**, *24*, 891–898. [CrossRef] [PubMed]

50. Freudlsperger, C.; Horn, D.; Weissfuss, S.; Weichert, W.; Weber, K.J.; Saure, D.; Sharma, S.; Dyckhoff, G.; Grabe, N.; Plinkert, P.; et al. Phosphorylation of AKT(Ser473) serves as an independent prognostic marker for radiosensitivity in advanced head and neck squamous cell carcinoma. *Int. J. Cancer* **2015**, *136*, 2775–2785. [CrossRef] [PubMed]

51. Okuzumi, T.; Fiedler, D.; Zhang, C.; Gray, D.C.; Aizenstein, B.; Hoffman, R.; Shokat, K.M. Inhibitor hijacking of Akt activation. *Nat. Chem. Biol.* **2009**, *5*, 484–493. [CrossRef] [PubMed]

52. Wisinski, K.B.; Tevaarwerk, A.J.; Burkard, M.E.; Rampurwala, M.; Eickhoff, J.; Bell, M.C.; Kolesar, J.M.; Flynn, C.; Liu, G. Phase I study of an AKT inhibitor (MK-2206) combined with lapatinib in adult solid tumors followed by dose expansion in advanced HER2+ breast cancer. *Clin. Cancer Res.* **2016**, *22*, 2659–2667. [CrossRef] [PubMed]

53. Winder, A.; Unno, K.; Yu, Y.; Lurain, J.; Kim, J.J. The allosteric AKT inhibitor, MK2206, decreases tumor growth and invasion in patient derived xenografts of endometrial cancer. *Cancer Biol. Ther.* **2017**, *18*, 958–964. [CrossRef] [PubMed]

54. Chandarlapaty, S.; Sawai, A.; Scaltriti, M.; Rodrik-Outmezguine, V.; Grbovic-Huezo, O.; Serra, V.; Majumder, P.K.; Baselga, J.; Rosen, N. AKT inhibition relieves feedback suppression of receptor tyrosine kinase expression and activity. *Cancer Cell* **2011**, *19*, 58–71. [CrossRef] [PubMed]

55. Lin, A.; Piao, H.L.; Zhuang, L.; Sarbassov dos, D.; Ma, L.; Gan, B. FoxO transcription factors promote AKT Ser473 phosphorylation and renal tumor growth in response to pharmacologic inhibition of the PI3K-AKT pathway. *Cancer Res.* **2014**, *74*, 1682–1693. [CrossRef] [PubMed]

![genes logo] **genes**

MDPI

Article

Lysine Acetylation Regulates Alanyl-tRNA Synthetase Activity in *Escherichia coli*

Takuya Umehara [1,2,*], Saori Kosono [1,3], Dieter Söll [4,5] and Koji Tamura [2,6]

1   Biotechnology Research Center, The University of Tokyo, Tokyo 113-8657, Japan; uskos@mail.ecc.u-tokyo.ac.jp
2   Department of Biological Science and Technology, Tokyo University of Science, Tokyo 125-8585, Japan; koji@rs.tus.ac.jp
3   Center for Sustainable Resource Science, RIKEN, Saitama 351-0198, Japan
4   Department of Molecular Biophysics and Biochemistry, Yale University, New Haven, CT 06520, USA; dieter.soll@yale.edu
5   Department of Chemistry, Yale University, New Haven, CT 06520, USA
6   Research Institute for Science and Technology, Tokyo University of Science, Chiba 278-8510, Japan
*   Correspondence: utumehara@mail.ecc.u-tokyo.ac.jp; Tel.: +81-3-5841-8030

Received: 14 August 2018; Accepted: 21 September 2018; Published: 28 September 2018

Abstract: Protein lysine acetylation is a widely conserved posttranslational modification in all three domains of life. Lysine acetylation frequently occurs in aminoacyl-tRNA synthetases (aaRSs) from many organisms. In this study, we determined the impact of the naturally occurring acetylation at lysine-73 (K73) in *Escherichia coli* class II alanyl-tRNA synthetase (AlaRS) on its alanylation activity. We prepared an AlaRS K73Ac variant in which N^ε-acetyl-L-lysine was incorporated at position 73 using an expanded genetic code system in *E. coli*. The AlaRS K73Ac variant showed low activity compared to the AlaRS wild type (WT). Nicotinamide treatment or CobB-deletion in an *E. coli* led to elevated acetylation levels of AlaRS K73Ac and strongly reduced alanylation activities. We assumed that alanylation by AlaRS is affected by K73 acetylation, and the modification is sensitive to CobB deacetylase in vivo. We also showed that *E. coli* expresses two CobB isoforms (CobB-L and CobB-S) in vivo. CobB-S displayed the deacetylase activity of the AlaRS K73Ac variant in vitro. Our results imply a potential regulatory role for lysine acetylation in controlling the activity of aaRSs and protein synthesis.

Keywords: alanyl-tRNA synthetase; class II aminoacyl-tRNA synthetase; expanded genetic code; lysine acetylation; posttranslational modification

1. Introduction

The recent development of mass spectrometry (MS)-based proteomic analysis enables the discovery of a wide variety of posttranslational modifications (PTMs) in prokaryotes [1]. Acetylation of lysine residues in proteins is distributed across all domains of life and is thought to be functionally important. Proteomic studies of acetylated protein (acetylome) indicated a large number of acetylated proteins in bacteria and archaea [2–7]. Since they also possess homologs of eukaryotic lysine acetyltransferases (KATs) and deacetylases (KDACs) [8], acetylation in bacteria and archaea are expected to be involved in the regulation of cellular processes as in eukaryotes. Recently, there has been growing interest in the role of acetylation [9–14]; however, the regulatory mechanisms and functional implications of prokaryotic acetylation are still largely unknown.

In *Escherichia coli*, acetylation is introduced to lysine residues by a KAT-dependent reaction utilizing acetyl-CoA and a non-enzymatic reaction utilizing acetyl-phosphate [15,16]. The non-enzymatic system is involved in global protein acetylation under carbon stress [17]. Moreover, the sirtuin-type KDAC

CobB is responsible for deacetylation of many proteins [18]. Acetylome analyses in *E. coli* have identified hundreds of acetylated proteins [7,15–17,19,20]; the most abundant group was metabolism-related proteins followed by translation-related proteins. Many of the translation-related proteins interact with RNA and nucleotides for their function. As the positive charge of lysine residues in those proteins is often crucial for their interaction with RNA and nucleotides, acetylation that cancels the charged lysine may affect function. We therefore searched for lysine acetylation found in aminoacyl-tRNA synthetases (aaRSs) (Table A1), which are essential enzymes to synthesize aminoacyl-tRNA in the first step of protein translation. Acetylation has been identified in all species of aaRSs and most of the acetylation sites seem to be located away from the catalytic and substrate-binding sites. However, some acetylations were detected at lysine residues, which are important for their aminoacylation activity and are expected to control the reaction.

Recently, it has been shown that naturally occurring acetylation of lysine residues in class I aaRSs, leucyl-tRNA synthetase (LeuRS), arginyl-tRNA synthetase (ArgRS), and tyrosyl-tRNA synthetase (TyrRS) in *E. coli* downregulates their enzymatic activities by using the expanded genetic code system with N^ε-acetyl-L-lysine (AcK). It was demonstrated that acetylation of K619 and K624 in LeuRS, which are located within and next to the conserved KMSKS motif involved in ATP recognition [21], and acetylation of K809, which interacts with leucine tRNA, impaired enzymatic activity [22]. They further elucidated that acetylation of K126 and K408 in ArgRS, which are involved in ATP binding and interaction with arginine tRNA, respectively, also decreases the enzymatic activity. Acetylation of K85, K235 and K238 in TyrRS reduced the enzyme activity; K235 and K238 are conserved lysine residues of the KMSKS motif and form the catalytic domain with K85 [23]. All of the acetylations in the three aaRSs are sensitive to CobB.

Although the impact of lysine acetylation on the class I aaRSs has been shown, the impact on class II enzymes, which have structurally different catalytic domains from those of class I [21], has not been examined. In this study, we demonstrate that one of the class II aaRSs, alanyl-tRNA synthetase (AlaRS) from *E. coli*, is regulated by acetylation at lysine-73 (K73). We further provide the first evidence that *E. coli* expresses two CobB isoforms in vivo and that its short variant (CobB-S) has deacetylase activity for K73 of AlaRS.

2. Materials and Methods

2.1. General

Custom oligonucleotide synthesis was ordered from Eurofins Genomics (Tokyo, Japan). The primer sequences used in this study are described in Table A2. DNA sequencing was performed by an ABI 3130xl (Applied Biosystems, Foster City, CA, USA) in the Research Equipment Center, Tokyo University of Science. Commercially available chemicals and enzymes were purchased from Wako Pure Chemical (Osaka, Japan) and Takara Bio (Shiga, Japan), respectively, unless otherwise noted. The plasmids and *E. coli* strains used in this study are listed in Tables A3 and A4, respectively.

2.2. Plasmid Constructions

The pT5C plasmid originated from pACYC184, in which a tetracycline resistant gene (tet^R) was replaced by the T5/*lac* promoter, C-terminal His-tagged sequence, the *rrnB* terminator, and the *lacI*^{q} gene (Figure A1a). The AlaRS gene (*alaS*; ACB03816) from *E. coli* DH10B (Invitrogen, Carlsbad, CA, USA) was inserted between the *Nde*I and *Xho*I site in pT5C, resulting in pT5C-Ec-alaS. The lysine codon (AAA) at position 74 of the *alaS* gene on pT5C-Ec-alaS was substituted to an alanine (GCG) or amber stop (TAG) codon by site-directed mutagenesis, resulting in pT5C-Ec-alaS-K74A and pT5C-Ec-alaS-K74amb, respectively. The discrepancy between K73 and the codon number is due to the posttranslational removal of the first methionine (Met) in cells [24].

pKTS-AcKRS1-PylT (Figure A1b), which was used for genetic incorporation of AcK, was constructed by inserting a *Methanosarcina mazei* pyrrolysine tRNA (tRNA^{Pyl}) expression cassette from pTECH-PylT into the *Nhe*I site of pKTS-AcKRS1, which overexpresses an N^{ε}-acetyl lysyl-tRNA synthetase (AcKRS) in *E. coli* [25].

For expression and promoter analyses, the pACYC184P was constructed from pACYC184, in which *tet*^{R} was replaced by the artificial transcription terminator (BBa_B1006), C-terminal His-tagged sequence, and the *rrn*B terminator (Figure A1c). CobB open reading frames (ORFs) (*cobB* gene; ACB02313) with the 1218, 417, and 30 nucleotide upstream sequences, *cobB-L* ORF, and *cobB-S* ORF (starts from the second Met codon of *cobB-L* ORF; Figure 5b) were inserted between the *Bgl*II and *Xho*I site in pACYC184P, resulting in pACYC184P-1218-cobB, pACYC184P-417-cobB, pACYC184P-30-cobB, pACYC184P-cobB-L, and pACYC184P-cobB-S.

For purification of the N-terminal His-tagged CobB protein, *cobB-L* and *cobB-S* ORFs were inserted between the *Nde*I and *Bam*HI site of pET15b (Novagen, Madison, WI, USA), resulting in pET15b-cobB-L and pET15b-cobB-S, respectively.

2.3. Construction of CobB-Deleted Escherichia coli Strain

To construct the CobB-deleted *E. coli* (ΔCobB) strain, the *cobB* gene was removed from the *E. coli* DH10B genome by using the λ-red recombination system [26,27]. The kanamycin resistant gene was next removed by FLP–FRT recombination. The gene deletion was verified by PCR.

2.4. Protein Expression and Purification

For preparation of the C-terminal His-tagged AlaRS wild type (WT) and its K73A mutant, *E. coli* DH10B transformed with pT5C-Ec-alaS or pT5C-Ec-alaS-K74A were grown in Luria-Bertani (LB) medium (1% Trypton, 0.5% yeast extract, and 1% NaCl) supplemented with 34 μg/mL chloramphenicol at 37 °C. When the OD_{600} of the culture reached 0.6, Isopropyl-β-D-1-thiogalactopyranoside (IPTG) was added to a final concentration of 0.2 mM and the cultivation was continued for 4 h. The cell pellet was resuspended with lysis buffer (20 mM Tris-HCl pH 8.0, 300 mM NaCl, and 10 mM imidazole) supplemented with 0.2 mg/mL lysozyme and 0.2% Triton X-100, and broken by sonication. After centrifugation (13,000× *g*, 4 °C, 20 min) to remove cell debris, the supernatant was charged onto a Ni-NTA agarose (Qiagen, Hilden, Germany) column equilibrated with lysis buffer. The column was washed with wash buffer (20 mM Tris-HCl pH 8.0, 300 mM NaCl, and 50 mM imidazole), and His-tagged AlaRS was eluted with elution buffer (20 mM Tris-HCl pH 8.0, 300 mM NaCl, and 250 mM imidazole). Fractions containing a homogeneous protein enzyme were pooled and dialyzed twice in 1 L of dialysis buffer (20 mM Tris-HCl pH 8.0, 10 mM $MgCl_2$, and 50 mM KCl) followed by concentration by Amicon Ultra Ultracel-30K (Merck Millipore, Billerica, MA, USA). Finally, the enzymes were stored in 10 mM Tris-HCl pH 8.0, 5 mM $MgCl_2$, 25 mM KCl, 2 mM dithiothreitol (DTT), and 50% glycerol.

For preparation of C-terminal His-tagged AcK-incorporated AlaRS (AlaRS K73Ac), *E. coli* DH10B or its ΔCobB strain co-transformed with pKTS-AcKRS1-PylT and pT5C-Ec-alaS-K74amb were grown in LB medium supplemented with 25 μg/mL kanamycin and 34 μg/mL chloramphenicol at 37 °C. Twenty millimolar of nicotinamide (NAM; Sigma-Aldrich, St. Louis, MO, USA), which is an inhibitor of CobB deacetylase, was added if necessary. When the OD_{600} of the culture reached 0.5, IPTG and AcK (Sigma-Aldrich) were added to a final concentration of 0.2 mM and 2 mM, respectively, and the cultivation was continued for 8 h. His-tag affinity purification was performed as described above.

For preparation of N-terminal His-tagged CobB protein used for the deacetylation assay, *E. coli* BL21(DE3) (Nippon Gene, Tokyo, Japan) transformed with pET15b-cobB-S was grown in LB medium supplemented with 100 μg/mL ampicillin at 37 °C. When the OD_{600} of the culture reached 0.6, IPTG was added to a final concentration of 0.2 mM. The cells were continuously cultivated for 4 h and then harvested. His-tag affinity purification was performed as described above. The purified enzyme was dialyzed in 1 L of dialysis buffer (20 mM Tris-HCl pH 8.0 and 100 mM NaCl) twice and concentrated by Amicon Ultra Ultracel-10K (Merck Millipore). Finally, the enzyme was stored in 10 mM Tris-HCl pH 8.0, 50 mM NaCl, 2 mM DTT, and 50% glycerol.

2.5. Western Blotting

After SDS-PAGE, proteins were electroblotted onto a Hybond-ECL membrane (GE Healthcare, Buckinghamshire, UK), which was blocked in Tris-buffered saline containing 0.1% Tween-20 (TBS-T) and 5% skimmed milk. Proteins on the membrane were probed with antibody in blocking buffer and then the membrane was washed with TBS-T. Rabbit polyclonal anti-AcK antibody (ImmuneChem, Burnaby, BC, Canada) and Anti-His-tag mAb-HRP-DirecT (Medical & Biological Laboratories, Aichi, Japan) were used at a 1/3000 and 1/10,000 dilution respectively as primary antibodies. Anti-Rabbit IgG HRP conjugate (Promega, Madison, WI, USA) was used as the secondary antibody. Proteins on the membrane were probed with the primary antibody in blocking buffer and then washed with TBS-T. Proteins were detected by Immobilon Western Chemilum HRP substrate (Merck Millipore) and visualized on a LAS 4000 (GE Healthcare).

2.6. Alanylation Assay

Alanine tRNA (tRNA^{Ala}) from *E. coli* was transcribed in vitro by T7 RNA polymerase and purified by 12% polyacrylamide gel with 7 M urea. The alanylation assay was performed at 37 °C in a reaction mixture (50 µL) containing 50 mM HEPES-NaOH pH 7.4, 10 mM MgCl$_2$, 30 mM KCl, 2 mM ATP, 10 µM [U-^{14}C] Ala (164 mCi/mmol; GE Healthcare), 5 µM transcribed tRNA^{Ala}, and 5 ng/µL (50 nM) His-tagged AlaRS or its variants. At time points of 2 min, 4 min, 8 min, and 10 min, 10 µL aliquots of the reaction mixture were spotted onto Whatman 3MM filter paper (GE Healthcare) and the reaction was immediately quenched with 5% trichloroacetic acid (TCA). After washing the filter paper with 5% TCA and drying, radioactivity on the filter paper was counted by a Beckman LS 6500 Scintillation Counter (Beckman Coulter, Brea, CA, USA).

2.7. Circular Dichroism Spectrometry Analysis

The circular dichroism (CD) spectra of AlaRS and its derivatives were recorded on a J-805 CD Spectrometer (JASCO Corporation, Tokyo, Japan). AlaRS WT, K73A, and K73Ac (NAM+) were respectively prepared at a concentration of 0.1 mg/mL (1.0 µM) in a buffer containing 5 mM Tris-HCl pH 7.5 and 5 mM MgCl$_2$, and scanned from 195 nm to 265 nm with a 50 nm/min speed.

2.8. Expression and Promoter Analysis for CobB-L and CobB-S

E. coli DH10B cells harboring pACYC184P-1218-cobB, pACYC184P-417-cobB, pACYC184P-30-cobB, pACYC184P-cobB-L, or pACYC184P-cobB-S were cultivated in LB medium at 37 °C overnight. The cells were broken in lysis buffer (50 mM Tris-HCl pH 8.0, 100 mM NaCl, 1 mM EDTA, 2% Triton X-100, 1% SDS, and 10% glycerol) by sonication. The protein concentration of the lysates was determined by an XL-Bradford (APRO SCIENCE, Tokushima, Japan). Ten micrograms of protein were subjected to 15% SDS-PAGE and Western blotting analysis using Anti-His-tag mAb-HRP-DirecT to detect the His-tagged CobB expression.

2.9. Deacetylation Assay

The assay was performed at 37 °C for 12 h in a reaction mixture (30 µL) containing 25 mM Tris-HCl pH 8.0, 100 mM NaCl, 27 mM KCl, 0.5 mM NAD$^+$ (Nacalai, Kyoto, Japan), 0.5 µM AlaRS K73Ac (ΔCobB), and 1.5 or 5 µM His-tagged CobB-S. The reaction was stopped by adding an equal volume of 2× sample buffer for SDS-PAGE. The resulting samples containing 0.2 pmol of AlaRS were separated by 7.5% SDS-PAGE and acetylation levels of the AlaRS K73Ac (ΔCobB) were determined by Western blotting analysis using anti-AcK antibody.

3. Results

3.1. Preparation of AlaRS K73Ac Variant by Expanded Genetic Code

The previously reported *E. coli* acetylome studies identified 15 lysine acetylation sites in AlaRS (Table A1). Of them, K73 is an essential residue in motif II of the enzyme active site (Figure 1a) [28] and interacts with the 3′-end of the cognate tRNA^{Ala} [29,30]. In addition, this residue is well-conserved among several bacteria and eukaryotes (Figure 1b). We therefore investigated the impact of acetylation at K73 of AlaRS using the genetic incorporation of AcK into position 73 of AlaRS. The expanded genetic code system is a technology to reassign non-canonical amino acids to one of the stop codons (usually the amber codon) by pairs of engineered aaRS and its cognate suppressor tRNA [31]. One such pair, AcKRS/tRNA^{Pyl} from *M. mazei*, can co-translationally introduce AcK at the amber codon, resulting in production of a protein with AcK at the specified sites [25]. Since the protein homogenously prepared by this system contains AcK, the protein with AcK is suitable to examine the impact of lysine acetylation.

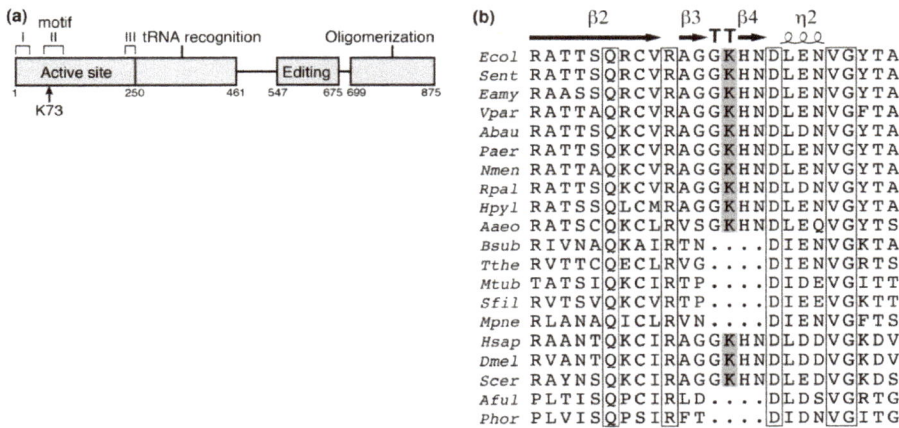

Figure 1. Schematic position of lysine-73 (K73) acetylation site in alanyl-tRNA synthetase (AlaRS). (**a**) Schematic representation of AlaRS with functional domains. K73 is located within the motif II, an essential constituent for the active site. (**b**) The alignment of sequences surrounding K73 in AlaRS. Parentheses indicate accession number. *Ecol*, *Escherichia coli* (ACB03816); *Sent*, *Salmonella enterica* (WP_065680553); *Eamy*, *Erwinia amylovora* (WP_004155915); *Vpar*, *Vibrio parahaemolyticus* (WP_086591229); *Abau*, *Acinetobacter baumannii* (ACC56466); *Paer*, *Pseudomonas aeruginosa* (WP_034012995); *Nmen*, *Neisseria meningitidis* (WP_010980961); *Rpal*, *Rhodopseudomonas palustris* (WP_044410751); *Hpyl*, *Helicobacter pylori* (WP_000354743); *Aaeo*, *Aquifex aeolicus* (WP_010880825); *Bsub*, *Bacillus subtilis* (WP_086344136); *Tthe*, *Thermus thermophilus* (WP_011228945); *Mtub*, *Mycobacterium tuberculosis* (WP_055366958); *Sfil*, *Streptomyces filamentosus* (WP_006123197); *Mpne*, *Mycoplasma pneumoniae* (WP_014574975); *Hsap*, *Homo sapiens* (NP_001596); *Dmel*, *Drosophila melanogaster* (NP_523511); *Scer*, *Saccharomyces cerevisiae* (EDV10897); *Aful*, *Archaeoglobus fulgidus* (WP_010879744); *Phor*, *Pyrococcus horikoshii* (WP_010884393); whole sequences were aligned by ClustalW [32] and ESPript 3 [33]. K73 in *E. coli* AlaRS is shown with the gray box and lies on the loop in the β-turn (TT). Strictly conserved residues among all organisms are shown with open boxes.

We prepared the AlaRS K73Ac variant using the AcKRS/tRNA^{Pyl} system and verified AcK-incorporation in AlaRS by Western blotting with the anti-AcK antibody (Figure 2a). Specific incorporation of AcK into position 73 of AlaRS was also verified by MALDI-TOF/TOF analysis (Figure A2). It has been reported that CobB is responsible for deacetylation against many proteins in *E. coli* [18]. In order to know whether acetylation of K73 was reversed by CobB, we examined the acetylation levels of AlaRS K73Ac purified from NAM-treated (NAM+) and untreated (NAM−) *E. coli*

DH10B. NAM is an inhibitor of sirtuin-type lysine deacetylases including CobB. AlaRS K73Ac (NAM+) showed higher acetylation signals than AlaRS K73Ac (NAM−), although AlaRS WT and its K73A mutant as negative controls did not give a signal of acetylation. The observation that acetylation of the AlaRS WT and K73A mutant could not be detected by Western blotting is thought to be due to the amount of the overexpressed enzymes being excess to a capability of endogenous lysine acetylation, resulting in reducing populations of acetylated AlaRS. We next examined the acetylation level of K73Ac purified from the ΔCobB strain. The acetylation level of K73Ac (ΔCobB) was almost the same as that of K73Ac (NAM+). Quantification of the band intensity showed that the acetylation levels of K73Ac (NAM+) and K73Ac (ΔCobB) were 3.3-fold and 4.0-fold higher than that of K73Ac (NAM−), respectively (Figure 2b). These results indicated that NAM treatment and CobB deletion enhanced the acetylation of AlaRS, suggesting that CobB can deacetylate K73Ac in vivo. It should be noted that lysine acetylation was not detected in the AlaRS WT and K73A mutant purified from the ΔCobB strain (Figure 2c). Taken together, these results support that the acetylation signal of the Western blot represented the K73 acetylation and CobB was responsible for the decreased acetylation of K73Ac.

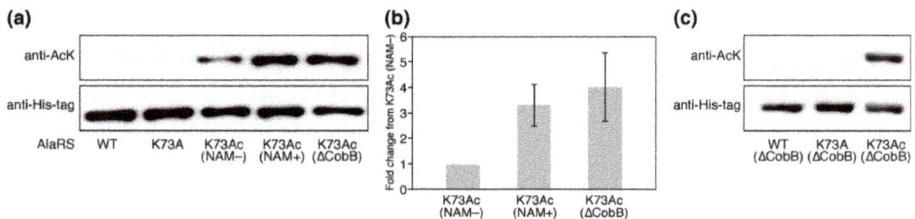

Figure 2. The verification of K73Ac-incorporation in AlaRS by Western blotting. (**a**) The acetylation levels of AlaRS variants were detected by Western blotting analysis using an anti- N^{ε}-acetyl-L-lysine (AcK) antibody. Western blotting analysis using an anti-His-tag antibody showed the loaded protein control. NAM indicates nicotinamide. The data is representative of three independent SDS-PAGE and Western blotting assays. (**b**) Quantification of the band intensities of three independent Western blotting assays. The relative acetylation levels of K73Ac (NAM+) and K73Ac (ΔCobB) to that of K73Ac (NAM−) were calculated. (**c**) Acetylation levels of the AlaRS wild-type (WT) and K73A mutant that were expressed under the same conditions as the K73Ac (ΔCobB) variant.

3.2. Alanylation Activity of AlaRS K73Ac Variants

Given that the K73 in AlaRS was important for the alanylation activity [29,30], we inferred that acetylation of the residue should influence the activity and consequently downregulate protein translation. To examine the activity of a series of AlaRS K73 variants, we performed alanylation assays in vitro (Figure 3). We confirmed that the K73A mutation abolished the alanylation activity of AlaRS, as reported previously [30]. AlaRS K73Ac (NAM−) showed decreased alanylation activity compared to the AlaRS WT. The alanylation activities of K73Ac (ΔCobB) and K73Ac (NAM+) variants, which showed higher levels of acetylation than K73Ac (NAM−), were more impaired. Considering that the acetylation signals reflect only K73 acetylation, these results indicated that K73 acetylation inhibits the alanylation activity of AlaRS.

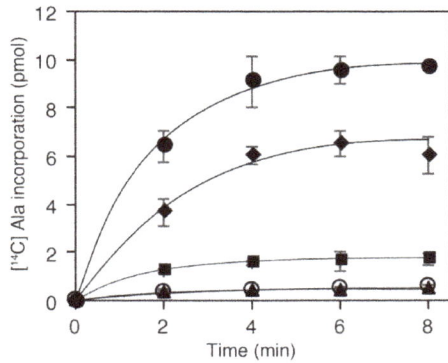

Figure 3. Alanylation activities of *E. coli* AlaRS K73Ac variants. In vitro transcribed *E. coli* tRNA^{Ala} was alanylated with AlaRS WT (filled circles), K73A (opened circles), K73Ac (NAM−; diamonds), K73Ac (NAM+; squares), and K73Ac (ΔCobB; triangles). The data is shown as the average of three independent assays with the standard deviations.

3.3. Circular Dichroism Spectrum of AlaRS K73Ac Variants

To exclude the possibility of a structural defect caused by genetic incorporation of AcK, we compared CD spectra of AlaRS WT, K73A and K73Ac (NAM+). Almost the same spectral curves were observed among the three proteins (Figure 4). The proportion of secondary structures contained in those proteins was calculated by the CAPITO web server [34]. WT, K73A and K73Ac (NAM+) contained 21%, 13%, and 22% α-helix, and 22%, 24%, and 29% β-sheet, respectively. This suggested that incorporation of AcK did not impair the protein structure of AlaRS.

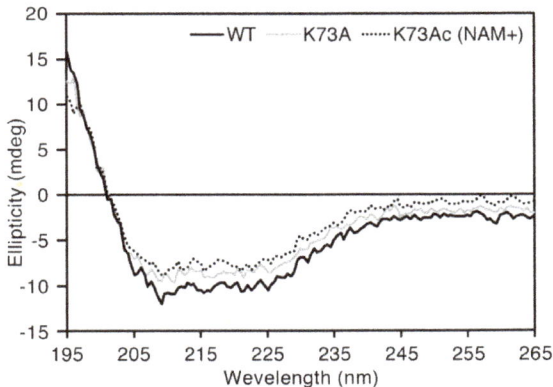

Figure 4. Circular dichroism (CD) spectral analysis. The black, gray, and dotted lines indicate AlaRS WT, K73A, and K73Ac (NAM+), respectively. Scanning was performed three times for each sample and the average was plotted.

3.4. Escherichia coli Expresses Two CobB Isoforms with Each Promoter

It has been reported that there are long and short isoforms of CobB (CobB-L and CobB-S, respectively) in *Salmonella enterica*, both of which are biologically active in the cell [35]. The existence of the two CobB isoforms has not been experimentally confirmed in *E. coli* yet. The *cobB* ORF in the public *E. coli* database encodes a 279-amino acid isoform (CobB-L), while a shorter 242-amino acid isoform (CobB-S) can be translated from the 112 nt-downstream second Met codon in the ORF. CobB-L contains a 37-amino acid N-terminal extension comprised of 15 basic and 13 hydrophobic residues, similar to CobB-L of *S. enterica* (Figure 5a). We expressed a C-terminal His-tagged CobB ORF with different upstream regions from pACYC184P plasmids in *E. coli* DH10B (Figure 5b). We detected two His-tagged protein bands by Western blotting, which corresponded to CobB-L and CobB-S from pACYC184P-1218-cobB and pACYC184P-417-cobB plasmids (Figure 5c). Meanwhile, CobB-S was only expressed from pACYC184P-30-cobB and pACYC184P-cobB-L. Neither form of CobB was detected with pACYC184P-cobB-S. These results indicated that the promoter for CobB-L expression existed between −417 and −30 within the *nagK* ORF and that for CobB-S expression existed between +1 and +111 within the *cobB* ORF. Using the online BPROM web tool (Softberry, Inc., Mount Kisco, NY, USA), we detected the candidates of the −10 and −35 consensus sequences between −347 and −319 for CobB-L and between +51 and +83 for CobB-S expression, respectively (Figure 5d). These results indicated that two isoforms of CobB-L and CobB-S exist in *E. coli* and they are directed from each promoter.

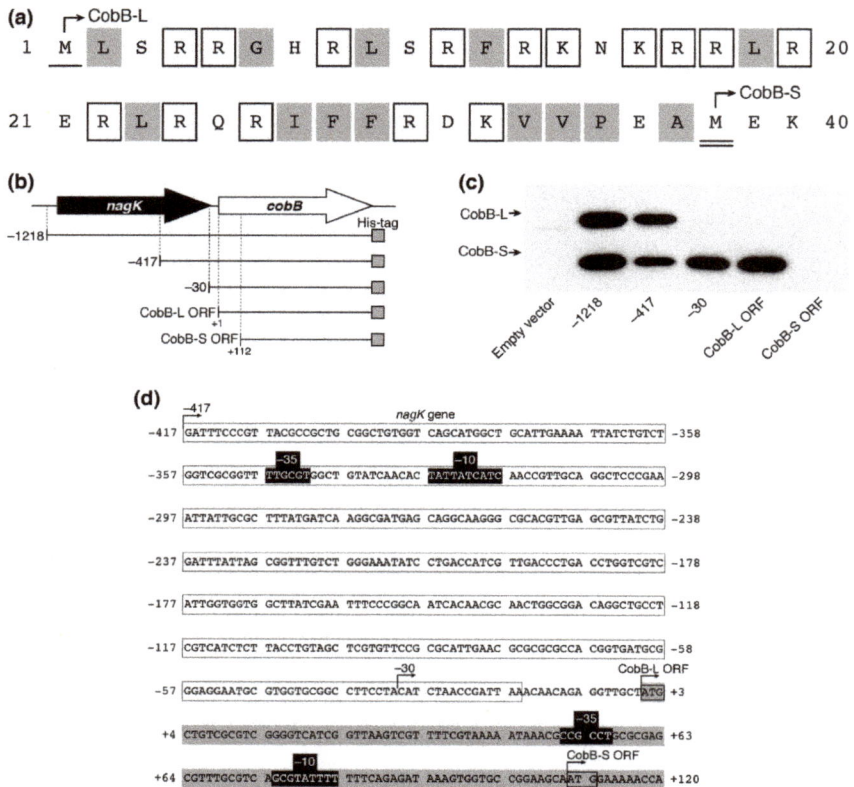

Figure 5. Expression of two CobB isoforms (CobB-L and CobB-S) in *E. coli*. (**a**) The primary sequence of N-terminal extension in CobB-L. The single and double underlines show the first methionine (Met) of CobB-L and CobB-S, respectively. The open and gray boxes indicate basic and hydrophobic amino acids, respectively (without Met). (**b**) Schematic representation of *cobB*-containing fragments in pACYC184P plasmid derivatives. (**c**) Expression of CobB isoforms in *E. coli*. Ten micrograms of lysate from *E. coli* DH10B harboring the pACYC184P derivatives was separated by 15% SDS-PAGE and analyzed by Western blotting using an anti-His-tag antibody to detect His-tagged CobB protein. (**d**) The nucleotide sequence of the *cobB* region. Open and gray boxes show the 3′-region of *nagK* and 5′-region of the *cobB* gene, respectively. The promoter containing −10 and −35 (black boxes) for CobB-L and CobB-S was predicted by BPROM of the Softberry web tool.

3.5. Deacetylation of K73 by CobB-S In Vitro

We attempted to produce N-terminal His-tagged CobB isoform proteins for deacetylation assays in vitro, but CobB-L could not be successfully obtained as soluble protein after several attempts. CobB-S contains the conserved Sir2 catalytic core and has been shown to have the deacetylase activity for a histone H4 acetylated peptide [36]. We therefore performed an in vitro deacetylation assay with CobB-S. After incubation of the His-tagged CobB-S and AlaRS K73Ac (ΔCobB) for 12 h, the residual acetylation was determined by Western blotting (Figure 6a). An NAD⁺-dependent decrease in acetylation was observed and it was enhanced by increasing concentrations of CobB-S; 27% and 39% decreases in acetylation were observed when 1.5 µM or 5.0 µM of CobB-S were used, respectively (Figure 6b). The deacetylase activity of CobB-S for AlaRS K73Ac seemed to be moderate, suggesting that CobB may require additional factors for full activity.

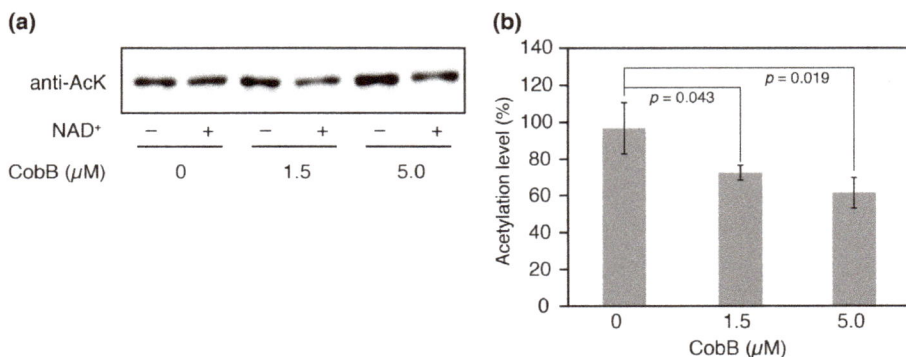

Figure 6. Deacetylation assay in vitro. (**a**) His-tagged AlaRS K73Ac (ΔCobB) (0.5 µM) and indicated concentrations of His-tagged CobB-S were reacted at 37 °C for 12 h. After the reaction, the remaining acetylation in AlaRS was determined by Western blotting with an anti-AcK antibody. This experiment was repeated three times. (**b**) The proportion of acetylation levels was calculated from the ratio of band intensities with NAD⁺ to those without NAD⁺ (negative control). The data is shown as the result of three independent reactions with the standard deviations. The p values were determined by Student's t-test.

4. Discussion

Serine phosphorylation in glutamyl-tRNA synthetase [37], lysine succinylation in methionyl-tRNA synthetase (MetRS) [38], and lysine acetylation in LeuRS, ArgRS [22] and TyrRS [23] have been reported as meaningful PTMs toward the aaRS function in *E. coli*. These modifications, with the exception of lysine succinylation in MetRS, downregulate their tRNA synthetase activities. The lysine succinylation of MetRS decreases the discrimination of cognate tRNA under stress conditions [38]. In this study, we newly found K73 acetylation that downregulates the activity of *E. coli* AlaRS which belongs to class II aaRSs.

K73 in AlaRS is known to be an essential residue that interacts with the 3′-end of tRNA^Ala [29]. Substitution of this residue to glutamine, asparagine, alanine or glutamate significantly reduces the alanylation activity but not the alanine activation, suggesting that the positive charge of the lysine plays an important role in the interaction [30]. The recently described crystal structure of AlaRS with the cognate tRNA^Ala from *Archaeoglobus fulgidus* elucidated the recognition mechanism of tRNA^Ala by AlaRS [39]. Although *A. fulgidus* AlaRS lacks the corresponding lysine residue in the primary sequence, the cytidine-75 in the tRNA^Ala of the complex is located in the vicinity of K73 of the *E. coli* enzyme when its structure is superposed on the complex structure. Based on the above-mentioned evidence, our results suggest that downregulation of AlaRS activity by K73 acetylation is due to the loss of the positive charge in the lysine residue, resulting in interference of the interaction between AlaRS and tRNA^{Ala}.

Our in vivo and in vitro studies indicate that the acetylation of K73 is reversed by CobB deacetylase. Systematic analysis to search for substrates of CobB has shown that CobB does not recognize a specific sequence, but tends to prefer aspartate (D), glutamate (E), alanine (A), glycine (G) and tyrosine (Y) on β-sheet or loop structures adjacent to an acetylated lysine [18]. The K73 exists in the AGGKHND sequence of the β-turn, which contains the favored amino acids of CobB (shown with underline). This observation supports our result that the acetylated K73 is a substrate of CobB. Our in vitro study showed the deacetylase activity of CobB-S for AlaRS K73Ac, but it seemed to be moderate. We consider the possibility that CobB-L may prefer to remove this acetylation or another factor like a chaperone might be required for the reaction.

We clarified the CobB-dependent deacetylation mechanism of K73 in AlaRS in vivo, while the mechanism of K73 acetylation remains elusive. The K73 acetylation has been detected at the stationary phase of growth without glucose [7,16] and at the early stationary phase when glucose is present [17]. In both cases, the intracellular level of acetyl-phosphate is likely increased by carbon overflow in the central metabolism pathway [16,40]. Thus, the acetylation of K73 is thought to be introduced by a non-enzymatic mechanism with acetyl-phosphate. This K73 acetylation has also been detected in *Vibrio parahaemolyticus*, mice, and humans [41,42]. The conserved K73 acetylation among different organisms may be of physiological significance, although it remains to be experimentally determined in other organisms.

Promoter analysis for CobB expression in *E. coli* was previously performed using an enhanced green fluorescent protein (EGFP)-based reporter system and real time quantitative PCR (RT-qPCR), showing that promoter activity for a *nagK-cobB* transcription was detected in the 5′-upstream region (−300 to +50) of *nagK* but no significant activity was detected in the upstream region (−400 to +151) of *cobB* [43]. Contrary to a previous report, we could find two promoter activities within the region of −417 to +111 of *cobB*: the one within *nagK* directs CobB-L expression and the other within *cobB* directs CobB-S expression (Figure 5d). These findings are similar to the case of *S. enterica*, in which the upstream and downstream promoters correspond to P2 and P3 promoters, respectively [35]. Since CobB-S contains the conserved catalytic core which is sufficient for the deacetylase activity [35,36], the extended N-terminal region composed of positively charged and hydrophobic amino acids may add an alternative function to CobB-L. It was proposed in *S. enterica* that the extended N-terminal region may be involved in interaction with other molecules such as nucleic acids and acetylated protein substrates, and cellular localization [35]. Our failure to purify N-terminal His-tagged CobB-L might be due to masking of the N-terminal His-tag by unknown interacting partners. Elucidation of the physiological significance of CobB-L and CobB-S is an intriguing challenge for the future studies.

As aaRSs are essential enzymes, severe inhibition of their activities causes growth retardation and cell death [44]. Therefore, it is thought that downregulation of AlaRS activity by K73 acetylation is likely transient and AlaRS should be rapidly reactivated by deacetylation when growth conditions are improved. Hence, our findings will provide insight into a new regulatory mechanism of translation by post-translational lysine acetylation of aaRSs. Future studies will shed light on the acetylation introduced into other aaRSs for a more comprehensive understanding of translation regulation.

Author Contributions: T.U. conceived and designed research; T.U. and S.K. performed research; T.U., S.K., D.S. and K.T. analyzed data; T.U. and S.K. wrote the paper; D.S. and K.T. supervised the work.

Funding: This research was supported by the Grants-in-Aid for Scientific Research (KAKENHI; Grant No. 15K18469 to T.U.) from the Ministry of Education, Science, Sports and Culture (MEXT), Japan. During a stay at Yale University, T.U. was supported by NIH grant GM22854 to D.S.

Acknowledgments: We thank Haruichi Asahara (New Engand BioLabs) for providing the plasmid containing the *E. coli* alanine tRNA gene for in vitro transcription, and Akiko Soma (Chiba University) for providing the plasmids pKD46, pKD13, and pCP20 to construct the *E. coli* knockout strain. We thank Takashi Nirasawa (Bruker, Japan) for the mass spectrometry analysis. We also thank Ryusaku Nemoto and Saki Kayamori for technical assistance.

Conflicts of Interest: The authors declare no conflict of interest.

Appendix

Table A1. Lysine acetylation sites in all aminoacyl-tRNA synthetases (aaRSs) from *E. coli*. The data was collected from the acetylomes performed by Kuhn et al. [16] and Schilling et al. [17].

aaRS	Acetyl Site	[16] Mascot Score	[17] Mascot Score	aaRS	Acetyl Site	[16] Mascot Score	[17] Mascot Score
AlaRS	55	45.4	58.7	GlnRS	160		71.2
	73	55.3	68.5		170	95.0	49.3
	278		85.2		195		93.0
	535	39.8	140.6		273	51.4	50.1
	591	53.8	71.2		399		41.8
	604		51.7		408	53.5	41.7
	647	70.8			419	57.9	
	721		59.0		515	34.3	65.4
	730	64.5	102.7	GluRS	102	41.1	
	743	57.6	97.9		215	50.9	65.0
	749	59.2	74.3		293		43.9
	763		53.9		345	107.9	77.7
	783	86.0	94.0		377	44.5	60.8
	819	54.8	101.2		423	70.4	
	873		51.8		459		44.3
ArgRS	112	53.4		GlyRS-α	3	38.1	42.5
	126	83.3	103.8		281	41.3	
	152	73.6		GlyRS-β	72	53.1	110.1
	198		50.6		86	72.4	73.5
	216	45.8	40.6		89		58.6
	295		44.5		112	75.6	
	306		54.3		141		54.1
	473		64.2		472		43.1
AsnRS	194		54.1		475	53.5	73.4
	199	51.9	42.3		592	55.1	99.6
	286	33.7	75.5		607	76.1	47.0
	294	53.1	67.1		624	43.5	59.1
	351	52.0	58.1		662	63.1	71.2
	357		66.1		674	52.4	46.3
	375		41.2	HisRS	53	51.5	72.0
AspRS	186	49.8	56.5		189		74.2
	283	65.2	80.6		370		47.3
	315		55.7		378	44.3	49.4
	332	57.1	54.9	IleRS	23	57.3	98.3
	353	58.5	78.1		112	61.9	82.3
	370	48.5			116	83.0	106.4
	399	35.8			183	56.6	66.6
	415	74.3	95.2		390		109.4
	513		73.1		398	90.2	96.7
CysRS	73	54.3	97.1		428	76.6	107.6
	76	50.8	90.9		436	57.1	58.1
	175	44.4			439	55.7	52.9
	282	53.4	59.4		605	49.1	
	310		48.4		619	108.3	104.3
					895		78.7
					913		50.0
					934	64.1	93.2

Table A1. *Cont.*

aaRS	Acetyl Site	[16] Mascot Score	[17] Mascot Score	aaRS	Acetyl Site	[16] Mascot Score	[17] Mascot Score
LeuRS	21	56.1	74.8	SerRS	17		54.2
	145	47.3			24	73.7	109.8
	182		54.7		29	49.6	102.6
	598	51.8	62.4		43	53.6	97.1
	619	53.2	81.4		77		42.2
	837		45.6		86		83.1
LysRS	82	112.9	109.9		115	68.7	
	114	88.0	123.2	ThrRS	122	87.8	95.8
	115	56.7	108.4		169	61.0	76.5
	156	57.9	132.8		200	52.3	
	335		91.8		226	61.8	76.3
LysRS (Inducible)	114	52.2	109.1		227	56.3	86.8
	115	52.0	109.2		286	56.9	
	156	119.0	116.1		314		68.6
	179	57.2	60.0		577	52.4	44.4
	322	76.7	95.4		599	52.6	58.6
	326		84.6		614	55.2	42.2
	335	37.9	103.6	TrpRS	173		47.2
MetRS	115		38.2		181	55.2	66.8
	121	48.6			214	57.6	76.6
	133	77.0	85.4		274		40.2
	143		38.2		307		46.9
	343	98.3	140.0	TyrRS	8	62.0	109.0
	403	50.0	54.8		85	76.7	131.3
	466		66.5		144	40.1	105.3
	498		85.4		235	55.2	62.5
	529	68.3	64.3		238	77.6	109.4
	597	68.1	124.5		416	55.1	
PheRS-α	34	47.1		ValRS	3		73.5
	75		71.0		178	52.7	45.9
	265	82.8			209		45.4
	324	31.9			248		35.2
PheRS-β	66	56.9			271	48.1	65.9
	104	70.9	86.8		291	52.9	101.4
	235		55.1		317	116.6	117.1
	371	66.6	85.0		334	70.9	68.6
	441	61.8	57.5		358		50.2
	496	42.1			593	65.9	96.3
	655	58.6	126.3		600		79.6
	780		42.9		633		41.2
ProRS	34	76.2	45.0		644	53.6	56.8
	126	78.2	91.0		861	76.9	110.1
	420		80.7		898		70.4
	495	41.8			909	108.8	104.5
	554	56.7	57.0		926	44.9	76.2
	568		78.9		930	48.3	80.0

Table A2. List of primers for PCR.

Name	Sequence	Purpose
pT5C-Ec-alaS-F	AGAGGAGAAATTACATATGAGCAAGAGCACCGCTG	Construction of pT5C-Ec-alaS
pT5C-Ec-alaS-R	TGGTGGTGGTGGTGCTCGAGTGCGGCTTGCAATTTC	
Ec-alaS-K74A-F	CTGCGTGCGTGCGGGTGGTGCGCACAACGACCTGGAAAAC	Construction of pT5C-Ec-alaS-K74A
Ec-alaS-K74A-R	GTTTTCCAGGTCGTTGTGCGCACCACCCGCACGCACGCAG	
Ec-alaS-K74amb-F	CTGCGTGCGTGCGGGTGGTTAGCACAACGACCTGGAAAAC	Construction of pT5C-Ec-alaS-K74amb
Ec-alaS-K74amb-R	GTTTTCCAGGTCGTTGTGCTAACCACCCGCACGCACGCAG	
pTECH-1960-Nhe-F	GGAAGCTAGCCTGTTGCCCGTCTCACTGGTGAAAAG	Construction of pKTS-AcKRS1-PylT
pTECH-1683-Nhe-R	GGAAGCTAGCCTGGCGAAAGGGGGATGTGCTG	
cobB-Nde-F	GTCACATATGCTGTCGCGTCGGGGTCATC	Construction of pET15b-cobB-L and pET15b-cobB-S
cobB-S-Nde-F	TGCCATATGGAAAAACCAAGAGTACTCGTACTGAC	
cobB-Bam-R	GACTGGATCCTTAGGCAATGCTTCCCGCTTTTAATC	
Ec-cobB-KO-F	GTGGTGCGGCCTTCCTACATCTAACCGATTAAACAACAGAG GTTGCTATGATTCCGGGGATCCGTCGACC	Disruption of *cobB* gene
Ec-cobB-KO-R	CCCCTTGCAGGCCTGATAAGCGTAGTGCATCAGGCAATGCT TCCCGCTTTTGTAGGCTGGAGCTGCTTCG	
cobB-1218-BglII-F	AGCAGATCTGTTCGGCAGCCTGTGTGGTGTG	Construction of pACYC184P-1218-cobB, -417-cobB, -30-cobB, -cobB-L, and -cobB-S
cobB-417-BglII-F	AGCAGATCTGATTTCCCGTTACGCCGCTG	
cobB-30-BglII-F	AGCAGATCTCATCTAACCGATTAAACAACAGAGGTTGC	
cobB-L-BglII-F	CAGAGATCTATGCTGTCGCGTCGGGGTC	
cobB-S-BglII-F	CAGAGATCTATGGAAAAACCAAGAGTACTCGTACTG	
cobB-XhoI-R	GCTCTCGAGGGCAATGCTTCCCGCTTTTAATCC	

Table A3. List of plasmids in this study. nts, nucleotides.

Plasmids	Description
pT5C-Ec-alaS	AlaRS wild type expression
pT5C-Ec-alaS-K74A	AlaRS K73A expression
pT5C-Ec-alaS-K74amb	AlaRS K73Ac expression
pKTS-AcKRS1-PylT	AcK incorporation system
pACYC184P	Empty vector for CobB expression analysis
pACYC184P-1218-cobB	*cobB* gene with 1218 nts upstream sequence
pACYC184P-417-cobB	*cobB* gene with 417 nts upstream sequence
pACYC184P-30-cobB	*cobB* gene with 30 nts upstream sequence
pACYC184P-cobB-L	*cobB* gene (long length) ORF
pACYC184P-cobB-S	*cobB* gene (short form) ORF
pET15b-cobB-L	CobB-L expression
pET15b-cobB-S	CobB-S expression

Table A4. *E. coli* strains in this study.

E. coli Strains	Genotype	Source
DH10B	F^-, *mcrA*, Δ(*mrr-hsdRMS-mcrBC*), *φ80lacZΔM15*, *ΔlacX74*, *recA1*, *endA1*, *araD139*, Δ(*ara-leu*)7697, *galU*, *galK*, λ^-, *rpsL*(Str^R), *nupG*	Invitrogen
BL21(DE3)	F^-, *dcm*, *ompT*, *hsdS*(r_B^- m_B^-), *gal*, *λ(DE3)*	Nippon Gene
ΔCobB	DH10B *cobB::frt*	This study

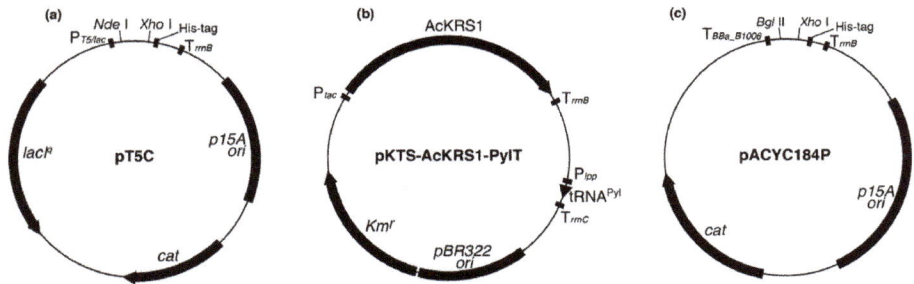

Figure A1. Plasmid maps of (**a**) pT5C, (**b**) pKTS-AcKRS1-PylT, and (**c**) pACYC184P. $P_{T5/lac}$, P_{lpp}, and P_{tac} indicate promoter. T_{rrnB} and T_{rrnC}, and T_{BBa_B1006} indicate transcription terminator. The artificial terminator BBa_B1006 blocks intrinsic promoter activity.

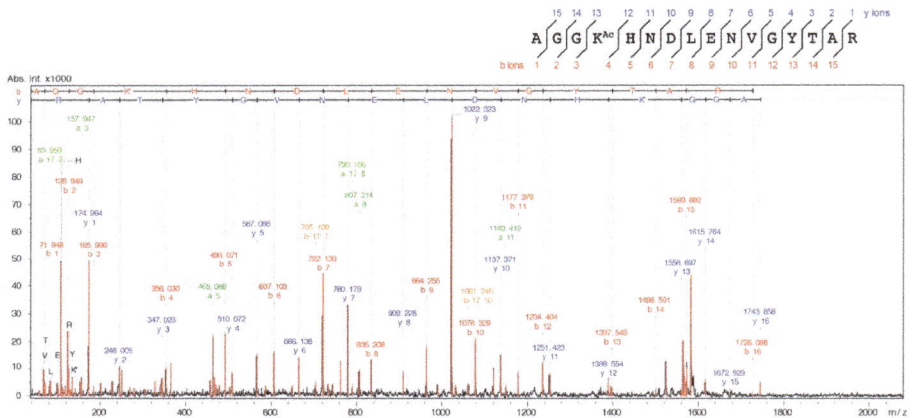

Figure A2. MS/MS ion spectrum of the K73Ac-acetylated peptide (71-AGGK^{Ac}HNDLENVGYTAR-86, *m/z* 1743.839). AlaRS K73Ac (ΔCobB) was reduced with 20 mM dithiothreitol (DTT) and subsequently alkylated with 30 mM iodoacetamide and digested with a sequence-grade trypsin (Promega). The proteolytic peptides were cleaned using a ZipTip C18 column (Merck Millipore) and analyzed by an autoflex speed TOF/TOF mass spectrometer (Bruker, Billerica, MA, USA).

References

1. Cain, J.A.; Solis, N.; Cordwell, S.J. Beyond gene expression: The impact of protein post-translational modifications in bacteria. *J. Proteom.* **2014**, *97*, 265–286. [CrossRef] [PubMed]

2. Ouidir, T.; Kentache, T.; Hardouin, J. Protein lysine acetylation in bacteria: Current state of the art. *Proteomics* **2016**, *16*, 301–309. [CrossRef] [PubMed]

3. Liu, L.; Wang, G.; Song, L.; Lv, B.; Liang, W. Acetylome analysis reveals the involvement of lysine acetylation in biosynthesis of antibiotics in *Bacillus amyloliquefaciens*. *Sci. Rep.* **2016**, *6*, 20108. [CrossRef] [PubMed]

4. Liu, J.; Wang, Q.; Jiang, X.; Yang, H.; Zhao, D.; Han, J.; Luo, Y.; Xiang, H. Systematic analysis of lysine acetylation in the halophilic archaeon *Haloferax mediterr*. *J. Proteom. Res.* **2017**, *16*, 3229–3241. [CrossRef] [PubMed]

5. Kosono, S.; Tamura, M.; Suzuki, S.; Kawamura, Y.; Yoshida, A.; Nishiyama, M.; Yoshida, M. Changes in the acetylome and succinylome of *Bacillus subtilis* in response to carbon source. *PLoS ONE* **2015**, *10*, e0131169. [CrossRef] [PubMed]

6. Kentache, T.; Jouenne, T.; De, E.; Hardouin, J. Proteomic characterization of N_α- and N_ε-acetylation in *Acinetobacter baumannii*. *J. Proteom.* **2016**, *144*, 148–158. [CrossRef] [PubMed]

7. Baeza, J.; Dowell, J.A.; Smallegan, M.J.; Fan, J.; Amador-Noguez, D.; Khan, Z.; Denu, J.M. Stoichiometry of site-specific lysine acetylation in an entire proteome. *J. Biol. Chem.* **2014**, *289*, 21326–21338. [CrossRef] [PubMed]

8. Hentchel, K.L.; Escalante-Semerena, J.C. Acylation of biomolecules in prokaryotes: A widespread strategy for the control of biological function and metabolic stress. *Microbiol. Mol. Biol. Rev.* **2015**, *79*, 321–346. [CrossRef] [PubMed]

9. Zhang, Q.F.; Gu, J.; Gong, P.; Wang, X.D.; Tu, S.; Bi, L.J.; Yu, Z.N.; Zhang, Z.P.; Cui, Z.Q.; Wei, H.P.; et al. Reversibly acetylated lysine residues play important roles in the enzymatic activity of *Escherichia coli* N-hydroxyarylamine O-acetyltransferase. *FEBS J.* **2013**, *280*, 1966–1979. [CrossRef] [PubMed]

10. Song, L.; Wang, G.; Malhotra, A.; Deutscher, M.P.; Liang, W. Reversible acetylation on Lys501 regulates the activity of RNase II. *Nucleic Acids Res.* **2016**, *44*, 1979–1988. [CrossRef] [PubMed]

11. Thao, S.; Chen, C.S.; Zhu, H.; Escalante-Semerena, J.C. N^ε-lysine acetylation of a bacterial transcription factor inhibits its DNA-binding activity. *PLoS ONE* **2010**, *5*, e15123. [CrossRef] [PubMed]

12. Liang, W.; Malhotra, A.; Deutscher, M.P. Acetylation regulates the stability of a bacterial protein: Growth stage-dependent modification of RNase R. *Mol. Cell* **2011**, *44*, 160–166. [CrossRef] [PubMed]

13. Ren, J.; Sang, Y.; Tan, Y.; Tao, J.; Ni, J.; Liu, S.; Fan, X.; Zhao, W.; Lu, J.; Wu, W.; et al. Acetylation of lysine 201 inhibits the DNA-binding ability of PhoP to regulate *Salmonella* virulence. *PLoS Pathog* **2016**, *12*, e1005458. [CrossRef] [PubMed]

14. Ramakrishnan, R.; Schuster, M.; Bourret, R.B. Acetylation at Lys-92 enhances signaling by the chemotaxis response regulator protein CheY. *Proc. Natl. Acad. Sci. USA* **1998**, *95*, 4918–4923. [CrossRef] [PubMed]

15. Weinert, B.T.; Iesmantavicius, V.; Wagner, S.A.; Scholz, C.; Gummesson, B.; Beli, P.; Nystrom, T.; Choudhary, C. Acetyl-phosphate is a critical determinant of lysine acetylation in *E. coli*. *Mol. Cell* **2013**, *51*, 265–272. [CrossRef] [PubMed]

16. Kuhn, M.L.; Zemaitaitis, B.; Hu, L.I.; Sahu, A.; Sorensen, D.; Minasov, G.; Lima, B.P.; Scholle, M.; Mrksich, M.; Anderson, W.F.; et al. Structural, kinetic and proteomic characterization of acetyl phosphate-dependent bacterial protein acetylation. *PLoS ONE* **2014**, *9*, e94816. [CrossRef] [PubMed]

17. Schilling, B.; Christensen, D.; Davis, R.; Sahu, A.K.; Hu, L.I.; Walker-Peddakotla, A.; Sorensen, D.J.; Zemaitaitis, B.; Gibson, B.W.; Wolfe, A.J. Protein acetylation dynamics in response to carbon overflow in *Escherichia coli*. *Mol. Microbiol.* **2015**, *98*, 847–863. [CrossRef] [PubMed]

18. AbouElfetouh, A.; Kuhn, M.L.; Hu, L.I.; Scholle, M.D.; Sorensen, D.J.; Sahu, A.K.; Becher, D.; Antelmann, H.; Mrksich, M.; Anderson, W.F.; et al. The *E. coli* sirtuin CobB shows no preference for enzymatic and nonenzymatic lysine acetylation substrate sites. *Microbiologyopen* **2015**, *4*, 66–83. [CrossRef] [PubMed]

19. Zhang, J.; Sprung, R.; Pei, J.; Tan, X.; Kim, S.; Zhu, H.; Liu, C.F.; Grishin, N.V.; Zhao, Y. Lysine acetylation is a highly abundant and evolutionarily conserved modification in *Escherichia coli*. *Mol. Cell. Proteom.* **2009**, *8*, 215–225. [CrossRef] [PubMed]

20. Zhang, K.; Zheng, S.; Yang, J.S.; Chen, Y.; Cheng, Z. Comprehensive profiling of protein lysine acetylation in *Escherichia coli*. *J. Proteom. Res.* **2013**, *12*, 844–851. [CrossRef] [PubMed]

21. Moras, D. Structural and functional relationships between aminoacyl-tRNA synthetases. *Trends Biochem. Sci.* **1992**, *17*, 159–164. [CrossRef]

22. Ye, Q.; Ji, Q.Q.; Yan, W.; Yang, F.; Wang, E.D. Acetylation of lysine ε-amino groups regulates aminoacyl-tRNA synthetase activity in *Escherichia coli*. *J. Biol. Chem.* **2017**, *292*, 10709–10722. [CrossRef] [PubMed]

23. Venkat, S.; Gregory, C.; Gan, Q.; Fan, C. Biochemical characterization of the lysine acetylation of tyrosyl-tRNA synthetase in *Escherichia coli*. *Chembiochem* **2017**, *18*, 1928–1934. [CrossRef] [PubMed]

24. Putney, S.D.; Sauer, R.T.; Schimmel, P.R. Purification and properties of alanine tRNA synthetase from *Escherichia coli* A tetramer of identical subunits. *J. Biol. Chem.* **1981**, *256*, 198–204. [PubMed]

25. Umehara, T.; Kim, J.; Lee, S.; Guo, L.T.; Söll, D.; Park, H.S. N-Acetyl lysyl-tRNA synthetases evolved by a CcdB-based selection possess N-acetyl lysine specificity in vitro and in vivo. *FEBS Lett.* **2012**, *586*, 729–733. [CrossRef] [PubMed]

26. Baba, T.; Ara, T.; Hasegawa, M.; Takai, Y.; Okumura, Y.; Baba, M.; Datsenko, K.A.; Tomita, M.; Wanner, B.L.; Mori, H. Construction of *Escherichia coli* K-12 in-frame, single-gene knockout mutants: The Keio collection. *Mol. Syst. Biol.* **2006**, *2*, 2006.0008. [CrossRef] [PubMed]

27. Datsenko, K.A.; Wanner, B.L. One-step inactivation of chromosomal genes in *Escherichia coli* K-12 using PCR products. *Proc. Natl. Acad. Sci. USA* **2000**, *97*, 6640–6645. [CrossRef] [PubMed]

28. Swairjo, M.A.; Otero, F.J.; Yang, X.L.; Lovato, M.A.; Skene, R.J.; McRee, D.E.; Ribas de Pouplana, L.; Schimmel, P. Alanyl-tRNA synthetase crystal structure and design for acceptor-stem recognition. *Mol. Cell* **2004**, *13*, 829–841. [CrossRef]

29. Hill, K.; Schimmel, P. Evidence that the 3′ end of a tRNA binds to a site in the adenylate synthesis domain of an aminoacyl-tRNA synthetase. *Biochemistry* **1989**, *28*, 2577–2586. [CrossRef] [PubMed]

30. Filley, S.J.; Hill, K.A. Amino acid substitutions at position 73 in motif 2 of *Escherichia coli* alanyl-tRNA synthetase. *Arch. Biochem. Biophys.* **1993**, *307*, 46–51. [CrossRef] [PubMed]

31. Dumas, A.; Lercher, L.; Spicer, C.D.; Davis, B.G. Designing logical codon reassignment—Expanding the chemistry in biology. *Chem. Sci.* **2015**, *6*, 50–69. [CrossRef] [PubMed]

32. Larkin, M.A.; Blackshields, G.; Brown, N.P.; Chenna, R.; McGettigan, P.A.; McWilliam, H.; Valentin, F.; Wallace, I.M.; Wilm, A.; Lopez, R.; et al. Clustal W and Clustal X version 2.0. *Bioinformatics* **2007**, *23*, 2947–2948. [CrossRef] [PubMed]

33. Robert, X.; Gouet, P. Deciphering key features in protein structures with the new ENDscript server. *Nucleic Acids Res.* **2014**, *42*, W320–W324. [CrossRef] [PubMed]

34. Wiedemann, C.; Bellstedt, P.; Gorlach, M. CAPITO—A web server-based analysis and plotting tool for circular dichroism data. *Bioinformatics* **2013**, *29*, 1750–1757. [CrossRef] [PubMed]

35. Tucker, A.C.; Escalante-Semerena, J.C. Biologically active isoforms of CobB sirtuin deacetylase in *Salmonella enterica* and *Erwinia amylovora*. *J. Bacteriol.* **2010**, *192*, 6200–6208. [CrossRef] [PubMed]

36. Zhao, K.; Chai, X.; Marmorstein, R. Structure and substrate binding properties of cobB, a Sir2 homolog protein deacetylase from *Escherichia coli*. *J. Mol. Biol.* **2004**, *337*, 731–741. [CrossRef] [PubMed]

37. Germain, E.; Castro-Roa, D.; Zenkin, N.; Gerdes, K. Molecular mechanism of bacterial persistence by HipA. *Mol. Cell.* **2013**, *52*, 248–254. [CrossRef] [PubMed]

38. Schwartz, M.H.; Waldbauer, J.R.; Zhang, L.; Pan, T. Global tRNA misacylation induced by anaerobiosis and antibiotic exposure broadly increases stress resistance in *Escherichia coli*. *Nucleic Acids Res.* **2016**, *44*, 10292–10303. [PubMed]

39. Naganuma, M.; Sekine, S.; Chong, Y.E.; Guo, M.; Yang, X.L.; Gamper, H.; Hou, Y.M.; Schimmel, P.; Yokoyama, S. The selective tRNA aminoacylation mechanism based on a single G•U pair. *Nature* **2014**, *510*, 507–511. [CrossRef] [PubMed]

40. Klein, A.H.; Shulla, A.; Reimann, S.A.; Keating, D.H.; Wolfe, A.J. The intracellular concentration of acetyl phosphate in *Escherichia coli* is sufficient for direct phosphorylation of two-component response regulators. *J. Bacteriol.* **2007**, *189*, 5574–5581. [CrossRef] [PubMed]

41. Pan, J.; Ye, Z.; Cheng, Z.; Peng, X.; Wen, L.; Zhao, F. Systematic analysis of the lysine acetylome in *Vibrio parahemolyticus*. *J. Proteom. Res.* **2014**, *13*, 3294–3302. [CrossRef] [PubMed]

42. Hornbeck, P.V.; Kornhauser, J.M.; Tkachev, S.; Zhang, B.; Skrzypek, E.; Murray, B.; Latham, V.; Sullivan, M. PhosphoSitePlus: A comprehensive resource for investigating the structure and function of experimentally determined post-translational modifications in man and mouse. *Nucleic Acids Res.* **2012**, *40*, D261–D270. [CrossRef] [PubMed]

43. Castano-Cerezo, S.; Bernal, V.; Blanco-Catala, J.; Iborra, J.L.; Canovas, M. cAMP-CRP co-ordinates the expression of the protein acetylation pathway with central metabolism in *Escherichia coli*. *Mol. Microbiol.* **2011**, *82*, 1110–1128. [CrossRef] [PubMed]

44. Ho, J.M.; Bakkalbasi, E.; Söll, D.; Miller, C.A. Drugging tRNA aminoacylation. *RNA Biol.* **2018**, *15*, 667–677. [CrossRef] [PubMed]

GCAT
TACG
GCAT

genes

MDPI

Review

Versatility of Synthetic tRNAs in Genetic Code Expansion

Kyle S. Hoffman [1], Ana Crnković [1] and Dieter Söll [1,2,*]

[1] Department of Molecular Biophysics and Biochemistry, Yale University, New Haven, CT 06520, USA; kyle.hoffman@yale.edu (K.S.H.); ana.crnkovic@yale.edu (A.C.)

[2] Department of Chemistry, Yale University, New Haven, CT 06520, USA

* Correspondence: dieter.soll@yale.edu; Tel.: +1-203-432-6200

Received: 16 October 2018; Accepted: 5 November 2018; Published: 7 November 2018

Abstract: Transfer RNA (tRNA) is a dynamic molecule used by all forms of life as a key component of the translation apparatus. Each tRNA is highly processed, structured, and modified, to accurately deliver amino acids to the ribosome for protein synthesis. The tRNA molecule is a critical component in synthetic biology methods for the synthesis of proteins designed to contain non-canonical amino acids (ncAAs). The multiple interactions and maturation requirements of a tRNA pose engineering challenges, but also offer tunable features. Major advances in the field of genetic code expansion have repeatedly demonstrated the central importance of suppressor tRNAs for efficient incorporation of ncAAs. Here we review the current status of two fundamentally different translation systems (TSs), selenocysteine (Sec)- and pyrrolysine (Pyl)-TSs. Idiosyncratic requirements of each of these TSs mandate how their tRNAs are adapted and dictate the techniques used to select or identify the best synthetic variants.

Keywords: genetic code expansion; transfer RNA; synthetic biology; non-canonical amino acids; selenocysteine

1. Introduction

Genetic code expansion (GCE) involves the engineering of protein synthesis machinery to site-specifically incorporate non-canonical amino acids (ncAAs) into a desired protein [1,2]. This is routinely done by assigning the ncAA to recoded stop or sense codons and delivering the ncAA to the ribosome via a suppressor transfer RNA (tRNA). The successful charging of an ncAA to the suppressor tRNA and incorporation at a defined codon requires an aminoacyl-tRNA synthetase (aaRS)•tRNA pair to function orthogonally (restricting interactions with host tRNAs, aaRSs, or canonical amino acids; Figure 1). Non-canonical amino acids endow proteins with unique chemical and physical properties that make them useful for a wide range of applications. They serve as affinity tags, imaging probes, environmental sensors, post-translational modifications, are used for protein crosslinking, conjugation, and altering pK_a or redox potential [3].

The most versatile aaRS for incorporating ncAAs is pyrrolysyl-tRNA synthetase (PylRS). Naturally, PylRS attaches pyrrolysine (Pyl), the 22nd genetically encoded amino acid, to its cognate tRNA^{Pyl}, a natural UAG suppressor. In archaea, PylRS is a single polypeptide chain; however, bacteria harbor a split protein where the C-terminal catalytic domain is only active in the presence of the N-terminal domain [4,5]. PylRS and its variants are polyspecific; to date they have facilitated the incorporation of over 100 ncAAs into proteins [6]. Moreover, PylRS•tRNA^{Pyl} pairs are used to engineer proteins with unique properties and functions in bacteria, viruses, insects, yeast, and animals [7–11].

Another valuable building block for protein engineering is the 21st amino acid, selenocysteine (Sec). Sec is a naturally occurring amino acid that resembles cysteine but has a selenol group instead of the thiol. Sec is found in the active site of redox enzymes of species that span all three domains of life, providing enhanced nucleophilic and reducing properties [12]. The site-specific incorporation of Sec can enhance enzyme activity when replacing cysteine (Cys), increase protein stability via diselenide bonds, and improve therapeutic peptides [13–15].

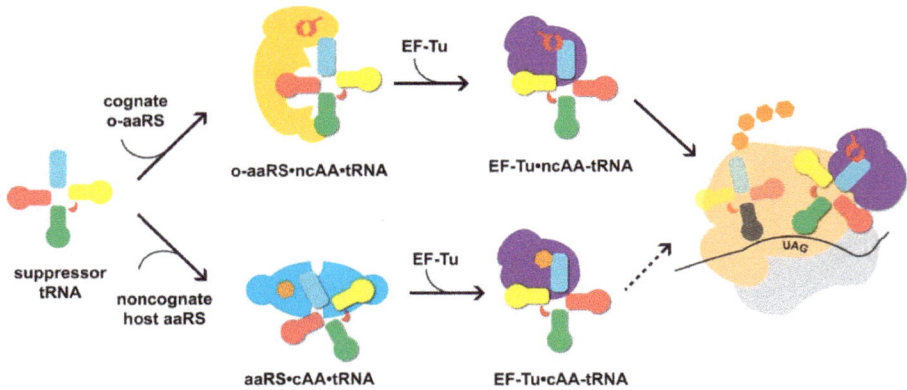

Figure 1. Suppressor transfer RNAs (tRNAs) interact with cognate orthogonal aminoacyl-tRNA synthetases (o-aaRSs) and the translational machinery of the host. For successful non-canonical amino acid (ncAA) incorporation, the suppressor tRNA needs to be recognized by its cognate o-aaRS and charged with the cognate ncAA (up). When not orthogonal, the tRNA can be erroneously recognized by an endogenous noncognate aaRS and aminoacylated with a canonical AA (cAA; down). The formation of cAA-tRNA can lead to cAA incorporation at the ribosome in response to UAG (depicted as a dotted arrow). Elements of the tRNA secondary structure are shown in light blue (acceptor stem), pink (D-arm), green (anticodon arm), red (variable loop), and yellow (T-arm). The o-aaRS is shown in yellow, noncognate, endogenous aaRS in cyan, elongation factor EF-Tu in purple, and the large and small ribosomal subunit in tan and light grey, respectively. NcAA is depicted as a red hexagonal shape, while the natural AAs are given in orange. The position of the UAG codon is indicated.

While PylRS directly ligates an ncAA onto tRNA^{Pyl}, there is no aaRS to form Sec-tRNA^{Sec}. Rather, Sec is biosynthesized in a tRNA-dependent manner (reviewed in [4]). In bacteria, this first involves the charging of serine (Ser) by seryl-tRNA synthetase (SerRS) to form Ser-tRNA^{Sec}, followed by the transfer of selenium from selenophosphate by selenocysteine synthase (SelA) for conversion to Sec-tRNA^{Sec} (Figure 2). In eukaryotes and archaea, Ser-tRNA^{Sec} is phosphorylated to form O-phosphoseryl-tRNA^{Sec} (Sep-tRNA^{Sec}) by Sep-tRNA kinase (PSTK) [16], to which the phosphate group is displaced with selenophosphate by Sep-tRNA:Sec-tRNA synthase (SepSecS) [17–19]. Sec-tRNA^{Sec} delivery to the ribosome is aided by a selenocysteine-specific elongation factor (SelB in bacteria or EFSec in eukaryotes) [20,21]. Furthermore, the Sec insertion sequence (SECIS), an RNA structure in selenoprotein mRNA, recruits the SelB/EFSec-bound Sec-tRNA^{Sec} to the ribosome for the recoding of a UGA stop codon [22,23] (Figure 2). Given the diverse set of interactions and different mechanisms for Sec incorporation versus PylRS-mediated ncAA incorporation, the task of improving each system requires very different considerations.

When refining Sec- and Pyl-orthogonal translation system (OTS) components for GCE, it is ideal to produce a high amount of the ncAA-tRNA while retaining orthogonality and limiting the effects on cellular fitness. Heterologous aaRS•tRNA pairs for the OTS of a particular host organism are often imported from a different domain of life, since tRNA identity elements and substrate recognition are dissimilar enough to function orthogonally [24]. Moreover, the malleable active site of PylRS allows straightforward directed evolution methods to identify new ncAA-activating variants; however, these variants are polyspecific [25], and mutations that decrease the orthogonality must be selected against. Selenocysteine-OTSs are often used in bacteria or mammalian cells that already have the Sec pathway. Therefore, Sec pathway components are removed to prevent interaction with the OTS. Recent work in *Escherichia coli* has focused on improving the Sec incorporation efficiency and discovering EF-Tu compatible tRNA^{Sec} variants for selenoprotein expression without the requirement for SECIS in the coding sequence [26–31].

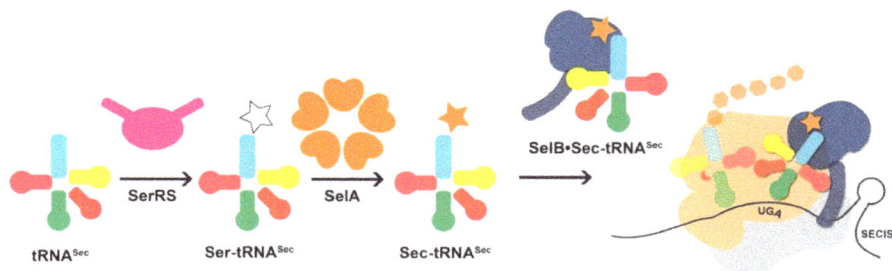

Figure 2. Idiosyncratic features of the natural Sec-incorporation pathway. tRNA^{Sec} is first misacylated with serine (white star) by seryl-tRNA synthetase (SerRS; purple). The intermediate, Ser-tRNA^{Sec} is a substrate for selenocysteine synthase (SelA) which converts the Ser moiety to Sec (orange star). Sec-tRNA^{Sec} is recognized by the Sec-specific elongation factor SelB (dark blue). In contrast to the general elongation factor EF-Tu, SelB approaches the ribosome bound to a Sec insertion sequence (SECIS), an RNA structure in its cognate mRNA. In this manner Sec-tRNA^{Sec} is directed to bind an upstream UGA codon and deliver Sec to the growing polypeptide chain.

The production of Sec-tRNA^{Sec} is naturally inefficient compared to canonical aminoacyl-tRNA formation; SerRS serylates tRNA^{Sec} 100-fold less efficiently than tRNA^{Ser} [32]. It is likely that this kinetic inefficiency of SerRS correlates with the low demand for Sec incorporation; there are a limited number of proteins requiring Sec. Thus, the most challenging aspect of Sec-OTS engineering is to achieve efficient serylation by SerRS, as well as complete conversion to Sec-tRNA^{Sec} to ensure limited amounts of Ser misincorporation during selenoprotein expression [28]. Similarly, Pyl-tRNA^{Pyl} formation is inefficient compared to other aaRSs and PylRS has a moderate level of catalytic activity [25,33]. It has been a candidate for the evolution of enzyme variants with increased catalytic turnover, as well as more desirable ncAA specificity [34,35].

As a result of increasing ncAA-tRNA concentrations, the cellular levels of the PylRS•tRNA^{Pyl} pairs and components of the Sec pathway must be manipulated to out-compete host tRNAs or release factors for the targeted codon, while maintaining cellular fitness. Furthermore, altering the stoichiometry of the Sec-OTS components is important for the efficiency and homogeneity of selenoprotein production [26,29]. Thus, in addition to mutagenesis approaches to improving OTS interactions, the expression levels of each individual component are critical.

Due to their interactions with various parts of the translation machinery, tRNAs are central to achieving highly efficient ncAA incorporation, and both Sec- and Pyl-OTSs can be significantly improved through tRNA engineering. This is often accomplished through rational design, structure-guided mutagenesis, and random mutagenesis. Current molecular biology techniques facilitate the construction of large libraries of mutants, while combining positive and negative selection has been a successful approach to finding better variants. Here, we discuss the aspects of tRNA biology that should be carefully considered prior to OTS engineering and review the recent developments of Sec- and Pyl-OTSs with a main focus on tRNA design.

2. Aspects of Heterologous tRNA Expression

2.1. Identity Elements and Recognition

The identity elements of tRNAs are nucleotides and their modifications, which function as substrate recognition determinants. These determinants are found throughout the tRNA molecule and are essential for interaction with enzymes for aminoacyl-tRNA formation, as well as elongation factors. Moreover, tRNA recognition involves anti-determinant nucleotides and modifications to prevent the binding and charging of non-cognate tRNAs. In some cases, a single nucleotide mutation can change the tRNA identity and allow aminoacylation by a non-cognate aaRS [36]. Similarly, modifications may also confer identity; for example, m^1G_{37} modification of tRNA^{Asp} in yeast is required to inhibit erroneous charging by ArgRS [37]. While some tRNAs (such as tRNA^{Asp} [38]), maintain their identity elements across all domains of life through divergent evolution, domain-specific idiosyncratic features required for aminoacylation are also present [39]. For this reason, aaRS•tRNA pairs can be transplanted from one domain of life to another and function orthogonally with respect to host aminoacylation.

Genetic code expansion designates a particular stop codon, or an "open" codon in genetically recoded organisms, for the insertion of an ncAA. Nonsense suppression is the most common way to insert ncAAs, since recoding is less detrimental to the proteome, given the low occurrence of stop codons. In this regard, the tRNAs of interest for GCE are typically those without identity elements in the anticodon, as their anticodons can be mutated to decode a stop codon of interest, while retaining aminoacylation capabilities. Conversely, if the active site of an aaRS is suitable for engineering ncAA substrate specificity, the anticodon binding domain can be evolved to recognize a nonsense suppressor tRNA [40–43].

The genetic code naturally expanded to include Sec and Pyl, through the recoding of UGA and UAG, respectively. However, Sec can be efficiently inserted at sense codons [44,45] and improving incorporation in a SECIS-independent manner is achieved through UAG suppression [26,28,29,31]. Anticodon mutations are sufficient to recode sense and stop codons with Sec and Pyl, since cognate SerRS and PylRS do not utilize identity elements in the anticodon loop of tRNA^{Sec} and tRNA^{Pyl}. Thus, ncAA insertion can be easily directed towards a codon of interest using tRNA^{Sec} and tRNA^{Pyl}, within the limitations of the host organism fitness and proteome perturbation.

2.2. Heterologous tRNA Modification and Maturation

Various factors influence the available pools of the aa-tRNA that can be used for peptide synthesis in the cell. These include amino acid and nutrient availability, tRNA expression and maturation (transcription, gene copy number, processing, and modifications), aaRS levels, and tRNA stability and degradation [46]. For GCE applications, the supply of ncAAs is controlled either by adding it in excess amounts to the growth medium or through metabolic engineering of the host organism (e.g., [47]). The biosynthesis and maturation of tRNA are more difficult processes to monitor and control. In *E. coli*, orthogonal tRNAs can be transcribed from "standard" constitutive and inducible promoters (e.g., *lpp*, *proK*, and P$_{BAD}$). To mimic the coding sequences of bacterial tRNAs, the naturally absent terminal CCA sequence is added to the 3'-end of the archaeal tRNA gene. In contrast, to ensure proper processing in eukaryotes, the 3'-CCA sequence of bacterial orthogonal tRNA genes is typically removed.

While archaeal tRNAs in principle are not orthogonal to eukaryotic aaRSs (one exception being tRNA^{Pyl}), bacterial tRNAs are utilized for GCE in eukaryotic hosts [1]. However, the normal transcription of tRNA genes in eukaryotic cells relies on RNA polymerase III, which recognizes A- and B-box promoter elements, present in the tRNA gene itself [48,49]. The majority of prokaryotic tRNAs lack such internal promoter sequences and the engineering of these o-tRNAs may lead to the artificial creation of A- and B-boxes in an o-tRNA variant (see below). To adapt the o-tRNAs of bacterial origin for transcription in yeast, two yeast Pol III promoters—the RPR1 promoter and the SNR52 promoter—have been shown to efficiently drive the expression of *E. coli* tRNAs [50]. Alternatively, a strong RNA polymerase II promoter with tandem tRNA repeats [51] or the yeast tRNA^{Arg} (used as a part of a dicistronic construct) fused upstream of the target tRNA [52] have also been developed.

Between 6.5% and 16.5% of tRNA nucleosides are post-transcriptionally modified, depending on the organisms [53], and over 100 different tRNA modifications have been identified (http://modomics. genesilico.pl/modifications/). Furthermore, tRNA processing is quite complex, sometimes involving intron splicing, trafficking to several subcellular locations [46], and even the ligation of two tRNA halves transcribed from different genes [54]. While the tRNAs used for GCE are orthogonal with respect to endogenous aaRSs, interactions with host modification and processing enzymes is required for function. The addition of tRNA modifications during biosynthesis is important for the stability [55], structure, and function of the molecule [56].

To ensure that an aberrant tRNA is not used for protein synthesis, tRNAs lacking certain modifications are targeted by nucleases for degradation. The nuclear surveillance turnover pathway ensures that a tRNA is properly modified during biosynthesis. For example, yeast pre-tRNA^{iMet} lacking m^{1}A$_{58}$ is polyadenylated by Trf4, which then triggers nuclease degradation by Rrp6 and the nuclear exosome [57,58]. The modifications m^{7}G and m^{5}C also play a role in tRNA stability. The rapid tRNA decay pathway (RTD) in yeast, involving 5'–3' exonucleases Rat1 and Xrn1, targets mature tRNA that lack the m^{7}G and m^{5}C modifications [59,60]. These nucleotides provide an additional level of tRNA regulation and can be manipulated (through mutagenesis or the deletion of nonessential tRNA modifying enzymes) to prevent RTD-targeting and increase tRNA abundance, or for targeted degradation to decrease the toxicity of a suppressor tRNA [61].

The modification of tRNA nucleotides also affects codon–anticodon interactions, binding at the ribosomal A site [62], and ultimately the suppression efficiency that is desired for GCE applications. For instance, natural *E. coli* suppressors depend on the isopentenylation of adenosine 37 for full activity [63,64]. A genetic approach to addressing this issue involves monitoring ncAA incorporation and reporter protein yields across *E. coli* or yeast strain collections containing deletions and/or the overexpression cassettes of metabolic genes. Recently it was shown that the yield and specificity of *O*-phosphoserine incorporation is significantly improved by the deletion of cysteine desulfurase and the overexpression of *E. coli* dimethylallyltransferase (MiaA) and pseudouridine synthase (TruB) [65]. Furthermore, a yeast study involving the removal of modifications by single gene deletions from U34, U35, A37, U47 and C48 in the anticodon stem-loop impairs nonsense suppression, with the strongest effect observed for U34 and A37. Interestingly, the overexpression of eEF1a rescues the activity of an ochre suppressor tRNA (*SUP4*) and other non-suppressor tRNAs that lack modifications [66]. Thus, when designing suppressor tRNAs for GCE, tRNA modifications must be maintained or compensated for, such that tRNA stability and ncAA incorporation is not compromised.

3. When Amino Acid Biosynthesis is o-tRNA-Dependent: Challenges in tRNA^{Sec} Engineering

The biosynthesis of Sec-tRNA^{Sec} and its delivery to the ribosome is complex compared to the canonical amino acid pathway and involves several interactions with different portions of tRNA^{Sec}. The major challenge in engineering tRNA^{Sec} for the more efficient incorporation of Sec is to improve serylation, while also having complete conversion of Ser-tRNA^{Sec} to Sec-tRNA^{Sec}. In addition to this, the requirement of a SECIS sequence directly after UGA necessitates an EF-Tu-mediated Sec insertion pathway for the design and expression of selenoproteins in bacteria.

3.1. tRNA^{Sec} Interactions

The first step in Sec biosynthesis is the charging of tRNA^{Sec} with Ser by SerRS (Figure 2). SerRS lacks an anticodon binding domain, and changes to the anticodon stem-loop do not affect aminoacylation [67]. Rather, SerRS recognizes a long variable arm, a G73 discriminator base, and identity elements in the acceptor and D stems [68–72], which are conserved between tRNA^{Ser} and tRNA^{Sec} (Figure 3). These elements contribute to the structural features and shape of the tRNA and are important for the backbone and sequence-specific interactions for recognition by SerRS [68]. Of these features, the variable arm is most critical for aminoacylation. SerRS possess an N-terminal helical extension that interacts with the variable arm of tRNA^{Ser} and tRNA^{Sec}, and properly orients the tRNA 3' end for aminoacylation [68,73,74]. The overall length of the variable arm is more important than the sequence; the insertion of only one or two nucleotides in the variable arm of tRNA^{Leu} and tRNA^{Tyr}, respectively, confers serylation activity and the deletion of a single base pair from the tRNA^{Sec} variable arm improves serylation 2–3 fold [32,70,75]. It is therefore not surprising that the variable arm accounts for the largest influence on the K_m/k_{cat} of aminoacylation [67].

Identity elements of the tRNA^{Sec} extend beyond aminoacylation and include features of SelA and SelB interactions. Whereas canonical tRNAs have a 12-base-pair amino acid acceptor branch (7/5; consisting of a seven-base-pair acceptor stem and a five-base-pair T stem) that is recognized by EF-Tu/eEF1a, tRNA^{Sec} has a longer 13-base-pair acceptor branch (8/5 or 9/4). The deletion of a base pair from the acceptor stem of *E. coli* tRNA^{Sec} to resemble that of canonical tRNA^{Ser} abolishes UGA read-through with Sec [45], likely due to the disruption of the complex formation of tRNA^{Sec} with SelA and SelB [32]. In addition to the effects of the acceptor stem length on SelA recognition, nucleotides in the D arm form a unique structure compared to tRNA^{Ser}, which is the basis of SelA-tRNA^{Sec} interaction [76].

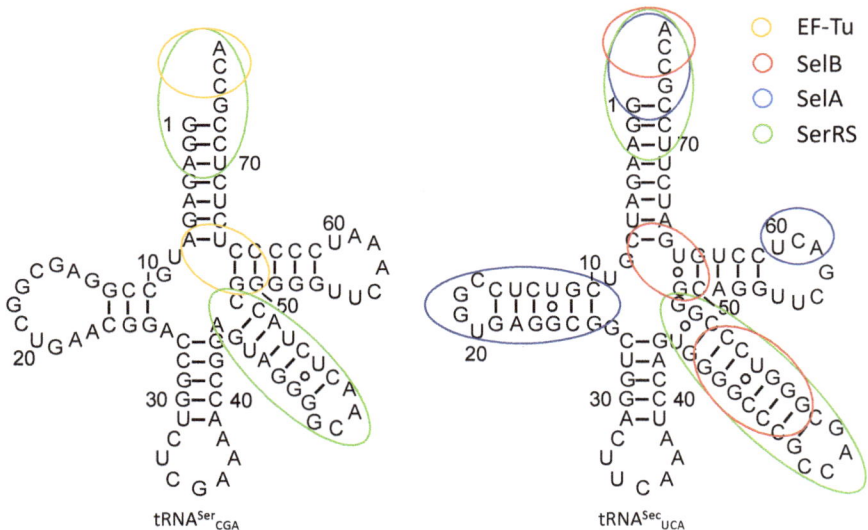

Figure 3. Secondary structure of *E. coli* tRNA$^{Ser}_{CGA}$ (**left**) and tRNA$^{Sec}_{UCA}$ (**right**). Identity elements required for accurate recognition by EF-Tu, SelB, SelA, and SerRS are given in orange, red, blue, and green, respectively.

Comparisons of SelB and EF-Tu complex structures show similarities of acceptor stem binding, but also unique domains and motifs that provide tRNA specificity. The N-terminal half of SelB consists of three domains, named D1, D2, and D3, that are analogous to those of EF-Tu [77]. D1 makes up the GTP-binding domain whereas, D2 and D3 consist of β-barrel-like and β-barrel structures for tRNA binding. Unique to SelB is a fourth domain (D4) comprised of four wing-helix motifs that recruit SelB to SECIS [78]. The structures of the SelB-Sec-tRNA^{Sec} complex obtained from single-particle cryo-electron microscopy depict how the linker region between D3 and D4 binds and distorts the variable arm of tRNA^{Sec}, while an extended loop of D3 interacts with the acceptor and T stems [79]. In conjunction with the positively-charged SelB binding pocket, which provides affinity for the selenol group of Sec, and the altered variable arm orientation of tRNA^{Sec} compared to tRNA^{Ser}, D3 and the linker between D3 and D4 of SelB provide Sec-tRNA^{Sec} specificity.

3.2. Converting tRNA^Sec Recognition from SelB to EF-Tu

The acceptor stem of the tRNA^Sec posed a challenge for engineering the EF-Tu-mediated Sec insertion. Although the binding specificity of tRNA^Sec can be switched from SelB to EF-Tu by shortening the acceptor stem [32], the eight-base-pair stem is important for the interaction with SelA. However, three base pairs in the T stem (49:65, 50:64, and 51:63) modulate the binding affinity of EF-Tu in a sequence-dependent manner [80]. In the same region, tRNA^Sec has different bases. Moreover, the last base pair of the acceptor stem and the first two base pairs of the T stem of tRNA^Sec are anti-determinants of EF-Tu complexed with GTP [81].

The first generation tRNA^Sec for EF-Tu recognition, named tRNA^UTu (U for Sec and Tu for EF-Tu), was designed using *E. coli* tRNA^Ser as a scaffold with the first seven base pairs of the *E. coli* tRNA^Sec acceptor stem [31]. The last base pair of the tRNA^UTu acceptor stem was transplanted from tRNA^Ser to eliminate the EF-Tu anti-determinant position. Serylation of tRNA^UTu was as efficient as canonical tRNA^Ser, however, the Ser to Sec conversion was hampered, which led to ~30% Ser misincorporation. Nonetheless, tRNA^UTu was successfully used to site-specifically incorporate Sec into selenoproteins of bacterial and human origin in a SECIS-independent manner.

3.3. Improving Ser-to-Sec Conversion

Complementary approaches were taken to address the incomplete conversion of Ser-tRNA^UTu to Sec-tRNA^UTu. *E. coli* tRNA^Sec was used as a scaffold for the random mutagenesis of the EF-Tu anti-determinant base pairs C7:G66, G49:U65, and C50:G64. A Sec-specific NMC-A β-lactamase reporter was selected as an efficient tRNA^Sec suppressor containing G7:C66, U49:G65, and C50:U64, which was named tRNA^SecUX [29]. In order to achieve nearly complete conversion of Ser to Sec, SelA expression was elevated, the tRNA^SecUX dosage was decreased, and PSTK was co-expressed to form a Sep-tRNA^SecUX intermediate, which would remain a substrate for SelA but not for EF-Tu prior to Sec conversion.

Other studies have built on tRNA^UTu to improve Ser to Sec conversion. Using the structure of *Aquifex aeolicus* SelA in a complex with *Thermus tengcongensis* tRNA^Sec, twenty-nine different tRNA^UTu variants were rationally designed to include tRNA^Sec features that interact with SelA, while maintaining those that are required for EF-Tu binding. *E. coli* FDH$_H$ was used as a Sec insertion reporter in a sensitive colorimetric assay to identify the best variant, named tRNA^UTuX, which differed from tRNA^UTu at 11 positions [28]. Kinetic assays confirmed that the serylation of tRNA^UTuX was comparable to tRNA^UTu and tRNA^Sec. Ser-to-Sec conversion was increased to 90%, reaching a similar conversion rate as *E. coli* tRNA^Sec. Furthermore, Fourier transform ion cyclotron resonance (FT-ICR) mass spectrometry analysis confirmed Sec insertion by tRNA^UTuX into the selenoprotein, Grx1, but did not detect a peak corresponding to Ser insertion. More recently, tRNA^UTu was used as a template for the generation of chimeric molecules to improve Sec incorporation and selenoprotein yields. It was found that a single base change of A59C in tRNA^UTu, generating a molecule named tRNA^UTu6, resulted in the highest expression levels of human GPx1 and nearly 90% Sec incorporation [27].

3.4. Different tRNA^Sec Structures for the Optimization of Selenoprotein Production

In a bioinformatic search for novel tRNA^Sec molecules, a group of tRNAs with unusual cloverleaf structures were identified, named allo-tRNA [82,83]. Certain allo-tRNA species had tRNA^Ser identities and functioned as efficient amber suppressors with Ser [82]. Allo-tRNAs also contain SelA identity elements, but have a 12-base-pair acceptor branch as opposed to the 13-pair branch present in most tRNA^Sec molecules. SelA from *Aeromonas salmonicida* subsp. *pectinolytica* 34mel (*As*) was coupled with allo-tRNA for selenoprotein expression, since its cognate tRNA^Sec also possesses a 12-base-pair acceptor branch [26]. Allo-tRNA nucleotides in the D stem and acceptor stem were mutated to include *As* tRNA^Sec identities. In addition, the stoichiometry of allo-tRNA to *As* SelA was altered to ensure the complete sequestration of the tRNA for Ser-to-Sec conversion while also maintaining non-toxic levels of *As* SelA. Further optimizations and metabolic engineering efforts created a Sec-OTS consisting of allo-tRNA^{UTu2D}, *As* SelD, *As* SelA, and *Treponema denticola* Trx1. Along with the high selenoprotein yields obtained with a Sec incorporation efficiency estimated at >90%, the stand-alone capabilities of this system make it ideal for use in other organisms [26].

4. Absolutely Orthogonal? Unique Features of tRNA^{Pyl}

Compared to the Sec system, the use of Pyl-OTS is comparatively less challenging, as its tRNA is orthogonal in the majority of model organisms used for GCE [84]; the enzyme is also orthogonal to both cellular tRNAs, as well as natural/canonical AAs [85]. Both bacterial and eukaryotic elongation factors accept tRNA^{Pyl}, and the AA-binding pocket can be separately adapted to accept some bulkier ncAAs [86]. Attempts to advance ncAA delivery by tRNA^{Pyl} engineering include those aiming to improve its compatibility with the cellular machinery of the host. In *E. coli*, tRNA^{Pyl} was evolved by targeting the EF-Tu-binding regions [87], although the optimizing mutations present in tRNA$^{Pyl}_{OPT}$ may be more suitable for the delivery of one particular ncAA and less for the other (e.g., N^{ε}-acetyllyine vs. 3-cyano-phenylalanine) [88]. The need to separately evolve an o-tRNA for a variety of "cognate" ncAAs or a variety of anticodons may require tunable binding by EF-Tu and the ribosome; while the stability of the EF-Tu•ncAA-tRNA complex reflects additive contributions by the ncAA and T-stem base pairs of the o-tRNA [80,89], the strength of codon–anticodon binding correlates with the nucleotide composition of the tRNA core [90]. For efficient expression in mammalian systems, a stabilizing mutation in the anticodon stem has been used (U29aC, Figure 4) [91,92]. By introducing elements conserved in human tRNAs, a better performing tRNA^{Pyl} was evolved. Mutations in the D-stem, D-loop, T-loop and the anticodon-stem U29aC proved to be indispensable for high activity [93]; compared to wild type tRNA^{Pyl}, the use of this variant in HEK293 cells improved the incorporation of two ncAAs, N^{ε}-carbobenzyloxy-lysine (Z-lysine) and N^{ε}-(tert-butoxycarbonyl)-lysine (Boc-lysine). Interestingly, a chimera between mitochondrial (mt) tRNA^{Ser} and *Methanosarcina mazei* tRNA^{Pyl} improved the insertion of Boc-lysine selectively (C15) [93]. Earlier attempts at using mttRNA^{Ser} in *E. coli* failed, due to the lack of orthogonality [94]. The improved activity of M15 and C15 variants in mammalian cell lines may have to do with the appearance of the B-box in the T-arm of the variants; prokaryotic o-tRNAs are usually placed under the external promoter, such as U6, but the endogenous tRNAs are transcribed from internal A- and B-box promoters [48].

One of the distinct features of the Pyl system is the minimal variable loop of tRNA^{Pyl}, which together with the T-loop forms a dipped surface [35,95] (Figure 4). From the crystal structure of the N-terminal domain of *M. mazei* PylRS it is evident that this minimalistic variable loop is a prerequisite for effective binding, as a larger variable loop would sterically clash with the N-domain [35]. In addition to *M. mazei*, *Desulfitobacterium hafniense* Pyl-OTS was employed in *E. coli*, either with its original N-terminal domain, or as a fusion with the recombinant (chimeric) N-domain of the archaeal system [96]. However, this system is not functional in mammalian cells [93]. As the N-terminal domain binds tRNA^{Pyl} with extremely high affinity [4], this element is likely to be an important contributor to (almost universal) Pyl-OTS orthogonality.

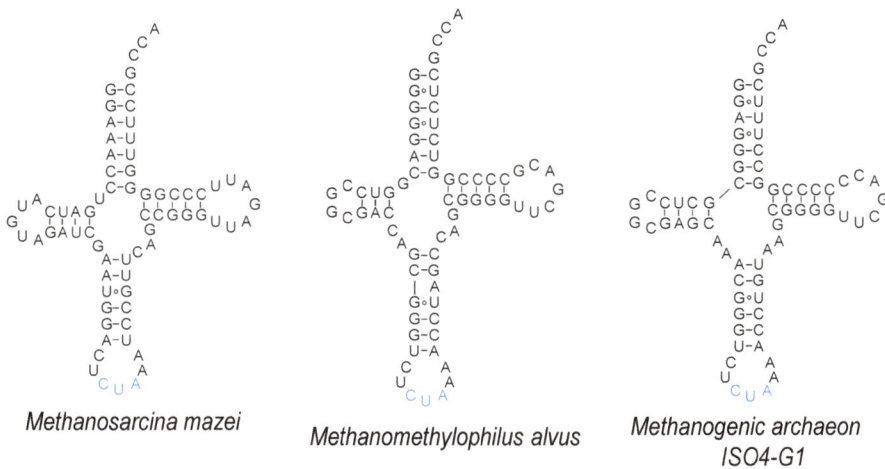

Figure 4. Structures of tRNA[Pyl] belonging to mutually orthogonal Pyl-orthogonal translation systems (OTSs) [97,98].

However, some organisms do not possess an equivalent to this N-domain [99], suggesting an alternative mode of recognition. This fact was recently exploited to develop mutually orthogonal Pyl-OTSs in *E. coli* [98] and mammalian cell lines [97,100]. Two PylRS enzymes that utilize the C-domain only (*Methanomethylophilus alvus* and methanogenic archaeon ISO4-G1) are highly active in *E. coli* [98]. Their cognate tRNAs retain some characteristic MmtRNA[Pyl] features (such as the identity of the discriminator base G73, or the minimalistic D-loop) but also diverge in the nucleotide composition of the acceptor stem and in the probable structure of the anticodon stem (Figure 4). Given that the *M. mazei* and *M. alvus*/G1 systems are not fully orthogonal, rational engineering was employed in order to generate MatRNA[Pyl] that would be recognized by MaPylRS and not MmPylRS. Variation of the nucleotide composition of the variable arm and/or its length allowed the generation of successful MatRNA[Pyl] variants. Given the malleability of the PylRS active site, orthogonality to other OTSs [26,101], together with high activity of Pyl-OTSs in the bacteria and cells of higher eukaryotes [6], it is foreseeable that this dual encoding system will be commonly used.

The creation of multiple, mutually orthogonal OTSs is inherently related to the number of liberated codons that can be targeted for ncAA incorporation. In addition to UAG-directed incorporation, Pyl-OTS was also employed for ncAA incorporation in response to rare arginine (AGG) codons in *E. coli*, alone [102] or in tandem with *Methanocaldococcus jannaschii* Tyr-OTS [26]. A similar strategy was attempted in *Mycoplasma capricolum*, which possesses only six arginine CGG codons that should, in theory, facilitate the reassignment (Arg-to-Pyl) [103]. However, upon mutation of the tRNA[Pyl] anticodon to CCG this almost universally orthogonal tRNA becomes a substrate for endogenous ArgRS. In conclusion, while the anticodon-blind recognition of PylRS allows the anticodon of tRNA[Pyl] to be mutated into any nucleotide triplet, synonymous anticodons (such as CCU and CCG) can be recognized by host aaRSs with very different affinity, causing one tRNA[Pyl] variant to lose its initial orthogonality.

5. Conclusions/Outlook

Improvements to OTSs have been emerging rapidly in recent years and are valuable for the accurate and efficient production of proteins containing ncAAs. The increasing amount of sequence data and bioinformatic/structural analyses reveal new molecules and novel mechanisms that help enhance each system. Moreover, advanced molecular cloning and directed evolution techniques help further shape the molecules that nature has provided into molecules that are better suited for the incorporation of ncAAs. tRNAs interact with each component of an OTS in the process of bringing the ncAA to the ribosome to insert a particular ncAA during peptide synthesis. For this reason, finding the best tRNA variant is critical for OTS developments. Our expanding knowledge of tRNA processing, maturation, and interaction mechanisms has guided tRNA engineering towards this goal. As we continue to learn more from nature and as technologies advance, it is conceivable that peptides with unique properties will be produced with significant industrial and medical implications.

Author Contributions: Conceptualization, K.S.H., A.C., and D.S.; Writing-Original Draft Preparation, K.S.H., A.C., and D.S.; Writing-Review and Editing, K.S.H., A.C., and D.S.; Visualization, K.S.H. and A.C.; Supervision, D.S.; Funding Acquisition, D.S.

Funding: Research in the Söll laboratory is supported by grants from the US National Institutes of Health (NIH) (R35GM122560), the US National Science Foundation (CHE-1740549) and from the Division of Chemical Sciences, Geosciences, and Biosciences, Office of Basic Energy Sciences of the Department of Energy (DE-FG02-98ER20311) to D.S.

Acknowledgments: We thank Oscar Vargas-Rodriguez, Hui Si Kwok, and Jeffery Tharp for critical discussion during the preparation of the manuscript.

Conflicts of Interest: The authors declare no conflict of interest.

References

1. Chin, J.W. Expanding and reprogramming the genetic code of cells and animals. *Annu. Rev. Biochem.* **2014**, *83*, 379–408. [CrossRef] [PubMed]
2. Mukai, T.; Lajoie, M.J.; Englert, M.; Soll, D. Rewriting the Genetic Code. *Annu. Rev. Microbiol.* **2017**, *71*, 557–577. [CrossRef] [PubMed]
3. Young, D.D.; Schultz, P.G. Playing with the molecules of life. *ACS Chem. Biol.* **2018**, *13*, 854–870. [CrossRef] [PubMed]
4. Herring, S.; Ambrogelly, A.; Gundllapalli, S.; O'Donoghue, P.; Polycarpo, C.R.; Söll, D. The amino-terminal domain of pyrrolysyl-tRNA synthetase is dispensable in vitro but required for in vivo activity. *FEBS Lett.* **2007**, *581*, 3197–3203. [CrossRef] [PubMed]
5. Jiang, R.; Krzycki, J.A. PylSn and the homologous N-terminal domain of pyrrolysyl-tRNA synthetase bind the tRNA that is essential for the genetic encoding of pyrrolysine. *J. Biol. Chem.* **2012**, *287*, 32738–32746. [CrossRef] [PubMed]
6. Wan, W.; Tharp, J.M.; Liu, W.R. Pyrrolysyl-tRNA synthetase: An ordinary enzyme but an outstanding genetic code expansion tool. *Biochim. Biophys. Acta* **2014**, *1844*, 1059–1070. [CrossRef] [PubMed]
7. Chin, J.W.; Cropp, T.A.; Anderson, J.C.; Mukherji, M.; Zhang, Z.; Schultz, P.G. An expanded eukaryotic genetic code. *Science* **2003**, *301*, 964–967. [CrossRef] [PubMed]
8. Greiss, S.; Chin, J.W. Expanding the genetic code of an animal. *J. Am. Chem. Soc.* **2011**, *133*, 14196–14199. [CrossRef] [PubMed]
9. Si, L.; Xu, H.; Zhou, X.; Zhang, Z.; Tian, Z.; Wang, Y.; Wu, Y.; Zhang, B.; Niu, Z.; Zhang, C.; et al. Generation of influenza A viruses as live but replication-incompetent virus vaccines. *Science* **2016**, *354*, 1170–1173. [CrossRef] [PubMed]
10. Bianco, A.; Townsley, F.M.; Greiss, S.; Lang, K.; Chin, J.W. Expanding the genetic code of *Drosophila melanogaster*. *Nat. Chem. Biol.* **2012**, *8*, 748–750. [CrossRef] [PubMed]
11. Liu, C.C.; Schultz, P.G. Adding new chemistries to the genetic code. *Annu. Rev. Biochem.* **2010**, *79*, 413–444. [CrossRef] [PubMed]
12. Reich, H.J.; Hondal, R.J. Why nature chose selenium. *ACS Chem. Biol.* **2016**, *11*, 821–841. [CrossRef] [PubMed]
13. Arai, K.; Takei, T.; Okumura, M.; Watanabe, S.; Amagai, Y.; Asahina, Y.; Moroder, L.; Hojo, H.; Inaba, K.; Iwaoka, M. Preparation of Selenoinsulin as a Long-Lasting Insulin Analogue. *Angew. Chem. Int. Ed. Engl.* **2017**, *56*, 5522–5526. [CrossRef] [PubMed]
14. Metanis, N.; Hilvert, D. Natural and synthetic selenoproteins. *Curr. Opin. Chem. Biol.* **2014**, *22*, 27–34. [CrossRef] [PubMed]
15. Shchedrina, V.A.; Novoselov, S.V.; Malinouski, M.Y.; Gladyshev, V.N. Identification and characterization of a selenoprotein family containing a diselenide bond in a redox motif. *Proc. Natl. Acad. Sci. USA* **2007**, *104*, 13919–13924. [CrossRef] [PubMed]

16. Carlson, B.A.; Xu, X.M.; Kryukov, G.V.; Rao, M.; Berry, M.J.; Gladyshev, V.N.; Hatfield, D.L. Identification and characterization of phosphoseryl-tRNA[Ser]Sec kinase. *Proc. Natl. Acad. Sci. USA* **2004**, *101*, 12848–12853. [CrossRef] [PubMed]

17. Xu, X.M.; Carlson, B.A.; Mix, H.; Zhang, Y.; Saira, K.; Glass, R.S.; Berry, M.J.; Gladyshev, V.N.; Hatfield, D.L. Biosynthesis of selenocysteine on its tRNA in eukaryotes. *PLoS Biol.* **2007**, *5*, e4. [CrossRef] [PubMed]

18. Palioura, S.; Sherrer, R.L.; Steitz, T.A.; Söll, D.; Simonovic, M. The human SepSecS-tRNA^Sec complex reveals the mechanism of selenocysteine formation. *Science* **2009**, *325*, 321–325. [CrossRef] [PubMed]

19. Yuan, J.; Palioura, S.; Salazar, J.C.; Su, D.; O'Donoghue, P.; Hohn, M.J.; Cardoso, A.M.; Whitman, W.B.; Soll, D. RNA-dependent conversion of phosphoserine forms selenocysteine in eukaryotes and archaea. *Proc. Natl. Acad. Sci. USA* **2006**, *103*, 18923–18927. [CrossRef] [PubMed]

20. Forchhammer, K.; Leinfelder, W.; Böck, A. Identification of a novel translation factor necessary for the incorporation of selenocysteine into protein. *Nature* **1989**, *342*, 453–456. [CrossRef] [PubMed]

21. Fagegaltier, D.; Hubert, N.; Yamada, K.; Mizutani, T.; Carbon, P.; Krol, A. Characterization of mSelB, a novel mammalian elongation factor for selenoprotein translation. *EMBO J.* **2000**, *19*, 4796–4805. [CrossRef] [PubMed]

22. Squires, J.E.; Berry, M.J. Eukaryotic selenoprotein synthesis: Mechanistic insight incorporating new factors and new functions for old factors. *IUBMB Life* **2008**, *60*, 232–235. [CrossRef] [PubMed]

23. Donovan, J.; Copeland, P.R. Threading the needle: Getting selenocysteine into proteins. *Antioxid. Redox Signal.* **2010**, *12*, 881–892. [CrossRef] [PubMed]

24. Giege, R.; Sissler, M.; Florentz, C. Universal rules and idiosyncratic features in tRNA identity. *Nucleic Acids Res.* **1998**, *26*, 5017–5035. [CrossRef] [PubMed]

25. Guo, L.T.; Wang, Y.S.; Nakamura, A.; Eiler, D.; Kavran, J.M.; Wong, M.; Kiessling, L.L.; Steitz, T.A.; O'Donoghue, P.; Söll, D. Polyspecific pyrrolysyl-tRNA synthetases from directed evolution. *Proc. Natl. Acad. Sci. USA* **2014**, *111*, 16724–16729. [CrossRef] [PubMed]

26. Proc Natl Acad Sci U S AFEBS LettOhtake, K.; Mukai, T.; Iraha, F.; Takahashi, M.; Haruna, K.-i.; Date, M.; Yokoyama, K.; Sakamoto, K. Engineering an automaturing transglutaminase with enhanced thermostability by genetic code expansion with two codon reassignments. *ACS Synth. Biol.* **2018**, *7*, 2170–2176. [CrossRef]

27. Fan, Z.; Song, J.; Guan, T.; Lv, X.; Wei, J. Efficient expression of glutathione peroxidase with chimeric tRNA in amber-less *Escherichia coli*. *ACS Synth. Biol.* **2018**, *7*, 249–257. [CrossRef] [PubMed]

28. Miller, C.; Bröcker, M.J.; Prat, L.; Ip, K.; Chirathivat, N.; Feiock, A.; Veszpremi, M.; Söll, D. A synthetic tRNA for EF-Tu mediated selenocysteine incorporation in vivo and in vitro. *FEBS Lett.* **2015**, *589*, 2194–2199. [CrossRef] [PubMed]

29. Thyer, R.; Robotham, S.A.; Brodbelt, J.S.; Ellington, A.D. Evolving tRNA^Sec for efficient canonical incorporation of selenocysteine. *J. Am. Chem. Soc.* **2015**, *137*, 46–49. [CrossRef] [PubMed]

30. Haruna, K.; Alkazemi, M.H.; Liu, Y.; Söll, D.; Englert, M. Engineering the elongation factor Tu for efficient selenoprotein synthesis. *Nucleic Acids Res.* **2014**, *42*, 9976–9983. [CrossRef] [PubMed]

31. Aldag, C.; Bröcker, M.J.; Hohn, M.J.; Prat, L.; Hammond, G.; Plummer, A.; Söll, D. Rewiring translation for elongation factor Tu-dependent selenocysteine incorporation. *Angew. Chem. Int. Ed. Engl.* **2013**, *52*, 1441–1445. [CrossRef] [PubMed]

32. Baron, C.; Böck, A. The length of the aminoacyl-acceptor stem of the selenocysteine-specific tRNA^Sec of *Escherichia coli* is the determinant for binding to elongation factors SELB or Tu. *J. Biol. Chem.* **1991**, *266*, 20375–20379. [PubMed]

33. Li, W.T.; Mahapatra, A.; Longstaff, D.G.; Bechtel, J.; Zhao, G.; Kang, P.T.; Chan, M.K.; Krzycki, J.A. Specificity of pyrrolysyl-tRNA synthetase for pyrrolysine and pyrrolysine analogs. *J. Mol. Biol.* **2009**, *385*, 1156–1164. [CrossRef] [PubMed]

34. Bryson, D.I.; Fan, C.; Guo, L.T.; Miller, C.; Söll, D.; Liu, D.R. Continuous directed evolution of aminoacyl-tRNA synthetases. *Nat. Chem. Biol.* **2017**, *13*, 1253–1260. [CrossRef] [PubMed]

35. Suzuki, T.; Miller, C.; Guo, L.T.; Ho, J.M.L.; Bryson, D.I.; Wang, Y.S.; Liu, D.R.; Söll, D. Crystal structures reveal an elusive functional domain of pyrrolysyl-tRNA synthetase. *Nat. Chem. Biol.* **2017**, *13*, 1261–1266. [CrossRef] [PubMed]

36. Auld, D.S.; Schimmel, P. Switching recognition of two tRNA synthetases with an amino acid swap in a designed peptide. *Science* **1995**, *267*, 1994–1996. [CrossRef] [PubMed]

37. Perret, V.; Garcia, A.; Grosjean, H.; Ebel, J.P.; Florentz, C.; Giege, R. Relaxation of a transfer RNA specificity by removal of modified nucleotides. *Nature* **1990**, *344*, 787–789. [CrossRef] [PubMed]
38. Becker, H.D.; Giege, R.; Kern, D. Identity of prokaryotic and eukaryotic tRNA^Asp for aminoacylation by aspartyl-tRNA synthetase from *Thermus thermophilus*. *Biochemistry* **1996**, *35*, 7447–7458. [CrossRef] [PubMed]
39. Lee, C.P.; RajBhandary, U.L. Mutants of *Escherichia coli* initiator tRNA that suppress amber codons in *Saccharomyces cerevisiae* and are aminoacylated with tyrosine by yeast extracts. *Proc. Natl. Acad. Sci. USA* **1991**, *88*, 11378–11382. [CrossRef] [PubMed]
40. Gan, R.; Perez, J.G.; Carlson, E.D.; Ntai, I.; Isaacs, F.J.; Kelleher, N.L.; Jewett, M.C. Translation system engineering in *Escherichia coli* enhances non-canonical amino acid incorporation into proteins. *Biotechnol. Bioeng.* **2017**, *114*, 1074–1086. [CrossRef] [PubMed]
41. Lee, S.; Oh, S.; Yang, A.; Kim, J.; Soll, D.; Lee, D.; Park, H.S. A facile strategy for selective incorporation of phosphoserine into histones. *Angew. Chem. Int. Ed. Engl.* **2013**, *52*, 5771–5775. [CrossRef] [PubMed]
42. Rogerson, D.T.; Sachdeva, A.; Wang, K.; Haq, T.; Kazlauskaite, A.; Hancock, S.M.; Huguenin-Dezot, N.; Muqit, M.M.; Fry, A.M.; Bayliss, R.; et al. Efficient genetic encoding of phosphoserine and its nonhydrolyzable analog. *Nat. Chem. Biol.* **2015**, *11*, 496–503. [CrossRef] [PubMed]
43. Chatterjee, A.; Xiao, H.; Schultz, P.G. Evolution of multiple, mutually orthogonal prolyl-tRNA synthetase/tRNA pairs for unnatural amino acid mutagenesis in *Escherichia coli*. *Proc. Natl. Acad. Sci. USA* **2012**, *109*, 14841–14846. [CrossRef] [PubMed]
44. Bröcker, M.J.; Ho, J.M.; Church, G.M.; Söll, D.; O'Donoghue, P. Recoding the genetic code with selenocysteine. *Angew. Chem. Int. Ed. Engl.* **2014**, *53*, 319–323. [CrossRef] [PubMed]
45. Baron, C.; Heider, J.; Böck, A. Mutagenesis of selC, the gene for the selenocysteine-inserting tRNA-species in *E. coli*: Effects on in vivo function. *Nucleic Acids Res.* **1990**, *18*, 6761–6766. [CrossRef] [PubMed]
46. Phizicky, E.M.; Hopper, A.K. tRNA biology charges to the front. *Genes Dev.* **2010**, *24*, 1832–1860. [CrossRef] [PubMed]
47. Steinfeld, J.B.; Aerni, H.R.; Rogulina, S.; Liu, Y.; Rinehart, J. Expanded cellular amino acid pools containing phosphoserine, phosphothreonine, and phosphotyrosine. *ACS Chem. Biol.* **2014**, *9*, 1104–1112. [CrossRef] [PubMed]
48. Galli, G.; Hofstetter, H.; Birnstiel, M.L. Two conserved sequence blocks within eukaryotic tRNA genes are major promoter elements. *Nature* **1981**, *294*, 626–631. [CrossRef] [PubMed]
49. Sharp, S.; DeFranco, D.; Dingermann, T.; Farrell, P.; Soll, D. Internal control regions for transcription of eukaryotic tRNA genes. *Proc. Natl. Acad. Sci. USA* **1981**, *78*, 6657–6661. [CrossRef] [PubMed]
50. Wang, Q.; Wang, L. Genetic incorporation of unnatural amino acids into proteins in yeast. *Methods Mol. Biol.* **2012**, *794*, 199–213. [CrossRef] [PubMed]
51. Chen, S.; Schultz, P.G.; Brock, A. An improved system for the generation and analysis of mutant proteins containing unnatural amino acids in *Saccharomyces cerevisiae*. *J. Mol. Biol.* **2007**, *371*, 112–122. [CrossRef] [PubMed]
52. Otter, C.A.; Straby, K.B. Transcription of eukaryotic genes with impaired internal promoters: The use of a yeast tRNA gene as promoter. *J. Biotechnol.* **1991**, *21*, 289–293. [CrossRef]
53. Machnicka, M.A.; Olchowik, A.; Grosjean, H.; Bujnicki, J.M. Distribution and frequencies of post-transcriptional modifications in tRNAs. *RNA Biol.* **2014**, *11*, 1619–1629. [CrossRef] [PubMed]
54. Randau, L.; Munch, R.; Hohn, M.J.; Jahn, D.; Söll, D. *Nanoarchaeum equitans* creates functional tRNAs from separate genes for their 5′- and 3′-halves. *Nature* **2005**, *433*, 537–541. [CrossRef] [PubMed]
55. Lorenz, C.; Lünse, E.C.; Mörl, M. tRNA Modifications: Impact on Structure and Thermal Adaptation. *Biomolecules* **2017**, *7*. [CrossRef] [PubMed]
56. Väre, Y.V.; Eruysal, R.E.; Narendran, A.; Sarachan, L.K.; Agris, F.P. Chemical and conformational diversity of modified nucleosides affects trna structure and function. *Biomolecules* **2017**, *7*. [CrossRef] [PubMed]
57. Kadaba, S.; Wang, X.; Anderson, J.T. Nuclear RNA surveillance in *Saccharomyces cerevisiae*: Trf4p-dependent polyadenylation of nascent hypomethylated tRNA and an aberrant form of 5S rRNA. *RNA* **2006**, *12*, 508–521. [CrossRef] [PubMed]
58. Kadaba, S.; Krueger, A.; Trice, T.; Krecic, A.M.; Hinnebusch, A.G.; Anderson, J. Nuclear surveillance and degradation of hypomodified initiator tRNA^Met in *S. cerevisiae*. *Genes Dev.* **2004**, *18*, 1227–1240. [CrossRef] [PubMed]

59. Dewe, J.M.; Whipple, J.M.; Chernyakov, I.; Jaramillo, L.N.; Phizicky, E.M. The yeast rapid tRNA decay pathway competes with elongation factor 1A for substrate tRNAs and acts on tRNAs lacking one or more of several modifications. *RNA* **2012**, *18*, 1886–1896. [CrossRef] [PubMed]

60. Alexandrov, A.; Chernyakov, I.; Gu, W.; Hiley, S.L.; Hughes, T.R.; Grayhack, E.J.; Phizicky, E.M. Rapid tRNA decay can result from lack of nonessential modifications. *Mol. Cell* **2006**, *21*, 87–96. [CrossRef] [PubMed]

61. Berg, M.D.; Hoffman, K.S.; Genereaux, J.; Mian, S.; Trussler, R.S.; Haniford, D.B.; O'Donoghue, P.; Brandl, C.J. Evolving mistranslating tRNAs through a phenotypically ambivalent intermediate in *Saccharomyces cerevisiae*. *Genetics* **2017**, *206*, 1865–1879. [CrossRef] [PubMed]

62. Grosjean, H.; Westhof, E. An integrated, structure- and energy-based view of the genetic code. *Nucleic Acids Res.* **2016**, *44*, 8020–8040. [CrossRef] [PubMed]

63. Gefter, M.L.; Russell, R.L. Role modifications in tyrosine transfer RNA: A modified base affecting ribosome binding. *J. Mol. Biol.* **1969**, *39*, 145–157. [CrossRef]

64. Laten, H.; Gorman, J.; Böck, R.M. Isopentenyladenosine deficient tRNA from an antisuppressor mutant of *Saccharomyces cerevisiae*. *Nucleic Acids Res.* **1978**, *5*, 4329–4342. [CrossRef] [PubMed]

65. Crnkovic, A.; Vargas-Rodriguez, O.; Merkuryev, A.; Söll, D. Effects of heterologous tRNA modifications on the production of proteins containing noncanonical amino acids. *Bioengineering* **2018**, *5*, 11. [CrossRef] [PubMed]

66. Klassen, R.; Schaffrath, R. Collaboration of tRNA modifications and elongation factor eEF1A in decoding and nonsense suppression. *Sci. Rep.* **2018**, *8*, 12749. [CrossRef] [PubMed]

67. Sampson, J.R.; Saks, M.E. Contributions of discrete tRNA^{Ser} domains to aminoacylation by *E. coli* seryl-tRNA synthetase: A kinetic analysis using model RNA substrates. *Nucleic Acids Res.* **1993**, *21*, 4467–4475. [CrossRef] [PubMed]

68. Biou, V.; Yaremchuk, A.; Tukalo, M.; Cusack, S. The 2.9 Å crystal structure of *T. thermophilus* seryl-tRNA synthetase complexed with tRNA^{Ser}. *Science* **1994**, *263*, 1404–1410. [CrossRef] [PubMed]

69. Normanly, J.; Ollick, T.; Abelson, J. Eight base changes are sufficient to convert a leucine-inserting tRNA into a serine-inserting tRNA. *Proc. Natl. Acad. Sci. USA* **1992**, *89*, 5680–5684. [CrossRef] [PubMed]

70. Himeno, H.; Hasegawa, T.; Ueda, T.; Watanabe, K.; Shimizu, M. Conversion of aminoacylation specificity from tRNA^{Tyr} to tRNA^{Ser} in vitro. *Nucleic Acids Res.* **1990**, *18*, 6815–6819. [CrossRef] [PubMed]

71. Rogers, M.J.; Söll, D. Discrimination between glutaminyl-tRNA synthetase and seryl-tRNA synthetase involves nucleotides in the acceptor helix of tRNA. *Proc. Natl. Acad. Sci. USA* **1988**, *85*, 6627–6631. [CrossRef] [PubMed]

72. Normanly, J.; Ogden, R.C.; Horvath, S.J.; Abelson, J. Changing the identity of a transfer RNA. *Nature* **1986**, *321*, 213–219. [CrossRef] [PubMed]

73. Itoh, Y.; Sekine, S.; Suetsugu, S.; Yokoyama, S. Tertiary structure of bacterial selenocysteine tRNA. *Nucleic Acids Res.* **2013**, *41*, 6729–6738. [CrossRef] [PubMed]

74. Wang, C.; Guo, Y.; Tian, Q.; Jia, Q.; Gao, Y.; Zhang, Q.; Zhou, C.; Xie, W. SerRS-tRNA^{Sec} complex structures reveal mechanism of the first step in selenocysteine biosynthesis. *Nucleic Acids Res.* **2015**, *43*, 10534–10545. [CrossRef] [PubMed]

75. Himeno, H.; Yoshida, S.; Soma, A.; Nishikawa, K. Only one nucleotide insertion to the long variable arm confers an efficient serine acceptor activity upon *Saccharomyces cerevisiae* tRNA^{Leu} in vitro. *J. Mol. Biol.* **1997**, *268*, 704–711. [CrossRef] [PubMed]

76. Itoh, Y.; Bröcker, M.J.; Sekine, S.; Hammond, G.; Suetsugu, S.; Söll, D.; Yokoyama, S. Decameric SelA•tRNA^{Sec} ring structure reveals mechanism of bacterial selenocysteine formation. *Science* **2013**, *340*, 75–78. [CrossRef] [PubMed]

77. Selmer, M.; Su, X.D. Crystal structure of an mRNA-binding fragment of *Moorella thermoacetica* elongation factor SelB. *EMBO J.* **2002**, *21*, 4145–4153. [CrossRef] [PubMed]

78. Kromayer, M.; Wilting, R.; Tormay, P.; Böck, A. Domain structure of the prokaryotic selenocysteine-specific elongation factor SelB. *J. Mol. Biol.* **1996**, *262*, 413–420. [CrossRef] [PubMed]

79. Fischer, N.; Neumann, P.; Bock, L.V.; Maracci, C.; Wang, Z.; Paleskava, A.; Konevega, A.L.; Schroder, G.F.; Grubmuller, H.; Ficner, R.; et al. The pathway to GTPase activation of elongation factor SelB on the ribosome. *Nature* **2016**, *540*, 80–85. [CrossRef] [PubMed]

80. Schrader, J.M.; Chapman, S.J.; Uhlenbeck, O.C. Understanding the sequence specificity of tRNA binding to elongation factor Tu using tRNA mutagenesis. *J. Mol. Biol.* **2009**, *386*, 1255–1264. [CrossRef] [PubMed]

81. Rudinger, J.; Hillenbrandt, R.; Sprinzl, M.; Giege, R. Antideterminants present in minihelix[Sec] hinder its recognition by prokaryotic elongation factor Tu. *EMBO J.* **1996**, *15*, 650–657. [CrossRef] [PubMed]

82. Mukai, T.; Vargas-Rodriguez, O.; Englert, M.; Tripp, H.J.; Ivanova, N.N.; Rubin, E.M.; Kyrpides, N.C.; Söll, D. Transfer RNAs with novel cloverleaf structures. *Nucleic Acids Res.* **2017**, *45*, 2776–2785. [CrossRef] [PubMed]

83. Mukai, T.; Englert, M.; Tripp, H.J.; Miller, C.; Ivanova, N.N.; Rubin, E.M.; Kyrpides, N.C.; Söll, D. Facile Recoding of Selenocysteine in Nature. *Angew. Chem. Int. Ed. Engl.* **2016**, *55*, 5337–5341. [CrossRef] [PubMed]

84. Tharp, J.M.; Ehnbom, A.; Liu, W.R. tRNA[Pyl]: Structure, function, and applications. *RNA Biol.* **2018**, *15*, 441–452. [CrossRef] [PubMed]

85. Polycarpo, C.; Ambrogelly, A.; Berube, A.; Winbush, S.M.; McCloskey, J.A.; Crain, P.F.; Wood, J.L.; Söll, D. An aminoacyl-tRNA synthetase that specifically activates pyrrolysine. *Proc. Natl. Acad. Sci. USA* **2004**, *101*, 12450–12454. [CrossRef] [PubMed]

86. Doi, Y.; Ohtsuki, T.; Shimizu, Y.; Ueda, T.; Sisido, M. Elongation factor Tu mutants expand amino acid tolerance of protein biosynthesis system. *J. Am. Chem. Soc.* **2007**, *129*, 14458–14462. [CrossRef] [PubMed]

87. Uhlenbeck, O.C.; Schrader, J.M. Evolutionary tuning impacts the design of bacterial tRNAs for the incorporation of unnatural amino acids by ribosomes. *Curr. Opin. Chem. Biol.* **2018**, *46*, 138–145. [CrossRef] [PubMed]

88. Fan, C.; Xiong, H.; Reynolds, N.M.; Söll, D. Rationally evolving tRNA[Pyl] for efficient incorporation of noncanonical amino acids. *Nucleic Acids Res.* **2015**, *43*, e156. [CrossRef] [PubMed]

89. LaRiviere, F.J.; Wolfson, A.D.; Uhlenbeck, O.C. Uniform binding of aminoacyl-tRNAs to elongation factor Tu by thermodynamic compensation. *Science* **2001**, *294*, 165–168. [CrossRef] [PubMed]

90. Shepotinovskaya, I.; Uhlenbeck, O.C. tRNA residues evolved to promote translational accuracy. *RNA* **2013**, *19*, 510–516. [CrossRef] [PubMed]

91. Chatterjee, A.; Sun, S.B.; Furman, J.L.; Xiao, H.; Schultz, P.G. A versatile platform for single- and multiple-unnatural amino acid mutagenesis in *Escherichia coli*. *Biochemistry* **2013**, *52*, 1828–1837. [CrossRef] [PubMed]

92. Schmied, W.H.; Elsässer, S.J.; Uttamapinant, C.; Chin, J.W. Efficient multisite unnatural amino acid incorporation in mammalian cells via optimized pyrrolysyl tRNA synthetase/tRNA expression and engineered eRF1. *J. Am. Chem. Soc.* **2014**, *136*, 15577–15583. [CrossRef] [PubMed]

93. Serfling, R.; Lorenz, C.; Etzel, M.; Schicht, G.; Bottke, T.; Mörl, M.; Coin, I. Designer tRNAs for efficient incorporation of non-canonical amino acids by the pyrrolysine system in mammalian cells. *Nucleic Acids Res.* **2018**, *46*, 1–10. [CrossRef] [PubMed]

94. Ambrogelly, A.; Gundllapalli, S.; Herring, S.; Polycarpo, C.; Frauer, C.; Söll, D. Pyrrolysine is not hardwired for cotranslational insertion at UAG codons. *Proc. Natl. Acad. Sci. USA* **2007**, *104*, 3141–3146. [CrossRef] [PubMed]

95. Nozawa, K.; O'Donoghue, P.; Gundllapalli, S.; Araiso, Y.; Ishitani, R.; Umehara, T.; Söll, D.; Nureki, O. Pyrrolysyl-tRNA synthetase-tRNA[Pyl] structure reveals the molecular basis of orthogonality. *Nature* **2009**, *457*, 1163–1167. [CrossRef] [PubMed]

96. Fladischer, P.; Weingartner, A.; Blamauer, J.; Darnhofer, B.; Birner-Gruenberger, R.; Kardashliev, T.; Ruff, A.J.; Schwaneberg, U.; Wiltschi, B. A semi-rationally engineered bacterial pyrrolysyl-tRNA synthetase genetically encodes phenyl azide chemistry. *Biotechnol. J.* **2018**, e1800125. [CrossRef] [PubMed]

97. Meineke, B.; Heimgärtner, J.; Lafranchi, L.; Elsässer, S.J. *Methanomethylophilus alvus* Mx1201 provides basis for mutual orthogonal pyrrolysyl tRNA/aminoacyl-tRNA synthetase pairs in mammalian cells. *ACS Chem. Biol.* **2018**. [CrossRef] [PubMed]

98. Willis, J.C.W.; Chin, J.W. Mutually orthogonal pyrrolysyl-tRNA synthetase/tRNA pairs. *Nat. Chem.* **2018**, *10*, 831–837. [CrossRef] [PubMed]

99. Borrel, G.; Gaci, N.; Peyret, P.; O'Toole, P.W.; Gribaldo, S.; Brugere, J.F. Unique characteristics of the pyrrolysine system in the 7th order of methanogens: Implications for the evolution of a genetic code expansion cassette. *Archaea* **2014**, *2014*, 374146. [CrossRef] [PubMed]

100. Beranek, V.; Willis, J.C.W.; Chin, J.W. An evolved *Methanomethylophilus alvus* pyrrolysyl-tRNA synthetase/tRNA pair is highly active and orthogonal in mammalian cells. *Biochemistry* **2018**. [CrossRef] [PubMed]

101. Venkat, S.; Sturges, J.; Stahman, A.; Gregory, C.; Gan, Q.; Fan, C. Genetically incorporating two distinct post-translational modifications into one protein simultaneously. *ACS Synth. Biol.* **2018**, *7*, 689–695. [CrossRef] [PubMed]

102. Mukai, T.; Yamaguchi, A.; Ohtake, K.; Takahashi, M.; Hayashi, A.; Iraha, F.; Kira, S.; Yanagisawa, T.; Yokoyama, S.; Hoshi, H.; et al. Reassignment of a rare sense codon to a non-canonical amino acid in *Escherichia coli*. *Nucleic Acids Res.* **2015**, *43*, 8111–8122. [CrossRef] [PubMed]

103. Krishnakumar, R.; Prat, L.; Aerni, H.R.; Ling, J.; Merryman, C.; Glass, J.I.; Rinehart, J.; Söll, D. Transfer RNA misidentification scrambles sense codon recoding. *Chembiochem* **2013**, *14*, 1967–1972. [CrossRef] [PubMed]

genes

MDPI

Article

Dissecting the Contribution of Release Factor Interactions to Amber Stop Codon Reassignment Efficiencies of the *Methanocaldococcus jannaschii* Orthogonal Pair

David G. Schwark, Margaret A. Schmitt and John D. Fisk *

Department of Chemistry, University of Colorado Denver, Campus Box 194, P.O. Box 173364, Denver, CO 80217-3364, USA; dschwark13@gmail.com (D.G.S.); margaret.schmitt@ucdenver.edu (M.A.S.)
* Correspondence: john.fisk@ucdenver.edu; Tel.: +1-303-315-7663

Received: 18 October 2018; Accepted: 5 November 2018; Published: 12 November 2018

Abstract: Non-canonical amino acids (ncAAs) are finding increasing use in basic biochemical studies and biomedical applications. The efficiency of ncAA incorporation is highly variable, as a result of competing system composition and codon context effects. The relative quantitative contribution of the multiple factors affecting incorporation efficiency are largely unknown. This manuscript describes the use of green fluorescent protein (GFP) reporters to quantify the efficiency of amber codon reassignment using the *Methanocaldococcus jannaschii* orthogonal pair system, commonly employed for ncAA incorporation, and quantify the contribution of release factor 1 (RF1) to the overall efficiency of amino acid incorporation. The efficiencies of amber codon reassignments were quantified at eight positions in GFP and evaluated in multiple combinations. The quantitative contribution of RF1 competition to reassignment efficiency was evaluated through comparisons of amber codon suppression efficiencies in normal and genomically recoded *Escherichia coli* strains. Measured amber stop codon reassignment efficiencies for eight single stop codon GFP variants ranged from 51 to 117% in *E. coli* DH10B and 76 to 104% in the RF1 deleted *E. coli* C321.ΔA.exp. Evaluation of efficiency changes in specific sequence contexts in the presence and absence of RF1 suggested that RF1 specifically interacts with +4 Cs and that the RF1 interactions contributed approximately half of the observed sequence context-dependent variation in measured reassignment efficiency. Evaluation of multisite suppression efficiencies suggests that increasing demand for translation system components limits multisite incorporation in cells with competing RF1.

Keywords: genetic code expansion; release factor 1; amber stop codon suppression; *M. jannaschii* orthogonal pair; fluorescence-based screen

1. Introduction

To a first approximation, the process of translation involves the direct reading of three nucleotide codons in mRNA by three nucleotide anticodons of tRNAs employing standard Watson-Crick base-pairing rules. This simple picture of translation is incomplete. System composition, molecular modifications, and sequence context effects contribute to fine tune the efficiency of decoding individual codons in different mRNA contexts. System composition includes factors such as the relative concentrations of cognate and near cognate tRNAs, and in the case of non-sense suppression, suppressor tRNAs and release factors. The efficiency of competition is further modulated by tRNA modifications that impact tRNA–protein interactions as well as alter the energy of codon–anticodon pairings. The efficiency of decoding a particular codon within a particular mRNA sequence is additionally dependent upon the sequence context of the codon within the mRNA. Factors including tRNA–ribosome interactions, extended tRNA interactions with the mRNA (outside of the codon), and tRNA–tRNA interactions on the ribosome have been invoked as mechanisms to explain observed differences in decoding the same codon at different positions within an mRNA. A clear picture of the relative quantitative importance of these and other contributing factors to the efficiency of translation has not emerged (reviewed in [1–3]). To productively engineer translation better, quantitative understanding of the contributions of the multiple factors affecting the efficiency of translation is needed.

Proteins containing biosynthetically incorporated non-canonical amino acids (ncAAs) are finding increasing use in basic biochemical studies and biomedical applications. The majority of ncAAs are incorporated into proteins in response to an amber stop (UAG) codon using an orthogonal tRNA/aminoacyl tRNA synthetase (aaRS) pair. The tRNA is modified to include a CUA anticodon to fully Watson-Crick base pair with the amber stop codon. The aaRS is engineered to recognize and aminoacylate the ncAA onto the orthogonal tRNA. Variants of only two orthogonal tRNA/aaRS pairs, the tyrosyl tRNA/aaRS pair from *Methanocaldococcus jannaschii* and the pyrrolysyl tRNA/aaRS pair from *Methanosarcina* species have been engineered to incorporate the majority of ncAAs [4,5].

Despite the common ancestry of the orthogonal systems, the efficiency of ncAA incorporation is highly variable. Orthogonal tRNA/aaRS pairs evolved to recognize different ncAAs exhibit broadly different efficiencies. Individual orthogonal pairs also exhibit varying efficiencies depending on the site of the UAG codon in the target protein. Variability across pairs is related to the differing enzymatic efficiencies of evolved variants to recognize different ncAAs. Variability in the expression of components and the cellular availability of different ncAAs further contributes to observed differences in the efficiency of incorporation for different ncAAs. Variability in the efficiency of incorporation based on the placement of a target codon within a gene sequence is the result of multiple differential competing interactions of tRNAs and release factors with the mRNA and the ribosome. The relative quantitative contributions of the multiple factors that determine the overall efficiency of incorporation of amino acids in response to amber stop codons are largely unknown.

Genetic code expansion via amber stop codon reassignment has been limited to incorporation of ncAAs at a single site because of competition with termination signals that curtail the amount of protein produced when multiple suppressions are attempted in a single protein. Recently, the Sakamoto, Wang, and Church laboratories each engineered genomic changes in *Escherichia coli* that mitigate the usual cytotoxic effect of deletion of the release factor that competes for decoding the amber stop signal [6–8]. The efficiency of incorporation of amino acids in response to amber stop codons is improved in these strains relative to strains that express release factor 1 (RF1); however, the contribution of release factor competition to the overall efficiency of amber stop codon reassignment has not been quantified.

To better understand the factors that contribute to the efficiency of reassigning amber stop codons in *E. coli* using the *M. jannaschii* tyrosyl tRNA/aaRS pair, the efficiency of suppression of individual and combinations of amber stop codons in green fluorescent protein (GFP) were quantified. To a first approximation, the system composition and tRNA modifications in each of these cases are expected to be identical. As a result, the suite of interactions directed by the mRNA sequence surrounding the targeted amber codon is expected to be a primary contributor to observed differences in the efficiency of stop codon suppression. The contribution of competition between the suppressing tRNA and release factors to amber codon reassignment efficiency was evaluated by quantifying the fluorescence of the GFP reporter variants in related *E. coli* strains with and without the competing release factor RF1. Evaluation of efficiency measurements of multiple reassignments within a single GFP gene enabled quantification of the contribution of competition and indicated that orthogonal pair machinery becomes a limiting factor during over-expression of genes containing multiple amber stop codons.

Systematic experimental manipulations of the translation system composition directed at quantitative evaluation of translational efficiency have not been reported. Manipulations with the purpose of improving protein production of mammalian genes in bacterial systems have qualitatively indicated that low abundance tRNAs can limit translational efficiency [9]. In the context of ncAA incorporation, multiple vector systems have been developed to improve the efficiency of ncAA incorporation through changes in the expression of orthogonal components, but systematic analyses of compositional changes between vector systems have not been performed [10–12]. Cell size, ribosome and tRNA numbers, and the ratio of ribosomes to tRNAs all vary with growth rate, making comparisons between experiments difficult. Experiments that report on codon-specific translational efficiencies have only recently become available (e.g., single molecule and ribosome profiling). The quantitative contribution of system composition variability to translational efficiency is largely unknown.

Beyond variability due to system composition, the sequence context in which a nonsense codon appears also modulates the efficiency of natural nonsense suppression [13,14]. Direct measurement of sequence context effects in sense codon reading are also evident from single molecule experiments and ribosome profiling, but systems that provide simple read-outs for sense codon reading are not available [2]. Related to context-specific reading differences, synonymous codons read by the same tRNA are decoded at different rates [15]. Both the efficiency of programed reading frame shifts and gain of function missense mutations are sequence context-dependent [16–18]. Evaluating the quantitative contributions of system composition and sequence context effects on the efficiency of nonsense suppression offers a window into the details of translation.

Context effects are of particular importance for genetic code expansion studies as the choice of substitution position within a target protein can have a large effect on the efficiency of ncAA introduction. In the case of amber suppressor tRNAs, the identity of the nucleotides immediately upstream and downstream of the amber codon in the mRNA have been shown to affect reading efficiency [19,20]. The bases in the vicinity of the amber stop codon are numbered such that the amber stop codon positions are +1, +2, +3, the 5′-nucleotides are assigned negative numbers and the 3′-nucleotides are assigned positive numbers. The magnitude and direction of context effects on amber stop codon reassignment with natural suppressor tRNAs depends on the specific suppressor tRNA employed. That context effects depend on the framework of the reassigning tRNA suggests that interactions outside of the codon-anticodon region contribute to observed modulations of reading efficiency. Evidence exists both for and against extended interactions of a tRNA with the mRNA sequence [19–21], indirect effects of tRNA-ribosome interactions [22,23], and termination factor-mRNA interactions [3,24] contributing to the context-depending efficiency of stop codon reassignment.

Data on multiple natural and engineered amber stop codon suppressors inserting different amino acids into the same seven test positions in a Lac repressor protein showed efficiencies spread over an approximately 10-fold range [25–28]. Different suppressor tRNAs showed different degrees of context effects and exhibited different patterns of relative increases and decreases in efficiency at the different positions [25,26]. Preferred codon contexts are evident through sequence analysis; however, the function of preferred contexts is not clear [29–34]. Although some codon orderings are not found in protein coding genes and different patterns of codon usage appear in highly and weakly expressed genes, simple rules regarding sequence preferences around individual codons are not evident [30,35]. Identifying signatures of codon context bias through bioinformatics analysis is confounded by the necessity of correcting for codon usage, tRNA abundances, GC content, and amino acid frequencies in proteins. Furthermore, the context of codons in naturally selected genes could be the result of evolutionary pressure to either increase or decrease codon reading efficiencies, resulting in confusion based upon examination of sequence alone [30,31,36,37].

Three recent works have presented data that can be analyzed to shed light on the context-specific reading of amber codons by the *M. jannaschii* and *Methanosarcina mazei* orthogonal tRNA/aaRS systems most commonly used in genetic code expansion [12,38,39]. Two of these studies were directed at generating improved systems for ncAA incorporation; however, the data presented can also be used to evaluate the effect of codon context on the efficiency of translation. The third study specifically addressed sequence context effects on the efficiency of amino acid incorporation by the *M. mazei* orthogonal tRNA/aaRS pair.

In a study describing the development of an improved vector system for the expression of ncAA-incorporating orthogonal tRNA/aaRS pairs, Young et al. measured the incorporation of 15 non-canonical amino acids at three amber stop codon positions in GFP [12]. The three sequence contexts in which suppression of the amber stop codon was evaluated showed an approximate 10-fold range of efficiencies, similar to that observed for natural amber suppressor tRNAs [25,40]. The ncAA incorporation efficiencies varied both with the particular evolved *M. jannaschii* aaRS used (and therefore, the identity of the amino acid) and with the position of the amber stop codon in the GFP reporter protein. Different patterns of relative efficiencies at the three sites were observed for different evolved aaRSs.

Pott et al. [39] selected optimized sequence contexts for the introduction of amino acids in response to an amber stop codon in an appended tag. Selection experiments were performed on libraries of N-terminal tag sequences containing an amber stop codon between two randomized codons (NNN-NNN-UAG-NNN-NNN). The nucleotide sequence contexts selected for improved ncAA incorporation using the *M. jannaschii* and *M. mazei* systems were similar, but not identical. The selected sequences increased the efficiency of amber stop codon suppression three-fold to five-fold relative to randomly chosen sequences. The preferred context for the *M. jannaschii* pair included a strong preference for adenosine (A) in the position immediately following the UAG stop codon, as had been previously observed for a number of natural amber stop codon suppression systems [24,25]. Additional strong preferences at the -5, -2, -1, $+6$ and $+7$ positions were also identified. Interestingly, the sequence contexts selected for efficient tyrosine and lysine sense codon reading were decidedly different than the contexts selected for amber stop codons. The different contexts for sense and nonsense codons suggests selection against release factor-mRNA interactions in the amber stop codon cases. The utility of the data presented for evaluating context effects on the orthogonal tRNA/aaRS systems was limited by the small number of selected sequences (fewer than 10 per selection) and an incomplete characterization of the efficiency of amber stop codon suppression for the selected sequences [39].

Codon context effects were explicitly examined for the *M. mazei* pyrrolysyl tRNA/aaRS orthogonal pair by Xu et al. employing a β-galactosidase assay to evaluate suppression efficiencies of a library containing randomized codons upstream and downstream of the targeted UAG codon (NNN-UAG-NNN) [38]. The overall consensus sequence for strong amber stop codon suppression identified was contained within the consensus sequence selected by the Pott et al. evaluation. The Pott et al. evaluation identified additional, stronger sequence effects based on the identity of the nucleotides at positions -6, -5, $+7$, $+8$ and $+9$, sites not analyzed by Xu et al. In the Xu et al. study [38], the efficiency of suppression across the evaluated sequence contexts varied over a 100-fold range, much larger than had been previously observed for natural suppressors. The relative importance of specific bases was difficult to interpret as many of the single base mutants were not reported. In cases where comparable single and double mutants were evaluated, the effects of the mutations on amber stop codon suppression efficiency appeared to be largely non-additive. Evaluation of the set of 18 single site mutants away from the consensus sequence indicated that the efficiency of suppression was exquisitely sensitive to the exact sequence.

Unraveling the mechanistic sources and quantitative contributions of the many factors affecting the overall efficiency of translation of a given codon is a largely open question and represents an important next step in quantitatively understanding the process of translation. The generation of high-quality quantitative efficiency measurements on a very large set of related systems in which individual variables are modulated would provide a necessary data set to begin to decipher base-level contributions to translational efficiency.

2. Materials and Methods

2.1. Cell Strains

DH10B: F⁻ mcrA Δ(*mrr-hsd*RMS-*mcr*BC) Φ80d*lacZ*ΔM15 Δ*lac*X74 *end*A1 *rec*A1 *deo*R Δ(*ara,leu*)7697 *ara*D139 *gal*U *gal*K *nup*G *rps*L λ⁻ (Thermo Fisher, Waltham, MA, USA).

C321.ΔA.exp: *E. coli* MG1655 Δ(*ybhB-bioAB*)::zeoR ΔprfA; all 321 UAG codons changed to UAA. C321.ΔA.exp was a gift from George Church (Addgene reference # 49018, Watertown, MA, USA) [41].

2.2. General Reagents and Materials

All restriction enzymes and DNA polymerases were purchased from New England Biolabs (Ipswich, MA, USA) and used according to the manufacturer's instructions. ATP was purchased from Fisher (BP413-25) (Waltham, MA, USA) and dNTPs were purchased form New England Biolabs (N0447S) (Ipswich, MA, USA). DNA isolation was performed using a Thermo Scientific GeneJET plasmid miniprep kit (K0503) (Thermo Scientific, Waltham, MA, USA) according to the manufacturer's protocols.

Luria broth (LB) liquid media (per liter: 10 g tryptone, 5 g yeast extract, 5 g NaCl) and LB agar plates with 15 g/L agar (TEKNova, A7777, Hollister, CA, USA) were used unless otherwise noted. Isopropyl-β-ᴅ-thiogalactoside (IPTG) was purchased from Gold Bio (I2481C5) (St. Louis, MO, USA). Spectinomycin (Enzo Life Science, BML-A281, Farmingdale, NY, USA) was used at 50 µg/mL to maintain the pUltra-based vectors harboring the tRNA and aaRS genes. Carbenicillin (Thermo Scientific, Waltham, MA, USA) was used at 50 µg/mL to maintain the vectors harboring the GFP reporter gene. All bacterial cultures were grown at 37 °C unless otherwise noted.

Electrocompetent stocks of all strains were prepared in-house according to the method of Sambrook and Russell [42]. Typical transformation efficiencies for electrocompetent cells produced in this way are 10^9 cfu/µg of supercoiled DNA. All transformations were recovered in SOC (20 g/L tryptone, 5 g/L yeast extract, 10 mM NaCl, 2.5 mM KCl, 10 mM MgCl$_2$, 20 mM glucose) for 1 h at 37 °C with shaking prior to transfer to media containing the appropriate antibiotics.

All oligonucleotides were purchased from Integrated DNA Technologies (Coralville, IA, USA). All DNA sequencing was performed by Genewiz (Plainfield, NJ, USA).

2.3. Construction of Amber Stop Codon-Containing GFP Reporter Variants

The complete sequence of the 100% fluorescent, wild-type GFP reporter plasmid has been reported (Supporting information in [43]). Amber stop codon containing variants were based on this vector.

Mutagenesis for the preparation of the GFP variants was achieved by using a protocol modified from the Stratagene QuikChange Multi-Site Directed Mutagenesis kit (San Diego, CA, USA). Briefly, 125 ng of mutagenic primer (one or two primers used per site mutated), 100 ng of template plasmid DNA, Q5 High Fidelity Polymerase Buffer (1X final concentration), dNTPs (200 µM final concentration), and Q5 DNA polymerase (0.5 units) were combined into a 200 µL PCR tube. Ultrapure water was added to adjust the final reaction volume to 25 µL. Reactions were then subjected to thermocycling in either a PeqStar (peqLab, Wilmington, DE, USA) or MjMini (Bio-Rad, Hercules, CA, USA) thermocycler. After initial denaturing at 98 °C for 1 min, 25 cycles of the following were performed: 98 °C for 20 s, 72 °C for 20 s, and 72 °C for 15 s/kb of template plasmid. After 25 cycles were completed, a final 5 min extension at 72 °C was performed, and the reactions were then cooled down to 4 °C. Following thermocycling, 1 µL (20 units) of DpnI restriction enzyme was added to the reaction tube and incubated for 2 h at 37 °C. Reactions were transformed into electrocompetent *E. coli* DH10B without cleanup.

The sequences of each mutagenic primer are provided in Appendix A, Table A1.

2.4. Vector for Expression of the Orthogonal M. jannaschii tRNA and aaRS

The vector containing the genes for expression of the *M. jannaschii* tyrosyl tRNA and aminoacyl tRNA synthetase has been described [44]. Expression of the aaRS is driven by a constitutive lpp promoter in the system used in this evaluation.

2.5. Fluorescence-Based Screen for Quantification of the Efficiency of Translation

The fluorescence-based screen provides a quantitative measure of the extent of amber stop codon suppression at each test position by bracketing the observed GFP signal for codon reassignment measurement between a "100% fluorescence" reference value produced by expressing superfolder GFP with a tyrosine UAC codon specifying the fluorophore and a "0% fluorescence" reference value established through expressing superfolder GFP with a non-tyrosine codon specifying the fluorophore. Both the 100% and 0% fluorescence reference systems include a plasmid expressing the *M. jannaschii* aaRS and tRNA to maintain a metabolic burden on the cells equivalent to that of the systems under evaluation.

The fluorescence-based screen for sense codon reassignment has been described [43,44]. Briefly, a given GFP reporter vector was co-transformed with the vector expressing the orthogonal translational components. After overnight growth, colonies were picked into 200 µL LB media in a 96-well plate. Cells were grown to at least mid-log phase (usually 8–10 h) with shaking at 37 °C. Cells were diluted 10-fold into media containing 1 mM IPTG to induce expression of the GFP reporter. Assays were performed in a Fluorotrac 200 clear bottom 96-well plate (Greiner 655096, Kremsmünster, Austria) and monitored in a BioTek Synergy H1 plate reader (BioTek, Winooski, VT, USA) at 37 °C with continuous double orbital shaking. The optical density (OD_{600}) and fluorescence of each well was measured every 15 min for at least 15 h; optical density was measured at 600 nm, and fluorescence was measured with an excitation at 485 nm and detection at 515 nm with an 8 nm band pass.

Calculation of the efficiency of codon reassignment has been described in detail in supplementary section S6 of Schmitt et al. [43]. Briefly, the relative fluorescence of each 200 µL culture was calculated by dividing the 0% reassignment control-blanked fluorescence for a reporter variant by the path length and blank corrected OD_{600} at each time point. The 100% relative fluorescence unit (reported as fluorescence units per unit optical density at 600 nm, RFU) value for amber codon reassignment efficiency was defined by taking an average of six separate colonies expressing wild type GFP and the orthogonal machinery. For each sample well, the relative fluorescence was calculated for each of the data points gathered between 8 and 12 h after induction of GFP with IPTG. These relative fluorescence values were averaged to determine the overall RFU for each sample well. A more detailed explanation of time point selection is provided in section S7 of Schmitt et al. Amber codon suppression efficiency for each UAG positional variant was calculated by dividing the overall RFU values from a single clone by the average 100% reference RFU value [43].

For a given GFP reporter vector, the reported amber codon reassignment efficiency is the average of at least six biological replicates. The majority of suppression efficiencies reported here are the average of between 12 and 18 biological replicates. The calculated efficiencies for only two reporter vectors, the two amber stop codon containing variants Y143/Y182 and Y106/Y143, in the C321.ΔA.exp cells are the average of fewer than six biological replicates (four and three, respectively).

3. Results and Discussion

In order to better understand the factors controlling the overall efficiency of amber stop codon suppression by the orthogonal *M. jannaschii* tyrosyl tRNA/aaRS pair, we employed a fluorescence-based screen to quantify amber stop codon reassignment efficiencies at eight different tyrosine positions in superfolder green fluorescent protein in two *E. coli* strains. To remove complicating effects resulting from differences in protein folding and performance between mutants and control proteins, the suppression of amber stop codons was accomplished with the tyrosine-incorporating orthogonal *M. jannaschii* tRNA/aaRS pair, and stop codons were placed at tyrosine positions in each GFP reporter variant. In this way, the protein produced by amber stop codon suppression with the *M. jannaschii* pair is phenotypically identical to wild type GFP controls, allowing fluorescence to be a direct measurement of amber stop codon reassignment efficiency. The mRNA sequence surrounding each amber stop codon was the only difference between the eight single amber codon GFP reporters. The identity and the expression levels of the orthogonal *M. jannaschii* tyrosyl tRNA and aaRS, the promoter driving the GFP reporter gene, and the cell lines used (i.e., system composition/cellular environment) were all consistent between the evaluated systems. Under these conditions, differences in decoding efficiency observed between amber stop codon positions should be predominantly due to codon context effects. The amber stop codon suppression efficiencies were evaluated in two *E. coli* strains, with and without the competing release factor RF1.

The hypothesis that the efficiency of suppression of multiple amber stop codons within a single gene is independent and additive was then evaluated for a series of related GFP variants. The measured amber stop codon suppression efficiencies for the single amber stop codon variants were used to predict the suppression efficiencies of GFP variants containing multiple amber stop codons, and the predicted and measured amber stop codon reassignment efficiencies for 14 multiple amber stop codon-containing GFP variants were compared. Comparison of reassignment efficiencies between the two strains enabled the isolation of the contribution of release factor interactions to the overall efficiency of translation.

3.1. Measurement of Reassignment Efficiencies for single UAG GFP Variants using the M. jannaschii Tyrosyl tRNA/aaRS Pair

In order for GFP fluorescence to accurately quantify successful incorporation of an amino acid in response to a stop codon, translation of the reporter variants in the absence of the amber stop codon reassignment machinery must result in non-fluorescent protein products. Singly, each of the nine tyrosine codons within the wild type GFP reporter gene (which served as the 100% fluorescent control) were mutated to the amber stop codon. GFP fluorescence was monitored in the absence of orthogonal translational machinery for each variant. No significant fluorescence from the GFP products for eight of the nine single amber stop codon variants was observed. One variant, Tyr 237 UAG, showed GFP fluorescence above that of the 100% reference wild type GFP. The observation of wild type levels of functional GFP in this system is unsurprising. Position 237 is the second to last residue in the GFP sequence, and stopping translation of the reporter variant so close to the C-terminus would not be expected to dramatically decrease function. The Tyr 237 UAG mutant was not utilized further.

Amber stop codon reassignment efficiency was measured with the *M. jannaschii* tyrosyl tRNA/aaRS pair for each of the eight GFP variants in *E. coli* DH10B cells. The measured incorporation efficiencies, reported as the cell density corrected fluorescence of the stop codon suppressing system divided by the cell density corrected fluorescence of a system in which wild type GFP was expressed, ranged from 51.2 ± 1.9% to 117.0 ± 2.6% (Table 1). The suppression efficiency reports on a combination of enzyme efficiency and relative expression level for the particular orthogonal system employed. The average reassignment efficiency across the eight positions measured is 87%, a value on the high end of the efficiencies measured for natural suppressors. The context-dependent variation of approximately two-fold observed for the *M. jannaschii* pair in this experimental evaluation is in line with that observed for natural suppressors. Typically, the efficiencies of more effective suppressors are less dependent on sequence context.

The identity of the nucleotide on the 3′-side of the UAG codon (i.e., +4 position) was the first recognized sequence context effect [28]. Multiple mechanistic rationales for the influence of a 3′-base on translational efficiency have been put forth, but the actual mechanisms remain unclear. An A or G immediately 3′ of the amber codon could potentially base pair with the universally conserved U33 of tRNAs. An alternative hypothesis for the preference of purines at the +4 position is that base stacking stabilizes the suppressor tRNA–mRNA interactions. Changing the nucleotide following the amber stop codon from C to A resulted in a 10-fold increase in the efficiency of suppression for the SupE amber suppressor [27,28]. The trend of purines rather than pyrimidines in the +4 position leading to improved translation efficiency was generally supported in a large study employing multiple suppressing tRNAs [27]. Context effects differed in direction and magnitude for the different suppressors in the same mRNA context. That context effects are different for different suppressor tRNAs suggests that the nucleotide in the +4 position of the mRNA is not the only factor involved in determining the efficiency of translation of a given codon [19,20]. An additional hypothesis, at least for the case of amber stop codon suppression, is that differential competition between interactions with release factors contributes to context effects [24].

The data gathered in our investigation generally support the hypothesis of +4 purines improving the efficiency of incorporation in response to an amber stop codon. The sequence contexts of the UAG codons evaluated include cases where the −1 bases are U, C and A and the +4 bases are C, A and G (Table 1). Three of the four most efficient amber suppressions with the orthogonal *M. jannaschii* tRNA/aaRS were observed in a context with a +4 A. In contrast, three of the four least efficient amber stop codon suppressions were in the context of a +4 C. A trend for the effect of the identity of the nucleotide at the −1 position on the efficiency of translation was not observed. This finding is in line with previous studies that did not identify clear patterns of context effects as a result of the nucleotides upstream of the amber stop codon (i.e., positions −1, −2, −3) [40].

Table 1. Amber stop codon reassignment efficiency for single amber green fluorescent protein (GFP) reporter variants.

Position of UAG Codon	Nucleotide Sequence Surrounding UAG [1]	Reassignment Efficiency DH10B	Reassignment Efficiency C321.ΔA.exp	*p*-Value (between Strains)
Tyr 66	CUG ACC *UAG* **G**GC GUC	116.9 ± 2.6%	97.5 ± 3.6%	1.93×10^{-11}
Tyr 74	UCC CGU *UAG* **C**CG GAC	51.2 ± 1.9%	75.5 ± 4.9%	2.18×10^{-12}
Tyr 92	GAA GGC *UAG* **G**UA CAG	83.6 ± 3.0%	76.4 ± 3.3%	4.05×10^{-6}
Tyr 106	GGG ACC *UAG* **A**AA ACC	89.7 ± 2.5%	81.5 ± 4.1%	1.34×10^{-5}
Tyr 143	CUC GAA *UAG* **A**AC UUC	94.9 ± 2.8%	101.9 ± 5.0%	0.000455
Tyr 151	AAC GUA *UAG* **A**UC ACG	95.9 ± 3.1%	97.8 ± 4.9%	0.243
Tyr 182	GAC CAC *UAG* **C**AG CAG	71.1 ± 2.5%	89.9 ± 5.8%	4.38×10^{-8}
Tyr 200	AAC CAC *UAG* **C**UG UCC	88.5 ± 4.0%	103.7 ± 5.5%	1.14×10^{-7}

[1] The targeted amber stop codon (UAG) is italicized. The sequence of the two codons 5′ and two codons 3′ of the targeted codon are shown. The nucleotide in the +4 position is shown in bold type.

3.2. Role of Ribosomal Release Factor 1 in Determining Amber Stop Codon Reassignment Efficiency

One of the predominant hypotheses regarding the source of context effects posits that differential interactions between the protein release factor and the mRNA contribute to observed sequence-dependent differences in the efficiency of amber suppression. Within the literature, studies both support and refute the role of release factor interactions in governing context effects [1,28]. In order to evaluate the contribution of release factor mRNA interactions in determining amber stop codon suppression efficiencies by the orthogonal *M. jannaschii* tRNA/aaRS pair, the set of amber stop codon GFP mutants were further evaluated in a genomically recoded *E. coli* strain, C321.ΔA.exp, in which all instances of the UAG codon in the genome were replaced, and the gene coding for ribosomal RF1 was removed [41] (Figure 1, Table 1). The single amber stop codon GFP variants were suppressed with efficiencies between 75.5 ± 4.9% and 103.7 ± 5.5% in the genomically recoded *E. coli* strain, compared to 51.2 ± 1.9% to 117.0 ± 2.6% in *E. coli* DH10Bs.

Utilization of near-cognate tRNAs to resolve issues of the ribosome stalling at the UAG codon in RF1 deleted cell lines has been observed [45]. In order to evaluate whether incorporation of non-tyrosine amino acids in response to amber codons could be contributing to the observed fluorescence when the orthogonal translation machinery and GFP reporters were expressed in C321.ΔA.exp, we examined two of the single site GFP reporters (positions 66 and 74) in concert with an inactive *M. jannaschii* tRNA variant. In these systems, the metabolic burden placed on the cells is as similar as possible to the systems under which codon reassignment is evaluated. In the position 74 case, any observable fluorescence above the intrinsic fluorescence of the medium could be indicative of background suppression. This statement assumes that position 74 in GFP is tolerant to mutation and able to fold and fluoresce with a different amino acid substituted for tyrosine. In the case of the position 66 mutant, only background incorporation of tyrosine would produce fluorescence. Significant fluorescence above background media, particularly within the time window over which reassignment efficiency is quantified, is not observed in these systems. This result suggests that, at least for the combination of orthogonal translation machinery, GFP reporters, and the cellular environment (e.g., media, temperature, antibiotic concentration) utilized in this study, suppression of amber codons by endogenous, near-cognate, non-tyrosine tRNAs is not a major contributor to the fluorescence observed in the RF1-deleted cell line. Further, an overestimation of the possible suppression by near-cognate tRNAs would contribute at most 0.006% to the calculated reassignment efficiency. A more elaborate discussion of these points is included in Appendix B.

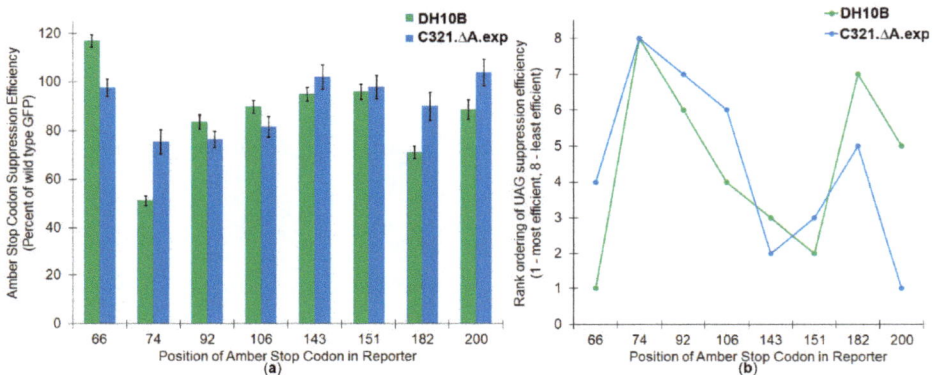

Figure 1. Side by side comparison of amber stop codon reassignment in two *Eschericia coli* strains. (**a**) Reassignment efficiencies as a percentage of wild type GFP at each of the eight single site UAG variants. (**b**) Rank ordering of suppression efficiencies of eight single site UAG positions from most efficient (placed at #1) to least efficient (placed at #8) in two *E. coli* strains. Green bars and dots represent data measured in DH10B cells; blue bars and dots represent data measured in C321.ΔA.exp.

The general trend of the relative efficiency of stop codon reassignment at the various positions was similar between the two cell lines. Figure 1b shows the rank ordering of the position of the most efficiently suppressed amber stop codon (1) to the position of the least efficiently suppressed amber stop codon (8) in each strain. While rank ordering was similar, there were significant differences in observed amber stop codon suppression efficiencies between the two cell lines. Three of the four least efficiently suppressed amber stop codon positions in *E. coli* DH10B cells (74, 182, and 200) were all significantly improved in the RF1 deleted cell line (+24.3, +18.8, and +15.2%, respectively). Three of the positions tested (66, 92, 106) showed decreases in absolute amber stop codon suppression efficiency upon removal of RF1 (−19.4, −7.2, and −8.2%, respectively).

A measure of the extent of the contribution of RF1 to the observed context effect can be derived from the deviation in measured values. The differences between measured stop codon suppression efficiencies in the two strains were statistically significant for seven of the eight single site variants evaluated (*p*-values less than 0.001, Table 1). These differences suggest that interactions with RF1 contribute to the variation in site to site reassignment efficiencies at all positions evaluated except position 151. The larger spread of measured efficiencies in the cells expressing RF1 (DH10B) suggests that both tRNA and release factor interactions are contributing to context effects. The extent of the context effect in DH10B is approximately 20% of the average value (i.e., the standard deviation of the average of reassignment efficiencies at the 8 positions). In contrast, the context effect in C321.ΔA.exp cells is approximately 11% of the average value. The difference in context effects between the two systems suggests that the contribution of release factor interactions accounts for about half of the observed context effect in DH10B cells.

The +4 sequence contexts for the set of codon positions that showed increased amber stop codon reassignment efficiencies upon removal of RF1 were consistent, having a +4 C and either U or C at the −1 position relative to the amber stop codon. Sequence consensus was not apparent across the positions that showed reduced or unchanged amber stop codon suppression efficiency upon RF1 removal. The consistency of the sequence contexts and the direction of changes in suppression efficiency suggest that RF1 does play a role in determining context effects. Our data suggests that RF1 has a preference for contexts with a +4 C.

3.3. Measurements of Amber Stop Codon Reassignment Efficiencies for Multiple UAG-Containing GFP Variants

The synergy/independence of the suppression of multiple amber stop codons in a single gene has not been subjected to widespread systematic analysis and could be useful in evaluating the contribution of local and global effects on context-dependent variations in codon translation. Recent manuscripts have evaluated the overall efficiency of ncAA incorporation in response to multiple amber codons in single genes in the C321.ΔA.exp cell line, but these data have not been evaluated in relation to deciphering the factors effecting amber codon reassignment efficiency [45,46]. The proposed mechanisms for mRNA sequence context effects on translation efficiency predominantly invoke local interactions and suggest that the effects of context on multiple incorporation events within a single gene would be additive and independent. To be clear, "additive and independent" interactions would imply that if, for example, two different amber stop codons are individually suppressed with 50% efficiency, then a double mutant containing the combination of the two amber stop codons would be expected to be produced at 25% overall efficiency. The measured single site suppression efficiencies were used to predict multisite suppression efficiencies based on an assumption of additivity and independence (Table 2).

Table 2. Predicted and measured UAG reassignment efficiencies for reporter variants containing multiple amber stop codons.

Position of UAG Codons	Predicted Reassignment Efficiency DH10B	Measured Reassignment Efficiency DH10B	Predicted Reassignment Efficiency C321.ΔA.exp	Measured Reassignment Efficiency C321.ΔA.exp
Y74, Y182	36.4 ± 1.8%	28.7 ± 1.2%	67.8 ± 6.2%	52.7 ± 3.2%
Y66, Y143	111.0 ± 4.1%	70.6 ± 3.2%	99.4 ± 6.1%	63.3 ± 4.6%
Y66, Y74	59.8 ± 2.6%	33.7 ± 0.9%	73.6 ± 5.5%	51.6 ± 6.5%
Y74, Y143	48.6 ± 2.3%	58.8 ± 1.4%	76.9 ± 6.3%	60.7 ± 8.1%
Y74, Y200	45.3 ± 2.7%	37.9 ± 1.8%	78.3 ± 6.6%	70.0 ± 7.3%
Y74, Y92	42.8 ± 2.2%	33.9 ± 1.7%	57.7 ± 4.5%	49.3 ± 5.0%
Y143, Y200	84.0 ± 4.5%	68.9 ± 3.7%	105.7 ± 7.6%	76.4 ± 2.1%
Y143, Y182	67.4 ± 3.1%	54.0 ± 2.3%	91.5 ± 7.4%	68.8 ± 1.8%
Y106, Y143	85.2 ± 3.5%	66.6 ± 2.4%	83.0 ± 5.9%	67.2 ± 0.4%
Y92, Y143	79.4 ± 3.7%	57.9 ± 2.2%	77.8 ± 5.1%	57.2 ± 6.3%
Y74, Y92, Y200	37.9 ± 2.6%	26.2 ± 0.9%	59.8 ± 5.6%	56.0 ± 2.8%
Y74, Y182, Y200	32.2 ± 2.2%	24.7 ± 0.8%	70.3 ± 7.4%	66.0 ± 1.5%
Y74, Y92, Y106, Y200	34.0 ± 2.5%	19.8 ± 0.3%	48.7 ± 5.2%	51.3 ± 1.5%
Y74, Y92, Y106, Y182, Y200	24.2 ± 2.0%	13.2 ± 0.3%	43.8 ± 5.5%	40.0 ± 1.5%

The amber stop codon suppression efficiencies for 14 different double, triple, quadruple and quintuple tyrosine to amber stop codon GFP mutants were measured in both *E. coli* DH10B and C321.ΔA.exp cells (Table 2). As expected, the overall UAG suppression efficiencies decrease with increasing numbers of UAG codons in the GFP reporter. In both cell lines there is a clear trend that the measured amber stop codon suppression efficiencies are consistently less than what would be predicted from the measured single site efficiencies (Figure 2). In the DH10B cell line, for all but one of the variants, the measured amber stop codon reassignment efficiencies were lower than the predicted efficiencies. In the RF1 deleted cell line, the measured amber stop codon reassignment efficiencies were consistently lower than the efficiencies predicted for non-interacting independent sites, but the measurements were more varied, reflecting greater variability in the growth of the C321.ΔA.exp cells. We analyzed the difference between the predicted and measured distributions with a two-factor analysis of variance (ANOVA) statistic. The measured efficiencies differ from those predicted using the assumption of additivity with *p*-values of 0.00054 (DH10B) and 0.00025 (C321.ΔA.exp). The observed trends suggest that some factor contributing to the efficiency of translation is exerting a synergistic as opposed to independent effect.

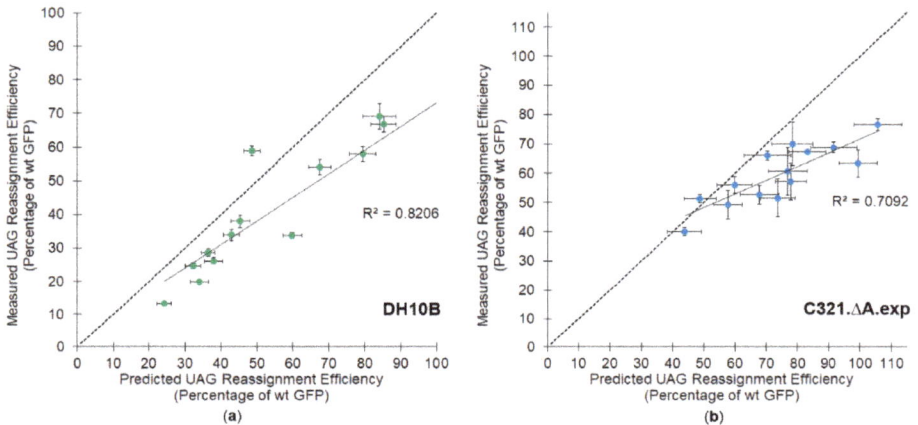

Figure 2. Measured versus predicted amber stop codon reassignment efficiencies for GFP reporter vectors containing two or more suppressible amber codons in (**a**) *E. coli* DH10B; (**b**) *E. coli* C321.ΔA.exp. The dotted line on the diagonal represents where the data would be expected to fall based on an assumption of additivity and independence for the efficiency of multisite amber stop codon reassignment. The full line represents the linear regression analysis of the data, with the R^2 value shown on each graph.

The most probable explanation for synergy involves variation in relative translational potential as a function of demand. As demand for orthogonal machinery (e.g., aminoacylated tRNA) increases with the number of suppressible amber codons in an mRNA, the reassignment efficiency at each individual site might be expected to decrease. Additional sources of synergistic interactions might involve differential ribosome drop off frequencies or mRNA degradation times related to overall translation efficiency [47,48].

To evaluate the contribution of competition between the available orthogonal tRNA and the release factor to the efficiency of amber codon reassignment, the average per site reassignment efficiencies were calculated for cases of increasing demand (1 vs. 2 vs. 3 amber codons per gene). The comparison between the average per site reassignment efficiency for the strains with and without release factor suggests that competition is occurring in the RF1-containing strain (DH10B), but is largely absent from the RF1 deleted strain (Table 3). The average per site reassignment efficiency drops from 87% for single reassignments per gene to 67% for genes requiring five reassignments in the RF1 containing cell line. In the RF1 deleted cell line, C321.ΔA.exp, the calculated per site reassignment efficiency is fairly constant at 85% for genes containing between one and five suppressible amber codons.

Perhaps unexpectedly, the average per site incorporation efficiency appears to plateau at approximately 67% for reporters with four and five UAG codons, rather than decreasing as was observed going from reporters with 1 UAG codon to 2 UAG codons. Our tentative explanation of this plateau is that an equilibrium is being reached between the combined kinetics of incorporation and tRNA recharging for the particular concentrations of *M. jannaschii* tRNA, *M. jannaschii* aaRS, RF1, and available tyrosine present in the cell. At lower codon demand (e.g., fewer suppressible stop codons per mRNA), the concentrations of aminoacylated tRNA and release factor in the cell favor orthogonal pair directed amber codon reassignment. As the number of suppressible stop codons per mRNA increases, the functional steady-state level of aminoacylated tRNA decreases, leading to lower per site reassignment efficiency. The drop-off is not expected to be linear; increased concentrations of non-aminoacylated tRNA increases the rate of re-aminoacylation. We expect that at some point a kinetic equilibrium is established with a particular, system-dependent per site efficiency. This explanation accounts for the little to no decrease in per site efficiency observed in C321.ΔA.exp as the number of UAG codons in each reporter vector increased. In these cells, aminoacylated tRNAs are not competing with RF1 for the amber stop codon. The available tRNAs are able to be aminoacylated, participate in translation, and be recharged for participation in another peptide bond-forming reaction without being kinetically outcompeted by RF1 during that course of events.

Table 3. Average measured *per codon* reassignment efficiency for UAG containing reporter variants.

Reporter Variant Class	Number of Systems Evaluated	Number of Codons Considered	Average Measured per Site Incorporation Efficiency DH10Bs [1]	Average Measured per Site Incorporation Efficiency C321.ΔA.exp
1 UAG codon	8	8	86.5%	90.5%
2 UAG codons	10	20	70.6%	78.4%
3 UAG codons	2	6	63.3%	84.7%
4 UAG codons	1	4	66.7%	84.6%
5 UAG codons	1	5	66.7%	83.3%

[1] The per site reassignment efficiencies were the averages of values determined for genes with equivalent numbers of amber codons. The per site reassignment was calculated as the nth root of the total reassignment efficiency, where n is equal to the number of amber codons in the gene.

3.4. Effect of Altering the Genetic Code on the Health of Amber Stop Codon Reassigning Systems

The extent to which cells tolerate reassignment of a stop codon to an amino acid has not been widely investigated. Missense errors are thought to be generally destabilizing, but translation errors introduced through defective aaRS editing or antibiotic treatment are broadly tolerated by cells. We examine the extent to which reassignment of the amber stop codon is tolerated by *E. coli* using the measured instantaneous doubling times of each reassignment system (Table 4). The requirement for cells to grow and divide in the presence of two antibiotics while replicating the DNA for both the orthogonal translation machinery vector and a GFP reporter vector slows the growth of both DH10B and C321.ΔA.exp. Allowing amber stop codon suppression by expressing the active *M. jannaschii* tRNA/aaRS pair exerts an additional deleterious effect on DH10B cells. Slowing of the instantaneous doubling times of C321.ΔA.exp beyond that which comes as a result of the antibiotic presence is not observed. Although C321.ΔA.exp cells exhibit longer doubling times than *E. coli* without the RF1-related genomic changes, amber codon reassignment does not have as much of an effect on the health of C321.ΔA.exp relative to the effect amber codon reassignment exerts on the health of DH10B.

Table 4. Instantaneous doubling times for amber stop codon suppressing systems.

Cellular Environment	Instantaneous Doubling Time (DH10B)	Instantaneous Doubling Time (C321.ΔA.exp)
Cells only (no antibiotic burden)	31.1 ± 1.2 min	40.0 ± 2.7 min
Inactive tRNA/aaRS and single site GFP reporter (2 antibiotics, no amber reassignment)	40.9 ± 0.9 min	46.1 ± 4.0 min
Amber suppressing tRNA/aaRS and single site GFP reporter (2 antibiotics, with amber reassignment)	54.1 ± 1.8 min	43.9 ± 1.7 min

aaRS: Aminoacyl tRNA synthetase.

4. Conclusions

We present a high-quality data set of amber stop codon suppression efficiency measurements employing a GFP fluorescence-based screen to quantify amber suppression by the orthogonal *M. jannaschii* tyrosyl tRNA/aaRS pair. Comparisons between sites of amber codon suppression within genes and between cells with and without competing release factor allow the quantification of some system composition-dependent contributions to orthogonal pair-directed nonsense suppression efficiency. Because transition is an incredibly complex process, no single set of measurements can define in an absolute way the contribution of particular factors to the overall process. The quantified values described here are a small step toward building a much larger set of quantitative measurements needed to describe the ways in which various system composition variables contribute to the overall activity of the translational apparatus. The data presented should help inform the use of the *M. jannaschii* pair in genetic code expansion applications. Measurements made using orthogonal machinery to incorporate natural amino acids are a useful tool for dissecting the workings of the natural translational apparatus.

Amber stop codon reassignment efficiencies for eight single stop codon GFP variants ranged from 51.2 ± 1.9 to 117.0 ± 2.6% in *E. coli* DH10B and 75.5 ± 4.9% to 103.7 ± 5.5% in the RF1 deleted *E. coli* C321.ΔA.exp. The relative ordering of reassignment efficiencies in the two strains was similar, suggesting that RF1 contributes to but is not a dominant factor for the observed context-dependent variation in translational efficiencies. Evaluation of reassignment efficiency changes in specific sequence contexts in the presence and absence of RF1 suggested that RF1 specifically interacts with +4 Cs. An estimate for the overall contribution of RF1 to sequence context-dependent variation in translational efficiency was calculated from the range of amber stop codon suppression efficiencies across positional GFP reporter variants. Relative sequence-dependent variation in efficiencies suggested that RF1 interactions are responsible for approximately half of the observed variation.

The amber stop codon suppression efficiencies for 14 different double, triple, quadruple and quintuple tyrosine to amber stop codon GFP mutants were measured in both *E. coli* DH10B and C321.ΔA.exp cells. The overall UAG reassignment efficiencies decrease with increasing numbers of UAG codons in the GFP reporters. In both cell lines there is a clear trend that the measured amber stop codon suppression efficiencies are consistently less than what would be predicted from measured single site efficiencies. The most probable explanation for long range synergistic interaction of suppression sites is that increasing demand for translational components reduces the per site suppression efficiency. Average per site reassignment efficiencies decreased and leveled off as the translation component demand increased in DH10B cells (i.e., in the presence of competing release factor). In C321.ΔA.exp cells without competing release factor, the observed per site suppression efficiency was consistent across the set of evaluated multisite mutants.

Author Contributions: Conceptualization, D.G.S. and J.D.F.; Funding acquisition, M.A.S. and J.D.F.; Investigation, D.G.S.; Methodology, D.G.S. and M.A.S.; Supervision, M.A.S. and J.D.F.; Writing—original draft, D.G.S.; Writing—review & editing, M.A.S. and J.D.F.

Funding: This research was funded by the National Science Foundation, grant number 1507055. The APC was funded by the University of Colorado Denver.

Acknowledgments: We thank George Church for the genetically recoded *E. coli* strain C321.ΔA.exp (Addgene reference # 49018).

Conflicts of Interest: The authors declare no conflict of interest. The funders had no role in the design of the study; in the collection, analyses, or interpretation of data; in the writing of the manuscript, or in the decision to publish the results.

Appendix A

Table A1. Sequences of mutagenic primers for generating amber stop codon-containing GFP reporter variants.

GFP Position to Be Mutagenized	Primer Sequence (5′ to 3′)
66	CACTCGTCACCACACTCACCTAGGGCGTCCAGTGCTTCTCCCG
74	TCCAGTGCTTCTCCCGTTAGCCGGACCACATGAAACGG
92	GCCATGCCCGAAGGCTAGGTACAGGAACGTACCATCTCCTTC
106	CTTCAAAGACGACGGGACCTAGAAAACCCGTGCCGAAGTC
143	CCTCGGACACAAACTCGAATAGAACTTCAACTCACACAACGTATACATC
151	CAACTTCAACTCACACAACGTATAGATCACGGCAGACAAACAGAAAAAC
182	CCAGCTCGCAGACCACTAGCAGCAGAACACCCCAATCG
200	GTCCTCTTACCAGACAACCACTAGCTGTCCACACAGTCCGTC
237	CGGCATGGACGAACTCTAGAAACACCACCACCACCAC

Appendix B

Evaluation of raw fluorescence data for systems with little to no expected amber codon reassignment is complicated by the observation that a fraction of the fluorescent components of the media are consumed as cells grow and divide. Various antibiotics also contribute to the background fluorescence of the media; utilizing the corrected fluorescence of non-fluorescent cells grown under different media conditions does not resolve the situation. Minor batch to batch variation in media composition contributes to the majority of differences observed in media fluorescence between experiments. The 0% fluorescent control evaluated in each plate with each batch of media is used to correct the raw fluorescence of systems in which codon reassignment takes place. These impact of these variations on the calculated reassignment efficiency is very low (less than 0.006%, discussed below).

The set of 3 green lines shows the raw fluorescence over time for blank media (open triangles, Δ), an amber GFP reporter co-expressed with inactive orthogonal translation machinery in C321.ΔA.exp (closed circles, ●) and DH10B (open circles, ○). As DH10B cells express RF1 and can terminate translation at UAG codons in the absence of an amber suppressing tRNA, one does not expect near-cognate suppression of the UAG codon by endogenous tRNAs. The raw fluorescence observed in both strains is similar as a function of time, which suggests that suppression of amber codons by near-cognate tRNAs is not significantly contributing to the fluorescence observed in reassignment systems expressing functional amber suppressing tRNAs and UAG GFP reporter variants.

The set of maroon and purple lines shows the raw fluorescence over time for blank media (open triangles, Δ) and either the GFP position 66 (closed maroon circles, ●) or GFP position 74 (closed purple circles, ●UAG variants co-expressed with inactive orthogonal translation machinery in C321.ΔA.exp cells. These two evaluations were performed on consecutive days using the same media, as noted by the similar raw fluorescence traces for blank media. The raw fluorescence versus time profiles for the 2 reporter variants are nearly identical. As noted above, in the case of the position 66 variant, only suppression by tyrosine would result in observed fluorescence, while in the case of the position 74 variant, suppression by any amino acid could lead to fluorescence (assuming that position 74 of GFP is tolerant to mutation). For these reasons, despite the absence of mass analysis, we do not suspect that suppression of the amber stop codons by near-cognate tRNA in the C321.ΔA.exp strain is contributing to our calculations of amber suppression efficiency.

Further, estimates of the relative contribution of possible suppression by near-cognate endogenous tRNAs would comprise very little of the overall reassignment efficiency as a percent of the 100% fluorescent wild type GFP. Raw fluorescence for the wild type GFP in these experiments reaches approximately 33,000 raw fluorescence units. 200 units of raw fluorescence (an estimate larger than would be expected by the representative raw profiles shown), would adjust the calculated reassignment efficiency by 0.006%, a value below the limit of detection of our in cell assay, which we have calculated to be 0.015% (Discussed in detail in [44]).

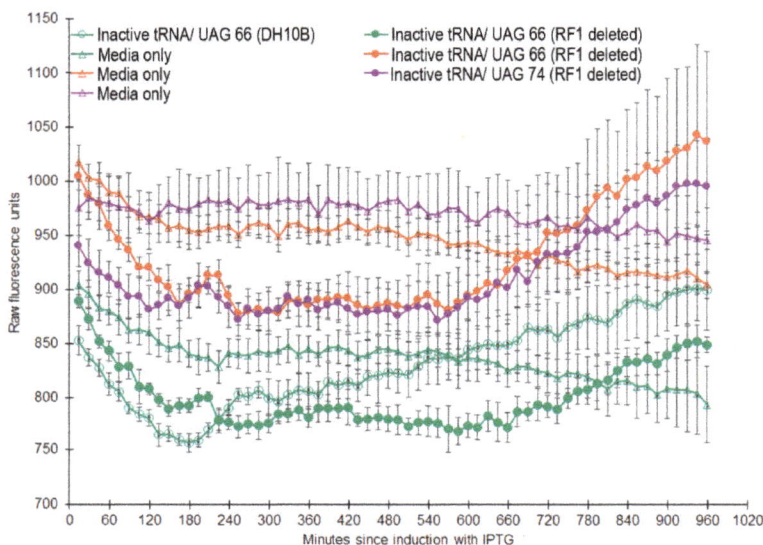

Figure A1. Raw fluorescence versus time profiles of single site UAG GFP reporter variants in the absence of an *M. jannaschii* tRNA capable of suppressing amber codons. Error bars represent the standard deviation across 6 biological replicates of each cell system (12 wells for media only). Data from experiments using two different media preparations are shown and reveal the contribution of media fluorescence to total fluorescence observed in non-reassigning systems.

References

1. Buckingham, R.H. Codon context. *Experientia* **1990**, *46*, 1126–1133. [CrossRef] [PubMed]
2. Brar, G.A. Beyond the triplet code: Context cues transform translation. *Cell* **2016**, *167*, 1681–1692. [CrossRef] [PubMed]
3. Tate, W.P.; Mannering, S.A. Three, four or more: The translational stop signal at length. *Mol. Microbiol.* **1996**, *21*, 213–219. [CrossRef] [PubMed]
4. Liu, C.C.; Schultz, P.G. Adding new chemistries to the genetic code. *Annu. Rev. Biochem.* **2010**, *79*, 413–444. [CrossRef] [PubMed]
5. Wan, W.; Tharp, J.M.; Liu, W.R. Pyrrolysyl-tRNA synthetase: An ordinary enzyme but an outstanding genetic code expansion tool. *Biochim. Biophys. Acta Proteins Proteom.* **2014**, *1844*, 1059–1070. [CrossRef] [PubMed]
6. Isaacs, F.J.; Carr, P.A.; Wang, H.H.; Lajoie, M.J.; Sterling, B.; Kraal, L.; Tolonen, A.C.; Gianoulis, T.A.; Goodman, D.B.; Reppas, N.B.; et al. Precise manipulation of chromosomes in vivo enables genome-wide codon replacement. *Science* **2011**, *333*, 348–353. [CrossRef] [PubMed]
7. Johnson, D.B.F.; Xu, J.F.; Shen, Z.X.; Takimoto, J.K.; Schultz, M.D.; Schmitz, R.J.; Xiang, Z.; Ecker, J.R.; Briggs, S.P.; Wang, L. RF1 knockout allows ribosomal incorporation of unnatural amino acids at multiple sites. *Nat. Chem. Biol.* **2011**, *7*, 779–786. [CrossRef] [PubMed]

8. Mukai, T.; Hayashi, A.; Iraha, F.; Sato, A.; Ohtake, K.; Yokoyama, S.; Sakamoto, K. Codon reassignment in the *Escherichia coli* genetic code. *Nucleic Acids Res.* **2010**, *38*, 8188–8195. [CrossRef] [PubMed]

9. Elena, C.; Ravasi, P.; Castelli, M.E.; Peiru, S.; Menzella, H.G. Expression of codon optimized genes in microbial systems: Current industrial applications and perspectives. *Front. Microbiol.* **2014**, *5*, 21. [CrossRef] [PubMed]

10. Chatterjee, A.; Sun, S.B.; Furman, J.L.; Xiao, H.; Schultz, P.G. A versatile platform for single- and multiple-unnatural amino acid mutagenesis in *Escherichia coli*. *Biochemistry* **2013**, *52*, 1828–1837. [CrossRef] [PubMed]

11. Ryu, Y.H.; Schultz, P.G. Efficient incorporation of unnatural amino acids into proteins in *Escherichia coli*. *Nat. Methods* **2006**, *3*, 263–265. [CrossRef] [PubMed]

12. Young, T.S.; Ahmad, I.; Yin, J.A.; Schultz, P.G. An enhanced system for unnatural amino acid mutagenesis in *E. coli*. *J. Mol. Biol.* **2010**, *395*, 361–374. [CrossRef] [PubMed]

13. Salser, W. Influence of reading context upon suppression of nonsense codons. *Mol. Gen. Genet.* **1969**, *105*, 125. [CrossRef] [PubMed]

14. Yahata, H.; Ocada, Y. Adjacent effect on suppression efficiency. II. Study on ochre and amber mutants of T4-phage lysozyme. *Mol. Gen. Genet.* **1970**, *106*, 208. [CrossRef] [PubMed]

15. Pedersen, S. *Escherichia coli* ribosomes translate in vivo with variable-rate. *EMBO J.* **1984**, *3*, 2895–2898. [CrossRef] [PubMed]

16. Murgola, E.J.; Pagel, F.T.; Hijazi, K.A. Codon context effects in missense suppression. *J. Mol. Biol.* **1984**, *175*, 19–27. [CrossRef]

17. Farabaugh, P.J.; Bjork, G.R. How translational accuracy influences reading frame maintenance. *EMBO J.* **1999**, *18*, 1427–1434. [CrossRef] [PubMed]

18. Kramer, E.B.; Farabaugh, P.J. The frequency of translational misreading errors in *E. coli* is largely determined by tRNA competition. *RNA* **2007**, *13*, 87–96. [CrossRef] [PubMed]

19. Yarus, M. Translational efficiency of transfer RNAs—Uses of an extended anticodon. *Science* **1982**, *218*, 646–652. [CrossRef] [PubMed]

20. Ayer, D.; Yarus, M. The context effect does not require a 4th base pair. *Science* **1986**, *231*, 393–395. [CrossRef] [PubMed]

21. Curran, J.F.; Poole, E.S.; Tate, W.P.; Gross, B.L. Selection of aminoacyl-tRNAs at sense codons: The size of the tRNA variable loop determines whether the immediate 3′ nucleotide to the codon has a context effect. *Nucleic Acids Res.* **1995**, *23*, 4104–4108. [CrossRef] [PubMed]

22. Murgola, E.J.; Hijazi, K.A.; Goringer, H.U.; Dahlberg, A.E. Mutant 16s ribosomal RNA: A codon-specific translational suppressor. *Proc. Natl. Acad. Sci. USA* **1988**, *85*, 4162–4165. [CrossRef] [PubMed]

23. Olejniczak, M.; Dale, T.; Fahlman, R.P.; Uhlenbeck, O.C. Idiosyncratic tuning of tRNAs to achieve uniform ribosome binding. *Nat. Struct. Mol. Biol.* **2005**, *12*, 788–793. [CrossRef] [PubMed]

24. Pedersen, W.T.; Curran, J.F. Effects of the nucleotide 3′ to an amber codon on ribosomal selection rates of suppressor tRNA and release factor-I. *J. Mol. Biol.* **1991**, *219*, 231–241. [CrossRef]

25. Kleina, L.G.; Masson, J.M.; Normanly, J.; Abelson, J.; Miller, J.H. Construction of *Escherichia coli* amber suppressor transfer-RNA genes. 2. Synthesis of additional transfer-RNA genes and improvement of suppressor efficiency. *J. Mol. Biol.* **1990**, *213*, 705–717. [CrossRef]

26. Normanly, J.; Kleina, L.G.; Masson, J.M.; Abelson, J.; Miller, J.H. Construction of *Escherichia coli* amber suppressor tRNAa genes. III. Determination of transfer-RNA specificity. *J. Mol. Biol.* **1990**, *213*, 719–726. [CrossRef]

27. Bossi, L. Context effects: Translation of UAG codon by suppressor tRNA is affected by the sequence following UAG in the message. *J. Mol. Biol.* **1983**, *164*, 73–87. [CrossRef]

28. Bossi, L.; Roth, J.R. The influence of codon context on genetic code translation. *Nature* **1980**, *286*, 123–127. [CrossRef] [PubMed]

29. Yarus, M.; Folley, L.S. Sense codons are found in specific contexts. *J. Mol. Biol.* **1985**, *182*, 529–540. [CrossRef]

30. Tats, A.; Tenson, T.; Remm, M. Preferred and avoided codon pairs in three domains of life. *BMC Genom.* **2008**, *9*, 463. [CrossRef] [PubMed]

31. Moura, G.R.; Pinheiro, M.; Freitas, A.; Oliveira, J.L.; Frommlet, J.C.; Carreto, L.; Soares, A.R.; Bezerra, A.R.; Santos, M.A.S. Species-specific codon context rules unveil non-neutrality effects of synonymous mutations. *PLoS ONE* **2011**, *6*, e26817. [CrossRef] [PubMed]

32. Behura, S.K.; Severson, D.W. Codon usage bias: Causative factors, quantification methods and genome-wide patterns: With emphasis on insect genomes. *Biol. Rev.* **2013**, *88*, 49–61. [CrossRef] [PubMed]

33. Behura, S.K.; Severson, D.W. Bicluster pattern of codon context usages between flavivirus and vector mosquito *Aedes aegypti*: Relevance to infection and transcriptional response of mosquito genes. *Mol. Genet. Genom.* **2014**, *289*, 885–894. [CrossRef] [PubMed]

34. Chevance, F.F.V.; Le Guyon, S.; Hughes, K.T. The effects of codon context on in vivo translation speed. *PLoS Genet.* **2014**, *10*, e1004392. [CrossRef] [PubMed]

35. Das, D.; Satapathy, S.S.; Buragohain, A.K.; Ray, S.K. Occurrence of all nucleotide combinations at the third and the first positions of two adjacent codons in open reading frames of bacteria. *Curr. Sci.* **2006**, *90*, 22–24.

36. Czech, A.; Fedyunin, I.; Zhang, G.; Ignatova, Z. Silent mutations in sight: Co-variations in tRNA abundance as a key to unravel consequences of silent mutations. *Mol. BioSyst.* **2010**, *6*, 1767–1772. [CrossRef] [PubMed]

37. Zhang, G.; Hubalewska, M.; Ignatova, Z. Transient ribosomal attenuation coordinates protein synthesis and co-translational folding. *Nat. Struct. Mol. Biol.* **2009**, *16*, 274–280. [CrossRef] [PubMed]

38. Xu, H.; Wang, Y.; Lu, J.Q.; Zhang, B.; Zhang, Z.W.; Si, L.L.; Wu, L.; Yao, T.Z.; Zhang, C.L.; Xiao, S.L.; et al. Re-exploration of the codon context effect on amber codon-guided incorporation of noncanonical amino acids in *Escherichia coli* by the blue–white screening assay. *ChemBioChem* **2016**, *17*, 1250–1256. [CrossRef] [PubMed]

39. Pott, M.; Schmidt, M.J.; Summerer, D. Evolved sequence contexts for highly efficient amber suppression with noncanonical amino acids. *ACS Chem. Biol.* **2014**, *9*, 2815–2822. [CrossRef] [PubMed]

40. Miller, J.H.; Albertini, A.M. Effects of surrounding sequence on the suppression of nonsense codons. *J. Mol. Biol.* **1983**, *164*, 59–71. [CrossRef]

41. Lajoie, M.J.; Rovner, A.J.; Goodman, D.B.; Aerni, H.-R.; Haimovich, A.D.; Kuznetsov, G.; Mercer, J.A.; Wang, H.H.; Carr, P.A.; Mosberg, J.A.; et al. Genomically recoded organisms expand biological functions. *Science* **2013**, *342*, 357–360. [CrossRef] [PubMed]

42. Sambrook, J.; Russell, D.W. *Molecular Cloning: A Laboratory Manual*, 3rd ed.; Cold Spring Harbor Laboratory Press: Cold Spring Harbor, NY, USA, 2001.

43. Schmitt, M.A.; Biddle, W.; Fisk, J.D. Mapping the plasticity of the *Escherichia coli* genetic code with orthogonal pair-directed sense codon reassignment. *Biochemistry* **2018**, *57*, 2762–2774. [CrossRef] [PubMed]

44. Biddle, W.; Schmitt, M.A.; Fisk, J.D. Evaluating sense codon reassignment with a simple fluorescence screen. *Biochemistry* **2015**, *54*, 7355–7364. [CrossRef] [PubMed]

45. Amiram, M.; Haimovich, A.D.; Fan, C.G.; Wang, Y.S.; Aerni, H.R.; Ntai, I.; Moonan, D.W.; Ma, N.J.; Rovner, A.J.; Hong, S.H.; et al. Evolution of translation machinery in recoded bacteria enables multi-site incorporation of nonstandard amino acids. *Nat. Biotechnol.* **2015**, *33*, 1272–1279. [CrossRef] [PubMed]

46. Zheng, Y.N.; Lajoie, M.J.; Italia, J.S.; Chin, M.A.; Church, G.M.; Chatterjee, A. Performance of optimized noncanonical amino acid mutagenesis systems in the absence of release factor 1. *Mol. BioSyst.* **2016**, *12*, 1746–1749. [CrossRef] [PubMed]

47. Deutscher, M.P. Degradation of RNA in bacteria: Comparison of mRNA and stable RNA. *Nucleic Acids Res.* **2006**, *34*, 659–666. [CrossRef] [PubMed]

48. Sin, C.; Chiarugi, D.; Valleriani, A. Quantitative assessment of ribosome drop-off in *E. coli. Nucleic Acids Res.* **2016**, *44*, 2528–2537. [CrossRef] [PubMed]

Review

Cyclic Peptides: Promising Scaffolds for Biopharmaceuticals

Donghyeok Gang, Do Wook Kim and Hee-Sung Park *

Department of Chemistry, Korea Advanced Institute of Science and Technology, 291 Daehak-ro, Yuseong-gu, Daejeon 34141, Korea; gangdh0154@kaist.ac.kr (D.G.); toyeca@kaist.ac.kr (D.W.K.)
* Correspondence: hspark@kaist.ac.kr; Tel.: +82-42-350-2813

Received: 22 October 2018; Accepted: 8 November 2018; Published: 16 November 2018

Abstract: To date, small molecules and macromolecules, including antibodies, have been the most pursued substances in drug screening and development efforts. Despite numerous favorable features as a drug, these molecules still have limitations and are not complementary in many regards. Recently, peptide-based chemical structures that lie between these two categories in terms of both structural and functional properties have gained increasing attention as potential alternatives. In particular, peptides in a circular form provide a promising scaffold for the development of a novel drug class owing to their adjustable and expandable ability to bind a wide range of target molecules. In this review, we discuss recent progress in methodologies for peptide cyclization and screening and use of bioactive cyclic peptides in various applications.

Keywords: cyclic peptides; biopharmaceuticals; mRNA display; yeast two hybrid

1. Introduction

Screening and identification of chemicals or molecules that can bind to or inhibit particular cellular targets is a critical step in biomedical and related research. For decades, natural compounds and man-made molecules have been searched for and exploited for this purpose. These molecules can be broadly classified into two categories: small molecules and macromolecules.

Small molecules have a molecular weight less than 1 kDa and are generally hydrophobic. This latter property enables small molecules to easily pass through the cell membrane and diffuse across intracellular space—and throughout the whole body—thus allowing for oral administration. Small molecules can be chemically synthesized, making mass production possible and quality control manageable. Despite such advantages and long tradition of use, small molecules still have some unresolved intrinsic limitations. Because of their small size and high hydrophobicity, small molecules can only interact with targets that have a rigid and hydrophobic patch or groove. In addition, small molecule binding regions or features are highly conserved among various cellular proteins that are closely or distantly related to target proteins [1]. For example, the ATP-binding domain has been a prime binding site for small molecules, but it is well conserved among protein kinases. Accordingly, many kinase inhibitors suffer from low selectivity, even after they have been screened and evolved for specific targets [2]. This low selectivity can inevitably cause many unexpected side effects.

Macromolecules have a wide range of molecular weights, from 5 kDa to 150 kDa [3,4]. Many are amino acid-based polypeptides, and some contain nucleic acids. Because of their large contact area and hydrophilic regions, which allow multiple tight interactions with a target, macromolecules usually have strong binding ability and are highly selective for their target. Such favorable characteristics, which small molecules lack, have expedited the development of macromolecule-based drugs; many macromolecules, mostly antibodies, have been approved and are now available commercially. However, macromolecules suffer from their own critical problems that have yet to be overcome.

For example, they are intrinsically unable to pass through cellular membranes; although their membrane transport can be improved by many means, such as conjugating to cell-penetrating peptides, success is limited and efficiency of transport is low [5,6]. Thus, their use is largely restricted to extracellular targets. Moreover, most macromolecules are currently produced using cell culture methods; thus, heterogeneity owing to post-translational modifications and other derivatizations are inevitable. Lastly, they are not orally available, and must generally be administered intravenously. Thus, macromolecules are easily degraded and neutralized in the bloodstream and can even cause severe immunogenicity [7].

Cyclic peptides have gained increasing attention in recent years as an alternative scaffold. Cyclic peptides are usually composed of 5 to 14 amino acids and have a molecular weight of about 500 to 2000 Da. The moderate size and diverse functional groups of peptides ensure that the contact area is large enough to provide high selectivity, and their potential to form multiple hydrogen bonds can lead to strong binding affinity. In addition, cyclization of peptides generates structural and functional features that are critical for their use as pharmaceutical agents [8]. The structural constraints provided by cyclization help to resist degradation by proteases in the blood, thereby increasing their serum stability [9]. Cyclization of peptides also facilitates passage through the cell membrane, thus broadening the potential use of cyclic peptides beyond extracellular targets to include intracellular targets [10]. Because of such favorable features, various cyclic peptides from natural sources and their derivatives have been exploited for biomedical and other purposes. Inspired by natural cyclic peptides, numerous approaches for designing man-made cyclic peptides with customized structural and functional properties have been in the spotlight in recent years. In this review, we will first discuss several representatives of natural cyclic peptides. Then, we will examine various methods that have been used for peptide cyclization, from ordinary chemical synthesis to enzymatic methods. Next, we will consider library construction and strategies for selecting active cyclic peptides. Lastly, potential applications of cyclic peptides in biomedical and other research areas will be described.

2. Natural Cyclic Peptides

Cyclosporine A, a natural cyclic peptide composed of 11 amino acids, has been used since the early 1980s as an immunosuppressive drug (Figure 1A). This cyclic peptide binds to cyclophilin in T-cells, forming a cyclosporine A-cyclophilin complex that inhibits the calcium/calmodulin-dependent protein phosphatase, calcineurin. Calcineurin induces dephosphorylation and nuclear translocation of the transcription factor NF-AT (nuclear factor of activated T cells), which promotes transcription of interlukin-2 and some cytokine genes and causes activation of T cells. Cyclosporin A inhibits this phosphatase pathway, thereby suppressing the activity of T-cells and repressing the immune system [11].

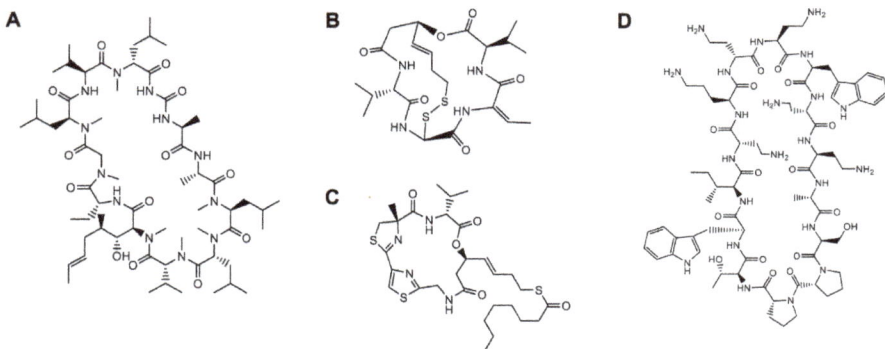

Figure 1. Natural cyclic peptides. (**A**) Cyclosporine A. (**B**) Romidepsin. (**C**) Largazole. (**D**) Murepavadin.

Romidepsin, a natural cyclic depsipeptide from *Chromobacterium violaceum* contains one or more ester bonds in place of amide bonds (Figure 1B). This cyclic peptide has a disulfide bond between two internal sulfhydryls that is reduced upon entry of the peptide into the cell. Reduction of this disulfide bond yields two thiol groups linked to long alkyl chains that can chelate divalent metal ions, especially zinc ions in the active site of class I histone deacetylase (HDAC), thus inhibiting HDAC activity [12]. Histone deacetylase, an essential epigenetic controller, is often overexpressed in many cancers. By inhibiting HDACs, romidepsin can restore the expression of genes involved in apoptosis, ultimately repressing the proliferation and differentiation of cancer cells [13]. Romidepsin was approved by the Food and Drug Administration (FDA) for cutaneous T-cell lymphoma in 2009 and for peripheral T-cell lymphomas in 2011. Largazole, a natural cyclic peptide from the marine cyanobacteria *Symploca* sp., has similar HDAC inhibiting activity (Figure 1C). Analogous to romidepsin, largazole is a kind of prodrug in which one aliphatic carboxylic acid-protected thiol group forms a thioester. The protecting group is cleaved by enzymes after translocation of largazole into the cytoplasm, and the exposed thiol group combines with a zinc ion cofactor in HDAC to inhibit its activity [14].

Murepavadin is a synthetic cyclic peptidomimetic derived from the natural peptide protegrin-1 (PG-1), containing two D-amino acids and 12 L-form proteogenic/nonproteogenic amino acids (Figure 1D). Because of its L-proline-D-proline scaffold, this cyclic peptide has a stabilized β-hairpin conformation. Murepavadin targets lipopolysaccharide transport protein D (LptD) of *Pseudomonas aeruginosa*, blocking its transport function and causing alterations in lipopolysaccharide in the outer membrane of bacteria, leading to cell death. On the basis of such antimicrobial activity [15,16], murepavadin is currently in Phase III clinical trials for treatment of nosocomial pneumonia.

3. Generation of Synthetic Cyclic Peptides

Inspired by pharmaceutical applications of natural cyclic peptides, researchers have developed several strategies for creating synthetic cyclic peptides with tailor-made functionalities. Cyclization of peptide has been achieved using various methods, including scaffold-based and enzyme-based approaches. Scaffold-based cyclization employs reactions between chemical compounds and specific functional groups of amino acids, whereas enzyme-based cyclization exploits naturally existing enzymes that can cyclize their own substrates.

3.1. Scaffold-Based Cyclization

Scaffold-based cyclization is one of the most frequently used methods because it can be applied to chemically or biologically synthesized peptides. In general, scaffold compounds such as organohalides (most frequently organobromides) selectively react with the sulfhydryl group of cysteine [17] (Figure 2A). One example of cysteine-specific scaffold-mediated cyclization includes modification of a peptide library displayed on bacteriophage [18]. In this approach, a library of genetically encoded peptides located between two cysteine residues can be transformed into circular forms by treating with the organobromide, tribromomethylbenzene (TBMB). These cyclic peptides can then be screened using the general phage-display method. Synthesis of double-bridged cyclic peptides using one or two kinds of organobromides has also recently been reported [19]. These di-bridged peptides were selected using the phage-display technique and the resulting peptide "hits" were found to have a binding affinity towards the target, plasma kallikrein, 10-times stronger than that of a previously screened bicyclic peptide.

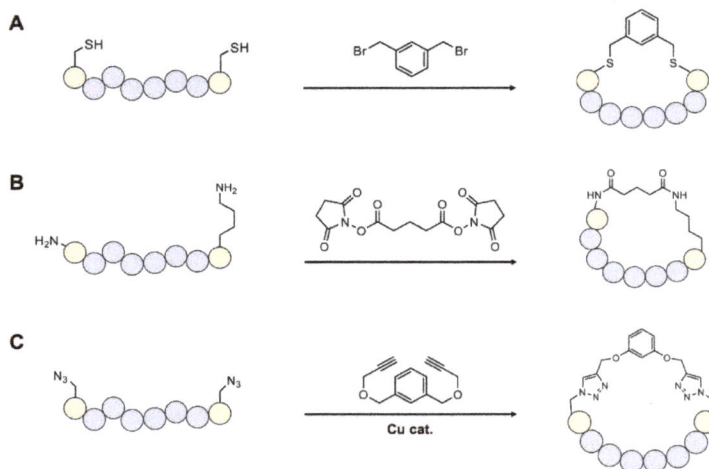

Figure 2. Scaffold-based cyclic peptide synthesis. (**A**) Organobromides selectively react with sulfhydryl group of cysteine. (**B**) *N*-hydroxysuccimide group reacts with the primary amine in p peptide and forms a stable amide bond. (**C**) Click reaction between azide and alkyne groups can be used for cyclization of chemically synthesized peptide.

Non-sulfhydryl groups, such as the primary amine of lysine or N-terminal amino group in a peptide, can be used for cyclization. *N*-hydroxysuccimide (NHS)-containing chemicals have long been used for this purpose (Figure 2B) [20], but this requires precise control of reaction conditions because NHS is easily hydrolyzed by water at physiological pH. Especially designed unnatural amino acids can be used for cyclization in peptides via a bio-orthogonal reaction. For example, if an azide-containing amino acid such as azidohomoalanine or azidophenylalanine exists in a peptide, a copper-mediated click reaction with an alkyne-bearing scaffold can lead to cyclization [21] (Figure 2C).

One of the greatest advantages of the scaffold-based method is the ease of controlling the cyclization process—and thus the size and hydrophilicity of the peptide—by changing the chemical structure of the scaffold. Multiple reaction sites, such as azide or alkyne groups, can be added to create further modifications, generating a multicyclic structure that is extremely rigid and stable under biological conditions. However, chemical scaffolds for cyclization may change the natural properties of peptides [22]. Moreover, in many peptide libraries, cysteines or lysines are present at multiple positions, potentially causing unpredictability and heterogeneity in the screening process.

3.2. Enzyme-Based Cyclization

Many natural cyclic peptides are synthesized in a ribosome- and mRNA-independent manner by specialized nonribosomal peptide synthetase (NRPS) enzymes using various non-proteogenic amino acids, including D-amino acids [23]. The NRPS system is composed of several modules, each of which is divided into at least three basic domains: adenylation (A domain), transfer (T domain), and condensation (C domain). The A domain recognizes proteogenic and non-proteogenic amino acids and adenylates its substrate using ATP. Aminoacyl-AMP (AA-AMP), produced by the A domain, reacts with an alkyl thiol in the T domain to form an AA-T domain. Then, the Nth module's C domain catalyzes formation of an amide bond between the N–1th and Nth amino acid. The last domain in the last module of NRPS is the termination (TE) domain, which determines the conformation of a synthesized peptide. If the thioester bond formed between the peptide and TE domain is cleaved by water or other nucleophiles, a linear peptide is released. However, if the thioester bond is attacked by an intramolecular nucleophile, such as an N-terminal amine, the primary amine of lysine, a serine hydroxyl group or a cysteine sulfhydryl group, a cyclic peptide is formed and released [24]. The chemical diversity of nonribosomal peptides can be expanded by changing the substrate specificity of the A domain or by replacing the whole A domain itself [25,26]. Such efforts have been successful in producing novel cyclic peptides, but the efficiency of the process is low [25]. To increase production efficiency, researchers developed a method that uses a synthetic non-native precursor peptide and thioester domain of NRPS (Figure 3A) [27,28]. Although the ratio of cyclic-to-linear forms of substrate peptide produced by this method was reduced, three different chemically prepared substrate peptides were successfully cyclized.

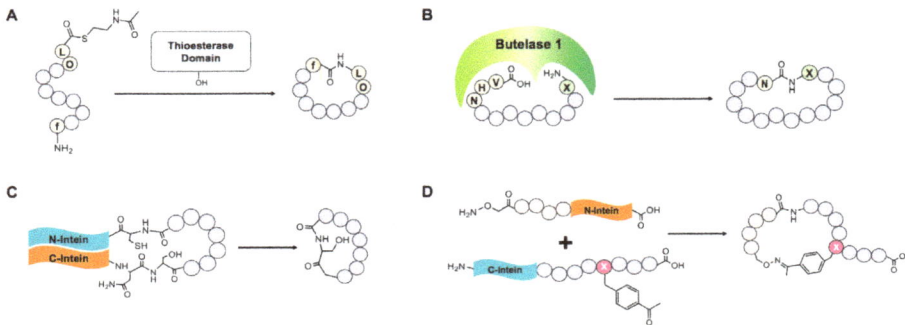

Figure 3. Enzyme-based cyclic peptide synthesis. (**A**) Cyclization of synthetic peptide using thioetsterase domain of nonribosomal peptide synthetase (NRPS). (**B**) Butelase 1-mediated cyclization requires C-terminal Asn-His-Val (NHV) tripeptide. After cyclization, only asparagine residue remains. (**C**) Split intein-mediated circular ligation for the synthesis of cyclic peptide. (**D**) Semi-synthetic method for peptide cyclization via oxime ligation using chemically synthesized N-intein and biologically prepared C-intein carrying the unnatural amino acid p-acetophenylalanine.

Another interesting example of cyclic peptide-containing non-proteogenic amino acids is cyanobactin [29]. Like the NRPS system, multiple genes are involved in the synthesis of cyanobactin. One major difference between nonribosomal peptides and cyanobactin is that the precursor of cyanobactin, composed of a leader peptide and core peptide, is encoded in the genome of the marine cyanobacteria and is translated by ribosomes. After translation of the precursor, enzymes in the same gene cluster recognize the leader peptide sequence and modify amino acids in the core region. The amino acid tolerance of these trunkamide-synthesizing enzymes was tested using a genetically encoded peptide library [30]. To determine whether non-native trunkamide peptides could be synthesized using this approach, researchers of this study screened the library in *Escherichia coli*, ultimately selecting and culturing 556 colonies for further analysis. This analysis identified 325 unique peptides containing 763 mutations, demonstrating successful synthesis of novel peptides. However, this method for preparing cyclic peptides using the cyanobactin synthesis pathway has not yet been widely adopted because of its low production yield and narrow substrate tolerance.

The Asn/Asp ligase butelase 1, an enzyme capable of catalyzing peptide cyclization, was recently discovered in *Clitoria ternatea* [31]. This enzyme recognizes asparagine or aspartate (Asx) near the C-terminus of a substrate, cleaves the amide bond after Asx, and ligates the N-terminus to the newly formed C-terminus (Figure 3B). Because butelase 1-mediated cyclization requires only Asn-His-Val (NHV) at the C-terminus, the substrate tolerance of this enzyme is extremely high. This advantage makes butelase 1 a good starting point for in vitro/in vivo peptide cyclization. Previously, reported enzymes that can be used for peptide cyclization, such as split inteins (see below), require ~10 amino acids or more for proper reaction [32]. Butelase 1 was shown to cyclize the peptides, kalata B1, conotoxin, thanatin and histatin-3, derived from plants, snails, insects and humans, respectively.

A protein splicing-based peptide cyclization method has also been developed [33]. This technique, called split intein-mediated circular ligation of peptide and proteins (SICLOPPS), catalyzes the ligation of two separate polypeptide chains, called exteins, into one. This is accomplished by transferring a peptide bond between the N-terminal intein domain and C-terminal extein, to the C-terminal intein domain and N-terminal extein. However, to promote conjugation of intramolecular N- and C-terminus, each N- and C-terminal intein domain must be fused at C- and N-termini of an extein (Figure 3C). Using an engineered split intein, these researchers synthesized pseudosterallin F, a plant-derived cyclic peptide tyrosinase inhibitor, in *E. coli*. The engineered split-intein approach has also been used to construct a cyclic peptide library, with the goal of identifying peptide inhibitors of various enzymes [34–36]. This peptide library was combined with a bacterial reverse two-hybrid system to screen for inhibitors of protein-protein interactions [37,38]. The split-intein system was further modified to synthesize lariat peptide forms by inhibiting downstream steps after transesterification by mutating Asn, which is responsible for cyclic peptide release [39]. A randomized lariat peptide library was subsequently screened using a yeast two-hybrid system to search for target protein-binding sequences [40]. An artificially designed intein, consensus fast DnaE intein (Cfa), was also recently developed [41]. As its name implies, this enzyme was evolved from consensus sequences of various DnaE inteins. A total of 105 DnaE inteins and 73 theoretically fast inteins were selected for multiple sequence alignment. After selection, the sequences of 73 inteins were used to generate Cfa intein, which has rapid and stable splicing activity. With additional mutations [42], Cfa intein was modified to cyclize eGFP without a dependence on extein sequence. The broad applicability of Cfa intein may be useful for constructing much larger cyclic peptide libraries.

3.3. Other Approaches for Peptide Cyclization

Attempts have been made to develop alternative methods for cyclization using endogenous reactive residues. Disulfide bonds are known to stabilize protein structures or protein complexes under oxidative conditions, such as found in the endoplasmic reticulum and extracellular regions [43]. Disulfide bond-mediated cyclization has been widely used to display cyclic peptides on the surface of bacteriophage. A method termed DNA-templated synthesis was developed to synthesize and cyclize peptides containing eight internal scaffold amino acids [44]. This method was used to prepare a library of cyclic tetrapeptides composed mostly of non-proteogenic amino acids. This system was recently expanded to 32 kinds of scaffold amino acids and applied to screen for inhibitory cyclic peptides targeting insulin-degrading enzyme (IDE) [45]. A cyclization method combining the intein system and unnatural amino acids was developed for generating monocyclic peptides (using GyrA intein) or bicyclic peptides (using split intein) [46,47]. One interesting feature of the GyrA-based method is that it is capable of producing side chain-to-tail cyclic peptides. A semi-synthetic method for peptide cyclization has also been reported [48]. In this method, the C-terminal domain of a split intein containing the unnatural amino acid p-acetylphenylalanine (AcF) and random mutations is expressed in *E. coli* using a genetic code expansion method, and the N-terminal domain of the split intein containing a hydroxylamine group and random mutations is chemically synthesized. When N- and C-terminal split inteins are mixed, a splicing reaction and oxime ligation occur between AcF and the hydroxylamine group, generating a cyclic peptide (Figure 3D).

4. Screening of Bio-Active Cyclic Peptides

An appropriate screening method is essential for identifying target-binding candidates from large cyclic peptide libraries in a reasonable length of time. Several screening methods have been developed; these can be categorized into in vitro and in vivo methods.

4.1. In Vitro Screening

Because there is no chance of interference by intracellular factors with in vitro methods, the primary selection criteria for these methods is binding affinity between the cyclic peptide and target protein. With successive screening steps, peptides with higher affinity can be selected. Using in vitro methods, researchers can also control all relevant variables necessary for appropriate screening, such as concentration of target protein, buffer, and salt conditions. Thus, in vitro methods usually offer a wide, adjustable dynamic range for screening active cyclic peptides [49].

4.1.1. Phage Display

One of the most frequently used in vitro peptide-screening methods is phage display [50]. In general, the phage-display technique uses filamentous bacteriophage to connect genotypes and phenotypes of displayed peptides. Other types of phage can also be used, including T7 phage and Qβ phage [51]. For a peptide to be displayed on a phage particle, it should be connected to the solvent-exposed region of a phage surface protein. The p3 protein of the filamentous bacteriophage, responsible for binding to the F pilus of the host, is most widely used because its N-terminus is exposed to a solvent. Peptides or proteins of various size can therefore be fused to p3 without affecting phage infection or production. Other phage surface proteins, such as p6, p7, p8 and p9, have also successfully been used for phage display [52–56]. After construction and transformation of a randomized peptide library, each peptide can be routinely displayed on the phage surface directly using a phage genome as a library carrier or by transfection of a helper phage in a phagemid system. This method is straightforward, but because the target molecule is often anchored on the hydrophobic surface of a plastic plate or magnetic beads through adsorption, the orientation of the target on the surface depends on the protein's surface properties. Methods for anchoring target molecules in a fixed orientation using biotin [57] or unnatural amino acids [58] have been established. In this method, library-displaying phages are mixed with an anchored target and incubated for an appropriate time. After binding, unbound phages are washed out and the remaining phages, which are presumably positive clones, are eluted. After repeating this series of steps ~3–5 times, selection-surviving peptides are isolated and identified. The maximum diversity of phage display is ~10^{10} because of limitations in transformation efficiency [59].

4.1.2. mRNA Display

Another powerful in vitro screening method is messenger RNA (mRNA) display. For synthesis of ribosomal polypeptide libraries, mRNA display depends on an in vitro translation system, such as *E. coli* S30 [60], rabbit reticulocyte lysates [61], or the PURE (protein synthesis using recombinant elements) system [62]. The mRNA used for peptide synthesis does not contain a stop codon for translation termination, and instead the translation inhibitor puromycin is conjugated at the 3' terminus via a flexible linker. When the peptidyl transfer reaction reaches a last codon, the ribosome stalls on the mRNA and puromycin enters the active site of the peptidyl transferase. The newly synthesized peptide is transferred to puromycin and covalently attached to its own mRNA, generating the linkage between genotype and phenotype and enabling in vitro screening of a large pool of potential peptide library. The basic workflow is depicted in Figure 4A. Many uncommon peptides, such as antibacterial peptides and highly hydrophobic or basic peptides, are usually difficult to express and display in a cellular system. However, mRNA display combined with reconstructed translation, as exemplified by the PURE system, is very useful for expressing such uncommon peptides [63]. The Suga group recently developed a modified flexible in vitro translation (FIT) system that allows for genetic incorporation of three non-proteogenic amino acids in addition to 20 basic proteogenic amino acids by dividing codon boxes for valine, glycine and arginine [64]. Using both a FIT-32nt system and flexizyme, an artificial AARS-like ribozyme [65], these researchers were able to incorporate eight different unnatural amino acids into linear and cyclic peptides. In theory, a maximum of 11 different non-proteogenic amino acids can be added to the natural repertoire in a one-pot reaction using this system. In addition, this in vitro display system ensures a higher maximum diversity (~10^{13}), since no transformation step is needed [66]. mRNA display has been used extensively to identify many active macrocyclic peptides [67]. For example, the anti-Akt2 macrocyclic peptide, Pakti-L1, isolated by this method was found to have strong inhibitory activity toward Akt2 (half maximum inhibitory concentration, $IC_{50} = 110$ nM), with only moderate activity towards Akt1 and Akt3d isoforms [68]. mRNA display has also led to the discovery of other macrocyclic peptides with low nanomolar IC_{50} values towards SIRT2 [69] and E6AP [70].

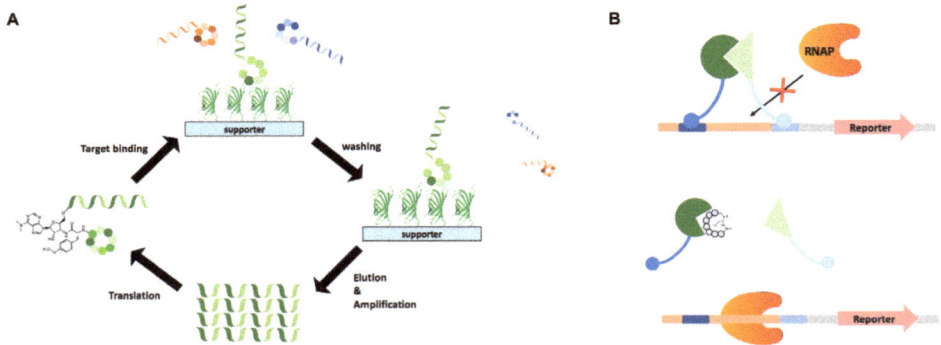

Figure 4. Screening of biologically active cyclic peptides. (**A**) Messenger RNA (mRNA) display-based cyclic peptide library screening. (**B**) Bacterial reverse two-hybrid system used for screening of active cyclic peptides. RNAP: RNA polymerase.

4.1.3. Solid-Phase Peptide Synthesis for Cyclic Peptide Screening

Total chemical synthesis of cyclic peptides is possible through solid-phase peptide synthesis (SPPS) [71]. Since SPPS does not depend on biological factors, amino acids with diverse chemical structures (such as those that are branched, elongated and N-methylated), can theoretically be used as reaction components. Synthesized peptides are anchored on a solid bead and can be modified further as necessary depending on the specific application. Cyclic peptides synthesized by SPPS are used for screening target proteins or molecules labeled with fluorescent dye or other reporters in an interaction-dependent manner. A cyclic peptide library containing specific chemical groups, such as 2-amino-8-hydroxyamino-8-oxooctanoic acid, as a "warhead" was synthesized by SPPS and screened against HDAC, leading to several hits with approximately nanomolar dissociation constant (K_d) values towards class 1 HDACs [72].

4.2. In Vivo Screening

In vivo screening methods rely on living organisms and their translation systems, and are affected by many conditions and factors, including culture conditions, copy number of the gene of interest and expression level of the target protein, among others. The genetic diversity with such screening methods is usually less than 10^{10}, because transformation of the library is mandatory. Despite such limitations, many in vivo screening methods have been widely used, since the selected peptides usually have low cytotoxicity. In addition, these methods ensure that the selected candidates inhibit the target, because the selections are often not only interaction-dependent but also activity-dependent.

4.2.1. Yeast Two-Hybrid

One of the most widely used in vivo screening methods is the yeast two-hybrid system. In this interaction-dependent selection method, the target protein is fused to a DNA binding domain and the cyclic peptide library is fused to a transcription-activating domain. When cyclic peptide (bait) interacts with target protein (prey), the resulting expression of a selective marker may be used to screen for cyclic peptide/protein interaction. This method has been used to identify the active cyclic peptides, L2 [39] and TG17 [40], which specifically interact with the bacterial SOS response regulator LexA and

Abl kinase, respectively. Interestingly, the cyclic peptide TG17 strongly binds to Abl kinase (K_d in the nanomolar range), but its IC_{50} value is quite high (>100 μM), illustrating a clear gap between target-binding ability and cellular inhibitory activity.

4.2.2. Bacterial Reverse Two-Hybrid

The bacterial reverse two-hybrid system (RTHS) is highly useful for screening modulators that inhibit target protein-protein interaction [73]. The RTHS utilizes a bacteriophage repression system [74], in which protein-protein interaction causes suppression of a selection marker, leading to cell death. If effective modulators are present in the cyclic peptide library, they block specific target interactions, causing the selection marker to become expressed and thus enabling cells that harbor the modulator to form colonies (Figure 4B). The RTHS combined with SICLOPPS has been used to screen for inhibitory cyclic peptides targeting specific homo- or heterodimeric protein-protein interactions in *E. coli* [37,38,75]. To date, most of the cyclic peptides screened by RTHS have had relatively weak binding ability (IC_{50} or K_d values of ~10 μM) compared with cyclic peptides screened using other affinity-dependent methods, such as phage display and mRNA display.

4.2.3. Protein-Fragment Complementation Assay

Protein-fragment complementation assays (PCAs) using various split reporters have been well documented and can be used for screening of active cyclic peptides. For example, in a split adenylate cyclase system, the interaction between two proteins of interest that are fused to each half of the cyclase molecule causes production of cAMP by the assembled adenylate cyclase, leading to expression of a special selection marker. When *pyrF* (orotidine 5′-phosphoate decarboxylase) is used as a marker, the expressed *pyrF* produces the toxic uracil analog 5-fluorouracil from the nontoxic compound 5-fluoroortic acid (5-FOA), ultimately leading to cell death. If an active cyclic peptide effectively inhibits this protein–protein interaction, expression of the *pyrF* gene is blocked; as a result, cells can form colonies in the presence 5-FOA [76]. However, with some rare exceptions, including luciferase-based PCAs, the assembly and disassembly of the components of many split reporters are not reversible and fragment pairs often form reporter complexes spontaneously without the induction of protein-protein interactions [77–79]. One such example is fluorescent protein-based PCAs, which often generate high background fluorescence owing to spontaneous assembly. Therefore, to date, most PCA methods have been used to detect protein–protein interactions, and not to screen for modulators of such interactions.

4.2.4. Other In Vivo Screening Approaches

In addition to targeting specific interactions between two proteins of interest, methods to screen for cyclic peptides that inhibit specific cellular enzymes have been established. DNA adenine methyltransferase (DAM methylase), which regulates the methylation state of bacterial chromosomes, is an important target for antibiotics development. To screen for inhibitors of the DAM methylase of *E. coli*, transposase-based approach was established [35]. If cyclic peptides in the library are able to inhibit DAM methylase, transposase can bind and transfer its unmethylated substrates including a chloramphenicol acetyltransferase gene and the cyclic peptide library SICLOPPS to the F plasmid. F plasmids containing transposons are transferred to transposase-lacking bacteria via conjugation, leading to colony formation on chloramphenicol-selective plates. After screening, the authors of this study obtained several active cyclic peptides. The selected cyclic peptide (SGWYVRNM) exhibited an inhibitory effect on DAM methylase comparable to that of the previously known small molecule inhibitor sinefungin, whereas a linearized peptide with the same sequence did not. The importance of cyclization is also illustrated by active cyclic peptides carrying the unnatural amino acid, p-benzoylphenylalanine, which shows inhibitory activity toward HIV protease only in circular form. An active cyclic peptide that inhibits the protein degradation pathway was also isolated [36]. In this application, green fluorescent protein (GFP) fused to an ssrA-tag sequence that is recognized for degradation by ClpXP protease was transformed into *E. coli* together with a cyclic peptide library

SICLOPPS. Active cyclic peptides with inhibitory activity against ClpXP protease lead to accumulation of the GFP-ssrA fusion protein, and thus emit a strong fluorescent signal. As expected, the cyclic form was much more potent than the linear form. Finally, active cyclic peptides with the ability to neutralize neural toxicity have been isolated using a cyclic peptide library SICLOPPS expressed in a yeast strain that mimics the neural toxicity caused by aggregation of α-synuclein [80]. Interestingly, the cyclic peptides selected were able to reduce the toxicity of α-synuclein in *Caenorhabditis elegans* dopaminergic neurons.

5. Active Cyclic Peptides as Inhibitors and Molecular Probes

As previously noted, screening for active cyclic peptides based on their target-binding ability does not ensure cellular inhibitory activity, emphasizing the importance of further derivatization and optimization in the development of efficient inhibitors from cyclic peptide hits. To date, there have been few successful cases of such derivatization and optimization from active cyclic peptide hits. In one such rare case, an in vivo method was used to screen a cyclic peptide library for an inhibitor that disrupts homodimeric interactions of aminoimidazole-4-carboxamide ribonucleotide (AICAR) transformylase, which catalyzes the last two steps of de novo purine synthesis [37]. This screen yielded the active cyclic peptide RYFNVC (peptide 1a), which exhibited a K_i value of 17 μM in circular form (Figure 5). An Ala-scanning analysis revealed that arginine and tyrosine played an important role in binding to the target protein [81]. Subsequent optimization of the hit led to more potent dipeptides, including Cpd14 (compound 14), which possessed the lowest K_i value (0.685 μM) and effectively inhibited proliferation of MCF-7 breast cancer cells. Cpd14 was also used to study metabolic disorders in a mouse model [82]. In high-fat-diet-fed mice treated with Cpd14, glucose levels in the blood rapidly decreased, an effect not observed in regular chow-fed mice. These results illustrate the potential of screened active cyclic peptides to serve as useful leads for multiple biomedical purposes.

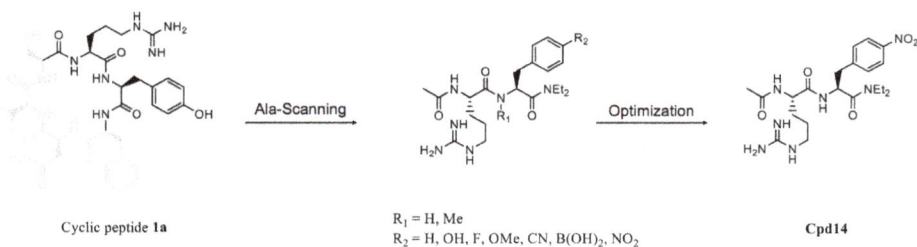

Figure 5. Optimization of active cyclic peptide. Ala-scanning of cyclic peptide 1a identified two essential amino acids, arginine and tyrosine, for target binding. Subsequent optimization led to a more potent dipeptide, Cpd14 (compound 14).

Most studies on cyclic peptides have focused on the possible role of these peptides as inhibitors. However, the high binding affinity and selectivity of cyclic peptides towards a target molecule can be expanded to include use as a molecular probe. Somatostatin, a naturally occurring cyclic peptide hormone, interacts with somatostatin receptors, which are overexpressed in many cancers, including gliomas, neuroendocrine tumors, breast cancer, and small-cell lung cancer [83]. Unfortunately, unmodified somatostatin cannot be used as a probe, owing to its low serum stability (half-life < 3 min) [84]. Accordingly, somatostatin was modified to develop cyclic peptides with higher resistance to in vivo proteases. The ^{111}In-labeled cyclic octapeptide octreotide, fCFwKTCT-ol (where lower-case letters represent D-amino acids) is the first FDA-approved peptide-based imaging agent for neuroendocrine tumors. Several ^{111}In-labeled octreotide variants and derivatives have subsequently been synthesized for possible clinical use [85]. The Arg-Gly-Asp (RGD) tripeptide also provides a good starting point for probe development, since it selectively binds to the $\alpha_v \beta_3$ receptor.

RGD-containing cyclic peptides have been labeled with various radioisotopes, including ^{18}F, ^{64}Cu, ^{68}Ga and ^{99m}Tc, to generate a new class of probes [86–89].

6. Conclusions

Because of their special functional features, including unique selectivity, versatility and structural stability, cyclic peptides have attracted increasing attention as a promising alternative to currently used small molecule and macromolecule pharmaceutical scaffolds. Many natural cyclic peptides have been modified further to develop biomedically useful compounds. Recent notable progress in peptide cyclization and screening technologies will expand the use of biologically or organically designed and synthesized cyclic peptides for therapeutic purposes. Rapid development of computational tools, including in silico-guided peptide library generation, will further accelerate cyclic peptide-based drug development. The selective and strong binding ability of cyclic peptides towards specific targets will find diverse important applications, including as excellent tools for probing specific proteins or metabolites in vivo. In addition, cyclic peptide-based chemical structures will find broader future use in various fields, including as building blocks for macromolecules.

Author Contributions: Conceptualization, D.G., D.W.K. and H.-S.P.; writing—original draft preparation, D.G., D.W.K. and H.-S.P.; writing—review and editing, H.-S.P.; visualization, D.G. and D.W.K.; supervision, H.-S.P.; funding acquisition, H.-S.P.

Funding: Research in the Park laboratory is supported by grants from the National Research Foundation of Korea (2014M3A6A4075060 and 2017R1A2B3011543) and from Samsung Science & Technology Foundation (SSTF-BA1702-09).

Acknowledgments: We apologize to all scientists in the field whose important work could not be cited in this report due to space limitations.

Conflicts of Interest: The authors declare no conflict of interest.

References

1. Gurevich, E.V.; Gurevich, V.V. Therapeutic potential of small molecules and engineered proteins. *Handb. Exp. Pharmacol.* **2014**, *219*, 1–12. [CrossRef] [PubMed]
2. Davis, M.I.; Hunt, J.P.; Herrgard, S.; Ciceri, P.; Wodicka, L.M.; Pallares, G.; Hocker, M.; Treiber, D.K.; Zarrinkar, P.P. Comprehensive analysis of kinase inhibitor selectivity. *Nat. Biotechnol.* **2011**, *29*, 1046–1051. [CrossRef] [PubMed]
3. Vazquez-Lombardi, R.; Phan, T.G.; Zimmermann, C.; Lowe, D.; Jermutus, L.; Christ, D. Challenges and opportunities for non-antibody scaffold drugs. *Drug Discov. Today* **2015**, *20*, 1271–1283. [CrossRef] [PubMed]
4. Singh, S.; Kumar, N.K.; Dwiwedi, P.; Charan, J.; Kaur, R.; Sidhu, P.; Chugh, V.K. Monoclonal antibodies: A review. *Curr. Clin. Pharmacol.* **2018**, *13*, 85–99. [CrossRef] [PubMed]
5. Marschall, A.L.; Frenzel, A.; Schirrmann, T.; Schungel, M.; Dubel, S. Targeting antibodies to the cytoplasm. *mAbs* **2011**, *3*, 3–16. [CrossRef] [PubMed]
6. Gu, Z.; Biswas, A.; Zhao, M.; Tang, Y. Tailoring nanocarriers for intracellular protein delivery. *Chem. Soc. Rev.* **2011**, *40*, 3638–3655. [CrossRef] [PubMed]
7. Kontermann, R.E. Strategies for extended serum half-life of protein therapeutics. *Curr. Opin. Biotechnol.* **2011**, *22*, 868–876. [CrossRef] [PubMed]
8. Lennard, K.R.; Tavassoli, A. Peptides come round: Using SICLOPPS libraries for early stage drug discovery. *Chemistry* **2014**, *20*, 10608–10614. [CrossRef] [PubMed]
9. Qian, Z.; Rhodes, C.A.; McCroskey, L.C.; Wen, J.; Appiah-Kubi, G.; Wang, D.J.; Guttridge, D.C.; Pei, D. Enhancing the cell permeability and metabolic stability of peptidyl drugs by reversible bicyclization. *Angew. Chem. Int. Ed.* **2017**, *56*, 1525–1529. [CrossRef] [PubMed]
10. Rezai, T.; Yu, B.; Millhauser, G.L.; Jacobson, M.P.; Lokey, R.S. Testing the conformational hypothesis of passive membrane permeability using synthetic cyclic peptide diastereomers. *J. Am. Chem. Soc.* **2006**, *128*, 2510–2511. [CrossRef] [PubMed]
11. Matsuda, S.; Koyasu, S. Mechanisms of action of cyclosporine. *Immunopharmacology* **2000**, *47*, 119–125. [CrossRef]

12. Furumai, R.; Matsuyama, A.; Kobashi, N.; Lee, K.H.; Nishiyama, M.; Nakajima, H.; Tanaka, A.; Komatsu, Y.; Nishino, N.; Yoshida, M.; et al. FK228 (depsipeptide) as a natural prodrug that inhibits class I histone deacetylases. *Cancer Res.* **2002**, *62*, 4916–4921. [PubMed]

13. Bose, P.; Dai, Y.; Grant, S. Histone deacetylase inhibitor (HDACI) mechanisms of action: Emerging insights. *Pharmacol. Ther.* **2014**, *143*, 323–336. [CrossRef] [PubMed]

14. Bowers, A.; West, N.; Taunton, J.; Schreiber, S.L.; Bradner, J.E.; Williams, R.M. Total synthesis and biological mode of action of largazole: A potent class I histone deacetylase inhibitor. *J. Am. Chem. Soc.* **2008**, *130*, 11219–11222. [CrossRef] [PubMed]

15. Martin-Loeches, I.; Dale, G.E.; Torres, A. Murepavadin: A new antibiotic class in the pipeline. *Expert Rev. Anti Infect. Ther.* **2018**, *16*, 259–268. [CrossRef] [PubMed]

16. Werneburg, M.; Zerbe, K.; Juhas, M.; Bigler, L.; Stalder, U.; Kaech, A.; Ziegler, U.; Obrecht, D.; Eberl, L.; Robinson, J.A. Inhibition of lipopolysaccharide transport to the outer membrane in Pseudomonas aeruginosa by peptidomimetic antibiotics. *ChemBioChem* **2012**, *13*, 1767–1775. [CrossRef] [PubMed]

17. Fadzen, C.M.; Wolfe, J.M.; Cho, C.F.; Chiocca, E.A.; Lawler, S.E.; Pentelute, B.L. Perfluoroarene-based peptide macrocycles to enhance penetration across the blood-brain barrier. *J. Am. Chem. Soc.* **2017**, *139*, 15628–15631. [CrossRef] [PubMed]

18. Heinis, C.; Rutherford, T.; Freund, S.; Winter, G. Phage-encoded combinatorial chemical libraries based on bicyclic peptides. *Nat. Chem. Biol.* **2009**, *5*, 502–507. [CrossRef] [PubMed]

19. Kale, S.S.; Villequey, C.; Kong, X.D.; Zorzi, A.; Deyle, K.; Heinis, C. Cyclization of peptides with two chemical bridges affords large scaffold diversities. *Nat. Chem.* **2018**, *10*, 715–723. [CrossRef] [PubMed]

20. Millward, S.W.; Takahashi, T.T.; Roberts, R.W. A general route for post-translational cyclization of mRNA display libraries. *J. Am. Chem. Soc.* **2005**, *127*, 14142–14143. [CrossRef] [PubMed]

21. Richelle, G.J.J.; Ori, S.; Hiemstra, H.; van Maarseveen, J.H.; Timmerman, P. General and facile route to isomerically pure tricyclic peptides based on templated tandem CLIPS/CuAAC cyclizations. *Angew. Chem. Int. Ed.* **2018**, *57*, 501–505. [CrossRef] [PubMed]

22. Chen, S.; Bertoldo, D.; Angelini, A.; Pojer, F.; Heinis, C. Peptide ligands stabilized by small molecules. *Angew. Chem. Int. Ed.* **2014**, *53*, 1602–1606. [CrossRef] [PubMed]

23. Finking, R.; Marahiel, M.A. Biosynthesis of nonribosomal peptides. *Annu. Rev. Microbiol.* **2004**, *58*, 453–488. [CrossRef] [PubMed]

24. Winn, M.; Fyans, J.K.; Zhuo, Y.; Micklefield, J. Recent advances in engineering nonribosomal peptide assembly lines. *Nat. Prod. Rep.* **2016**, *33*, 317–347. [CrossRef] [PubMed]

25. Eppelmann, K.; Stachelhaus, T.; Marahiel, M.A. Exploitation of the selectivity-conferring code of nonribosomal peptide synthetases for the rational design of novel peptide antibiotics. *Biochemistry* **2002**, *41*, 9718–9726. [CrossRef] [PubMed]

26. Stachelhaus, T.; Schneider, A.; Marahiel, M.A. Rational design of peptide antibiotics by targeted replacement of bacterial and fungal domains. *Science* **1995**, *269*, 69–72. [CrossRef] [PubMed]

27. Lin, H.; Thayer, D.A.; Wong, C.H.; Walsh, C.T. Macrolactamization of glycosylated peptide thioesters by the thioesterase domain of tyrocidine synthetase. *Chem. Biol.* **2004**, *11*, 1635–1642. [CrossRef] [PubMed]

28. Kohli, R.M.; Takagi, J.; Walsh, C.T. The thioesterase domain from a nonribosomal peptide synthetase as a cyclization catalyst for integrin binding peptides. *Proc. Natl. Acad. Sci. USA* **2002**, *99*, 1247–1252. [CrossRef] [PubMed]

29. Jaspars, M. The origins of cyanobactin chemistry and biology. *Chem. Commun.* **2014**, *50*, 10174–10176. [CrossRef] [PubMed]

30. Ruffner, D.E.; Schmidt, E.W.; Heemstra, J.R. Assessing the combinatorial potential of the RiPP cyanobactin tru pathway. *ACS Synth. Biol.* **2015**, *4*, 482–492. [CrossRef] [PubMed]

31. Nguyen, G.K.; Wang, S.; Qiu, Y.; Hemu, X.; Lian, Y.; Tam, J.P. Butelase 1 is an Asx-specific ligase enabling peptide macrocyclization and synthesis. *Nat. Chem. Biol.* **2014**, *10*, 732–738. [CrossRef] [PubMed]

32. Aranko, A.S.; Oeemig, J.S.; Zhou, D.; Kajander, T.; Wlodawer, A.; Iwai, H. Structure-based engineering and comparison of novel split inteins for protein ligation. *Mol. Biosyst.* **2014**, *10*, 1023–1034. [CrossRef] [PubMed]

33. Scott, C.P.; Abel-Santos, E.; Wall, M.; Wahnon, D.C.; Benkovic, S.J. Production of cyclic peptides and proteins in vivo. *Proc. Natl. Acad. Sci. USA* **1999**, *96*, 13638–13643. [CrossRef] [PubMed]

34. Young, T.S.; Young, D.D.; Ahmad, I.; Louis, J.M.; Benkovic, S.J.; Schultz, P.G. Evolution of cyclic peptide protease inhibitors. *Proc. Natl. Acad. Sci. USA* **2011**, *108*, 11052–11056. [CrossRef] [PubMed]

35. Naumann, T.A.; Tavassoli, A.; Benkovic, S.J. Genetic selection of cyclic peptide Dam methyltransferase inhibitors. *ChemBioChem* **2008**, *9*, 194–197. [CrossRef] [PubMed]

36. Cheng, L.; Naumann, T.A.; Horswill, A.R.; Hong, S.J.; Venters, B.J.; Tomsho, J.W.; Benkovic, S.J.; Keiler, K.C. Discovery of antibacterial cyclic peptides that inhibit the ClpXP protease. *Protein Sci.* **2007**, *16*, 1535–1542. [CrossRef] [PubMed]

37. Tavassoli, A.; Benkovic, S.J. Genetically selected cyclic-peptide inhibitors of AICAR transformylase homodimerization. *Angew. Chem. Int. Ed.* **2005**, *44*, 2760–2763. [CrossRef] [PubMed]

38. Tavassoli, A.; Lu, Q.; Gam, J.; Pan, H.; Benkovic, S.J.; Cohen, S.N. Inhibition of HIV budding by a genetically selected cyclic peptide targeting the Gag-TSG101 interaction. *ACS Chem. Biol.* **2008**, *3*, 757–764. [CrossRef] [PubMed]

39. Barreto, K.; Bharathikumar, V.M.; Ricardo, A.; DeCoteau, J.F.; Luo, Y.; Geyer, C.R. A genetic screen for isolating "lariat" Peptide inhibitors of protein function. *Chem. Biol.* **2009**, *16*, 1148–1157. [CrossRef] [PubMed]

40. Bharathikumar, V.M.; Barreto, K.; Decoteau, J.F.; Geyer, C.R. Allosteric lariat peptide inhibitors of Abl kinase. *ChemBioChem* **2013**, *14*, 2119–2125. [CrossRef] [PubMed]

41. Stevens, A.J.; Brown, Z.Z.; Shah, N.H.; Sekar, G.; Cowburn, D.; Muir, T.W. Design of a split intein with exceptional protein splicing activity. *J. Am. Chem. Soc.* **2016**, *138*, 2162–2165. [CrossRef] [PubMed]

42. Stevens, A.J.; Sekar, G.; Shah, N.H.; Mostafavi, A.Z.; Cowburn, D.; Muir, T.W. A promiscuous split intein with expanded protein engineering applications. *Proc. Natl. Acad. Sci. USA* **2017**, *114*, 8538–8543. [CrossRef] [PubMed]

43. Fass, D. Disulfide bonding in protein biophysics. *Annu. Rev. Biophys.* **2012**, *41*, 63–79. [CrossRef] [PubMed]

44. Tse, B.N.; Snyder, T.M.; Shen, Y.; Liu, D.R. Translation of DNA into a library of 13,000 synthetic small-molecule macrocycles suitable for in vitro selection. *J. Am. Chem. Soc.* **2008**, *130*, 15611–15626. [CrossRef] [PubMed]

45. Usanov, D.L.; Chan, A.I.; Maianti, J.P.; Liu, D.R. Second-generation DNA-templated macrocycle libraries for the discovery of bioactive small molecules. *Nat. Chem.* **2018**, *10*, 704–714. [CrossRef] [PubMed]

46. Frost, J.R.; Wu, Z.; Lam, Y.C.; Owens, A.E.; Fasan, R. Side-chain-to-tail cyclization of ribosomally derived peptides promoted by aryl and alkyl amino-functionalized unnatural amino acids. *Org. Biomol. Chem.* **2016**, *14*, 5803–5812. [CrossRef] [PubMed]

47. Bionda, N.; Fasan, R. Ribosomal synthesis of natural-product-like bicyclic peptides in *Escherichia coli*. *ChemBioChem* **2015**, *16*, 2011–2016. [CrossRef] [PubMed]

48. Palei, S.; Mootz, H.D. Cyclic peptides made by linking synthetic and genetically encoded fragments. *ChemBioChem* **2016**, *17*, 378–382. [CrossRef] [PubMed]

49. Li, W.; Caberoy, N.B. New perspective for phage display as an efficient and versatile technology of functional proteomics. *Appl. Microbiol. Biotechnol.* **2010**, *85*, 909–919. [CrossRef] [PubMed]

50. Nixon, A.E.; Sexton, D.J.; Ladner, R.C. Drugs derived from phage display: From candidate identification to clinical practice. *mAbs* **2014**, *6*, 73–85. [CrossRef] [PubMed]

51. Skamel, C.; Aller, S.G.; Bopda Waffo, A. In vitro evolution and affinity-maturation with Coliphage Qβ display. *PLoS ONE* **2014**, *9*, e113069. [CrossRef] [PubMed]

52. Scott, J.K.; Smith, G.P. Searching for peptide ligands with an epitope library. *Science* **1990**, *249*, 386–390. [CrossRef] [PubMed]

53. Felici, F.; Castagnoli, L.; Musacchio, A.; Jappelli, R.; Cesareni, G. Selection of antibody ligands from a large library of oligopeptides expressed on a multivalent exposition vector. *J. Mol. Biol.* **1991**, *222*, 301–310. [CrossRef]

54. Chappel, J.A.; He, M.; Kang, A.S. Modulation of antibody display on M13 filamentous phage. *J. Immunol. Methods* **1998**, *221*, 25–34. [CrossRef]

55. Gao, C.; Mao, S.; Kaufmann, G.; Wirsching, P.; Lerner, R.A.; Janda, K.D. A method for the generation of combinatorial antibody libraries using pIX phage display. *Proc. Natl. Acad. Sci. USA* **2002**, *99*, 12612–12616. [CrossRef] [PubMed]

56. Hufton, S.E.; Moerkerk, P.T.; Meulemans, E.V.; de Bruine, A.; Arends, J.W.; Hoogenboom, H.R. Phage display of cDNA repertoires: The pVI display system and its applications for the selection of immunogenic ligands. *J. Immunol. Methods* **1999**, *231*, 39–51. [CrossRef]

57. Seo, M.H.; Lee, T.S.; Kim, E.; Cho, Y.L.; Park, H.S.; Yoon, T.Y.; Kim, H.S. Efficient single-molecule fluorescence resonance energy transfer analysis by site-specific dual-labeling of protein using an unnatural amino acid. *Anal. Chem.* **2011**, *83*, 8849–8854. [CrossRef] [PubMed]

58. Seo, M.H.; Han, J.; Jin, Z.; Lee, D.W.; Park, H.S.; Kim, H.S. Controlled and oriented immobilization of protein by site-specific incorporation of unnatural amino acid. *Anal. Chem.* **2011**, *83*, 2841–2845. [CrossRef] [PubMed]

59. Hanahan, D.; Jessee, J.; Bloom, F.R. Plasmid transformation of *Escherichia coli* and other bacteria. *Methods Enzymol.* **1991**, *204*, 63–113. [PubMed]

60. Yang, L.M.; Wang, J.L.; Kang, L.; Gao, S.; Liu, Y.H.; Hu, T.M. Construction and analysis of high-complexity ribosome display random peptide libraries. *PLoS ONE* **2008**, *3*, e2092. [CrossRef] [PubMed]

61. Wang, R.; Cotten, S.W.; Liu, R. mRNA display using covalent coupling of mRNA to translated proteins. *Methods Mol. Biol.* **2012**, *805*, 87–100. [CrossRef] [PubMed]

62. Nagumo, Y.; Fujiwara, K.; Horisawa, K.; Yanagawa, H.; Doi, N. PURE mRNA display for in vitro selection of single-chain antibodies. *J. Biochem.* **2016**, *159*, 519–526. [CrossRef] [PubMed]

63. Rogers, J.M.; Kwon, S.; Dawson, S.J.; Mandal, P.K.; Suga, H.; Huc, I. Ribosomal synthesis and folding of peptide-helical aromatic foldamer hybrids. *Nat. Chem.* **2018**, *10*, 405–412. [CrossRef] [PubMed]

64. Iwane, Y.; Hitomi, A.; Murakami, H.; Katoh, T.; Goto, Y.; Suga, H. Expanding the amino acid repertoire of ribosomal polypeptide synthesis via the artificial division of codon boxes. *Nat. Chem.* **2016**, *8*, 317–325. [CrossRef] [PubMed]

65. Ohuchi, M.; Murakami, H.; Suga, H. The flexizyme system: A highly flexible tRNA aminoacylation tool for the translation apparatus. *Curr. Opin. Chem. Biol.* **2007**, *11*, 537–542. [CrossRef] [PubMed]

66. Takahashi, T.T.; Austin, R.J.; Roberts, R.W. mRNA display: Ligand discovery, interaction analysis and beyond. *Trends Biochem. Sci.* **2003**, *28*, 159–165. [CrossRef]

67. Josephson, K.; Ricardo, A.; Szostak, J.W. mRNA display: From basic principles to macrocycle drug discovery. *Drug Discov. Today* **2014**, *19*, 388–399. [CrossRef] [PubMed]

68. Hayashi, Y.; Morimoto, J.; Suga, H. In vitro selection of anti-Akt2 thioether-macrocyclic peptides leading to isoform-selective inhibitors. *ACS Chem. Biol.* **2012**, *7*, 607–613. [CrossRef] [PubMed]

69. Morimoto, J.; Hayashi, Y.; Suga, H. Discovery of macrocyclic peptides armed with a mechanism-based warhead: Isoform-selective inhibition of human deacetylase SIRT2. *Angew. Chem. Int. Ed.* **2012**, *51*, 3423–3427. [CrossRef] [PubMed]

70. Yamagishi, Y.; Shoji, I.; Miyagawa, S.; Kawakami, T.; Katoh, T.; Goto, Y.; Suga, H. Natural product-like macrocyclic N-methyl-peptide inhibitors against a ubiquitin ligase uncovered from a ribosome-expressed de novo library. *Chem. Biol.* **2011**, *18*, 1562–1570. [CrossRef] [PubMed]

71. Jaradat, D.M.M. Thirteen decades of peptide synthesis: Key developments in solid phase peptide synthesis and amide bond formation utilized in peptide ligation. *Amino Acids* **2018**, *50*, 39–68. [CrossRef] [PubMed]

72. Olsen, C.A.; Ghadiri, M.R. Discovery of potent and selective histone deacetylase inhibitors via focused combinatorial libraries of cyclic alpha3beta-tetrapeptides. *J. Med. Chem.* **2009**, *52*, 7836–7846. [CrossRef] [PubMed]

73. Horswill, A.R.; Savinov, S.N.; Benkovic, S.J. A systematic method for identifying small-molecule modulators of protein-protein interactions. *Proc. Natl. Acad. Sci. USA* **2004**, *101*, 15591–15596. [CrossRef] [PubMed]

74. Di Lallo, G.; Castagnoli, L.; Ghelardini, P.; Paolozzi, L. A two-hybrid system based on chimeric operator recognition for studying protein homo/heterodimerization in *Escherichia coli*. *Microbiology* **2001**, *147*, 1651–1656. [CrossRef] [PubMed]

75. Osher, E.L.; Castillo, F.; Elumalai, N.; Waring, M.J.; Pairaudeau, G.; Tavassoli, A. A genetically selected cyclic peptide inhibitor of BCL6 homodimerization. *Bioorg. Med. Chem.* **2018**, *26*, 3034–3038. [CrossRef] [PubMed]

76. Kjelstrup, S.; Hansen, P.M.; Thomsen, L.E.; Hansen, P.R.; Lobner-Olesen, A. Cyclic peptide inhibitors of the β-sliding clamp in *Staphylococcus aureus*. *PLoS ONE* **2013**, *8*, e72273. [CrossRef] [PubMed]

77. Hu, C.D.; Chinenov, Y.; Kerppola, T.K. Visualization of interactions among bZIP and Rel family proteins in living cells using bimolecular fluorescence complementation. *Mol. Cell* **2002**, *9*, 789–798. [CrossRef]

78. Magliery, T.J.; Wilson, C.G.; Pan, W.; Mishler, D.; Ghosh, I.; Hamilton, A.D.; Regan, L. Detecting protein-protein interactions with a green fluorescent protein fragment reassembly trap: Scope and mechanism. *J. Am. Chem. Soc.* **2005**, *127*, 146–157. [CrossRef] [PubMed]

79. Nyfeler, B.; Michnick, S.W.; Hauri, H.P. Capturing protein interactions in the secretory pathway of living cells. *Proc. Natl. Acad. Sci. USA* **2005**, *102*, 6350–6355. [CrossRef] [PubMed]

80. Kritzer, J.A.; Hamamichi, S.; McCaffery, J.M.; Santagata, S.; Naumann, T.A.; Caldwell, K.A.; Caldwell, G.A.; Lindquist, S. Rapid selection of cyclic peptides that reduce α-synuclein toxicity in yeast and animal models. *Nat. Chem. Biol.* **2009**, *5*, 655–663. [CrossRef] [PubMed]

81. Spurr, I.B.; Birts, C.N.; Cuda, F.; Benkovic, S.J.; Blaydes, J.P.; Tavassoli, A. Targeting tumour proliferation with a small-molecule inhibitor of AICAR transformylase homodimerization. *ChemBioChem* **2012**, *13*, 1628–1634. [CrossRef] [PubMed]

82. Asby, D.J.; Cuda, F.; Beyaert, M.; Houghton, F.D.; Cagampang, F.R.; Tavassoli, A. AMPK activation via modulation of de novo purine biosynthesis with an inhibitor of ATIC homodimerization. *Chem. Biol.* **2015**, *22*, 838–848. [CrossRef] [PubMed]

83. Kwekkeboom, D.J.; Krenning, E.P. Somatostatin receptor imaging. *Semin. Nucl. Med.* **2002**, *32*, 84–91. [CrossRef] [PubMed]

84. Benuck, M.; Marks, N. Differences in the degradation of hypothalamic releasing factors by rat and human serum. *Life Sci.* **1976**, *19*, 1271–1276. [CrossRef]

85. Rufini, V.; Calcagni, M.L.; Baum, R.P. Imaging of neuroendocrine tumors. *Semin. Nucl. Med.* **2006**, *36*, 228–247. [CrossRef] [PubMed]

86. Chen, X.; Park, R.; Shahinian, A.H.; Tohme, M.; Khankaldyyan, V.; Bozorgzadeh, M.H.; Bading, J.R.; Moats, R.; Laug, W.E.; Conti, P.S. ^{18}F-labeled RGD peptide: Initial evaluation for imaging brain tumor angiogenesis. *Nucl. Med. Biol.* **2004**, *31*, 179–189. [CrossRef] [PubMed]

87. Liu, Z.; Niu, G.; Shi, J.; Liu, S.; Wang, F.; Liu, S.; Chen, X. ^{68}Ga-labeled cyclic RGD dimers with Gly$_3$ and PEG$_4$ linkers: Promising agents for tumor integrin αvβ$_3$ PET imaging. *Eur. J. Nucl. Med. Mol. Imaging* **2009**, *36*, 947–957. [CrossRef] [PubMed]

88. Shi, J.; Wang, L.; Kim, Y.S.; Zhai, S.; Liu, Z.; Chen, X.; Liu, S. Improving tumor uptake and excretion kinetics of ^{99m}Tc-labeled cyclic arginine-glycine-aspartic (RGD) dimers with triglycine linkers. *J. Med. Chem.* **2008**, *51*, 7980–7990. [CrossRef] [PubMed]

89. Shi, J.; Kim, Y.S.; Zhai, S.; Liu, Z.; Chen, X.; Liu, S. Improving tumor uptake and pharmacokinetics of ^{64}Cu-labeled cyclic RGD peptide dimers with Gly$_3$ and PEG$_4$ linkers. *Bioconjug. Chem.* **2009**, *20*, 750–759. [CrossRef] [PubMed]

![genes logo] *genes*

MDPI

Article

Acceptor Stem Differences Contribute to Species-Specific Use of Yeast and Human tRNA^{Ser}

Matthew D. Berg *, Julie Genereaux, Yanrui Zhu, Safee Mian, Gregory B. Gloor and Christopher J. Brandl *

Department of Biochemistry, The University of Western Ontario, London, ON N6A 5C1, Canada;
jgenerea@uwo.ca (J.G.); yzhu633@uwo.ca (Y.Z.); mmian6@uwo.ca (S.M.); ggloor@uwo.ca (G.B.G.)
* Correspondence: mberg2@uwo.ca (M.D.B.); cbrandl@uwo.ca (C.J.B.)

Received: 12 October 2018; Accepted: 3 December 2018; Published: 7 December 2018

Abstract: The molecular mechanisms of translation are highly conserved in all organisms indicative of a single evolutionary origin. This includes the molecular interactions of tRNAs with their cognate aminoacyl-tRNA synthetase, which must be precise to ensure the specificity of the process. For many tRNAs, the anticodon is a major component of the specificity. This is not the case for the aminoacylation of alanine and serine to their cognate tRNAs. Rather, aminoacylation relies on other features of the tRNA. For tRNA^{Ser}, a key specificity feature is the variable arm, which is positioned between the anticodon arm and the T-arm. The variable arm is conserved from yeast to human. This work was initiated to determine if the structure/function of tRNA^{Ser} has been conserved from *Saccharomyces cerevisiae* to human. We did this by detecting mistranslation in yeast cells with tRNA^{Ser} derivatives having the UGA anticodon converted to UGG for proline. Despite being nearly identical in everything except the acceptor stem, human tRNA^{Ser} is less active than yeast tRNA^{Ser}. A chimeric tRNA with the human acceptor stem and other sequences from the yeast molecule acts similarly to the human tRNA^{Ser}. The 3:70 base pair in the acceptor stem (C:G in yeast and A:U in humans) is a prime determinant of the specificity. Consistent with the functional difference of yeast and human tRNA^{Ser} resulting from subtle changes in the specificity of their respective SerRS enzymes, the functionality of the human and chimeric tRNA^{Ser}$_{UGG}$ molecules was enhanced when human SerRS was introduced into yeast. Residues in motif 2 of the aminoacylation domain of SerRS likely participated in the species-specific differences. Trp290 in yeast SerRS (Arg313 in humans) found in motif 2 is proximal to base 70 in models of the tRNA-synthetase interaction. Altering this motif 2 sequence of hSerRS to the yeast sequence decreases the activity of the human enzyme with human tRNA^{Ser}, supporting the coadaptation of motif 2 loop–acceptor stem interactions.

Keywords: tRNA^{Ser}; mistranslation; anticodon; functional conservation

1. Introduction

A fundamental property of all cells is the conversion of DNA sequence into protein via an RNA intermediate. The genetic code, the mechanism of translation, and the machinery involved are highly conserved in all domains of life. Fidelity during translation is maintained at two steps. First, specific aminoacyl-tRNA synthetases (aaRS) are responsible for ligating the correct amino acid onto their cognate tRNAs. Secondly, decoding at the ribosome ensures the correct pairing between a tRNA anticodon and the mRNA codon. Errors at either step can result in the incorporation of the wrong amino acid into the growing polypeptide chain, known as mistranslation [1].

Each aaRS selectively recognizes its cognate tRNAs through individual bases and base pairs, known as identity elements. In two dimensions, the canonical tRNA structure consists of an acceptor stem, where the amino acid is ligated, and three stem-loop structures (reviewed in Reference [2]).

From 5′ to 3′ these are the D-arm, the anticodon stem-loop that interacts with the mRNA at the ribosome, and the T-arm, which contributes to the interaction with elongation factor thermo unstable (EF-Tu) to recruit the tRNA to the ribosome. Serine, leucine, and bacterial tyrosine tRNAs have an addition arm, i.e., the variable arm, between the anticodon and T-arm. In three dimensions, all tRNAs fold into a L-shape (reviewed in Reference [3]). In this structure, the acceptor stem and T-arm interact to form one part of the L and the anticodon stem and D-arm interact to form the other. Many aaRS enzymes span the entire length of the tRNA with both the anticodon and acceptor stem being major determinants for recognition [4–6]. In contrast, the anticodon is not a determinant for seryl- and alanyl-tRNA synthetases [6]. For the alanyl-tRNA synthetase (AlaRS), the major identity element is a G3:U70 base pair in the acceptor stem [7–10]. A major determinant for seryl-tRNA synthetase (SerRS) is the variable arm [11–13].

All aaRSs catalyze the attachment of an amino acid onto the 3′ end of a tRNA in a two-step reaction. In the first step, an amino acid is recognized and activated with ATP to form an aminoacyl-adenylate. In the second step, the activated amino acid is transferred to the terminal 3′ adenosine of the tRNA (reviewed in Reference [14]). The aaRSs are divided into two classes based on their catalytic domains [14–16]. Class I aaRS enzymes contain a Rossmann fold with two consensus motifs (HIGH and KMSKS) that interact with ATP [15,17,18]. In class II enzymes, a unique seven sheet antiparallel beta fold [19] containing the highly degenerate motif 2 (FRXE) and motif 3 (GXGXGF[D/E]R) are responsible for binding ATP [15,20,21].

Class II enzymes are composed of a catalytic domain, often an editing domain and, in all members except AlaRS and SerRS, an anticodon binding domain [22]. Like other class II enzymes, SerRS approaches the tRNA acceptor stem from the major groove side [12,23]. In the catalytic site, beta sheets form the base of a pocket with two sides composed of large hairpins [22]. A conserved phenylalanine near the end of motif 2 stacks with the adenine ring of ATP, while a conserved arginine interacts with the alpha-phosphate [12]. The hairpin loop between two beta-strands of motif 2 contacts the upper bases of the acceptor stem [12,22–25]. The main recognition element of SerRS is the unique long variable arm, which is bound by an N-terminal tRNA binding domain comprising two alpha helices in all organisms except methanogens [12,26]. The enzyme functions as a homo-dimer, where the tRNA binding domain of one subunit binds the variable arm and positions the 3′ end of the tRNA in the active site of the other subunit [12,27,28]. Additional in vitro and in vivo experiments have identified other weak identity elements in the acceptor stem of bacterial tRNA^{Ser} [29,30].

Translation is a fundamental process where both the mechanism and the molecules involved in steps such as aminoacylation of tRNAs are conserved. This is evident from the ability of heterologous aaRS to complement deletions of native synthetase genes. For example, the human LysRS rescues a deletion of *Escherichia coli* LysRS [31], and the human AlaRS can replace the native yeast enzyme [32]. More recently, complementation of MetRS in yeast by the human enzyme was used to investigate disease-related variants [33]. However, coevolution of tRNA and synthetase pairs often results in decreased or abolished tRNA function when expressed in non-native organisms. For example, Burke et al. [34] found that human tRNA^{Pro} is not aminoacylated by *E. coli* ProRS due to differences in the acceptor stem and D-arm of the tRNA. Edwards et al. [35] used a tyrosine tRNA suppressor from *E. coli* to suppress stop codons in yeast and found it was not aminoacylated by yeast TyrRS but rather was aminoacylated with leucine. In addition, approaches to encode noncanonical amino acids at stop codons rely on orthogonal tRNA:aaRS pairs, often transplanted from other organisms, that do not interact with endogenous tRNAs or aaRSs (reviewed in Reference [36]).

This work was initiated to determine if the structure/function of tRNA^{Ser} has been conserved from *Saccharomyces cerevisiae* to human. We analyzed human tRNA^{Ser} function in *S. cerevisiae* by detecting mistranslation and suppression with tRNA^{Ser} derivatives having the UGA anticodon converted to UGG for proline. We found that despite the conservation of tRNA^{Ser}, differences in the acceptor stem result in reduced functionality of the human tRNA in yeast. The 3:70 base pair in the acceptor stem (C:G in yeast and A:U in humans) is a prime determinant of the specificity. Toxicity of the humanized tRNAs increases when human SerRS is introduced into yeast, suggesting a role for aminoacylation in the differential function.

2. Materials and Methods

2.1. Yeast Strains

Yeast strains are derivatives of the wild-type haploid strain BY4742 (Table S1; Reference [37]). The *tti2* disruption strain complemented by *tti2-L187P* on a *LEU2* centromeric plasmid (CY7020) has been previously described [10]. The *met22Δ* strain (CY7640) was derived from a spore colony of the yeast magic marker strain in the BY4743 diploid background [38].

Yeast strains were grown in yeast peptone media containing 2% glucose or synthetic media supplemented with nitrogenous bases and amino acids at 30 °C. Selections for CY7020-derived strains were on yeast peptone dextrose (YPD) plus 5% (*v/v*) ethanol. Cells were grown in dropout medium to stationary phase, 10-fold serially diluted, and spotted onto the indicated medium.

Growth curves were generated by diluting saturated cultures to OD_{600} ~ 0.1 in minimal media and incubating at 30 °C. OD_{600} was measured every 15 min using a BioTek Epoch 2 (BioTek, Winooski, VT, USA) microplate spectrophotometer for 24 h.

2.2. Yeast Transformation

Approximately 10^6 cells were pelleted and washed twice in an equivalent volume of 100 mM lithium acetate, 1.0 mM ethylenediaminetetraacetic acid, and 10 mM TrisHCl (pH 7.5; LiAc) and suspended in 200 µl of LiAc. Five milligrams of denatured calf thymus DNA and ~1.0 µg of plasmid DNA was incubated with 100 µl of cells for 15 min at 30 °C. Then, 1.0 mL of 40% polyethylene glycol 4000 was added, and the cells were incubated 15 min at 30 °C. Dimethyl sulfoxide was added to 7%, and the cells were incubated at 42 °C for 15 min. Cells were pelleted, plated on selective medium, and grown at 30 °C.

2.3. Plasmid Constructs

SUP17 [*Sc*tRNA^{Ser}$_{UGA}$] including ~300 bp upstream and downstream and either the wild-type anticodon (UGA; pCB3076) or the proline anticodon (UGG; pCB3082) in the *URA3* centromeric plasmid YCplac33 have been previously described [39]. The G26A derivative of pCB3082 (pCB4023) and A4G derivative of pCB3082 (pCB4097) were identified by selection [39]. DNA containing the sequence encoding human tRNA^{Ser} (chr6.trna172-SerUGA) with proline anticodon UGG inserted directly in place of *SUP17* (*Hs*tRNA^{Ser}$_{UGG}$; see Figure S1) was purchased from Invitrogen (Pleasanton, CA, USA). This was digested with *Pst*I and *Eco*RI and inserted into YCplac33 (pCB4062). The *SUP17*-human acceptor stem chimera (*ch*tRNA^{Ser}$_{UGG}$) with proline anticodon was constructed by two-step PCR using the outside primers UG5953 and UG5954, internal primers VD5993 and VD5994 to alter the 3′ end of the acceptor stem, VD5578 and VD5579 to alter the 5′ end of the acceptor stem, and pCB4062 as template. The molecule was cloned as a *Hind*III-*Eco*RI fragment into YCplac33 (pCB4122). (See Table S2 for oligonucleotides.) The A4G:U69C derivative of pCB3082 (pCB4145) was similarly constructed using pCB4097 as template and the inside primers VF8687 and VF8688. The A3C:T70G derivative of the chimeric tRNA^{Ser}$_{UGG}$ (pCB4336) was similarly constructed using outside primers UG5953 and UG5954, inside primers WE0866 and WE0867 to alter the 3′ end, WE0868 and WE0869 to alter the 5′ end, and pCB4122 as the template.

The cloned coding region of human seryl-tRNA synthetase (SARS) was purchased from NovoPro Biosciences Inc. (Cat# 715151-2, Shanghai, China; RefSeq NM_006513). This was amplified by PCR using oligonucleotides VD5232 and VD5803 and cloned as a *Not*I-*Eco*RI fragment into YCplac111 containing a *DED1* promoter and myc^9-tag (pCB4329; Hoffman et al. [40]) and YEplac181 containing a *DED1* promoter and myc-tag (pCB4355). Note that codon 435, which codes for C in the purchased clone, was converted by two-step PCR to an R codon to resemble those more commonly found in mammalian SerRS. A derivative of SARS with mutations T312A and R313W was constructed by ligating an *NcoI-EcoRI* fragment synthesized by PCR with oligonucleotides WF1846 and VD5803 into pCB4355 to give pCB4357. *S. cerevisiae SES1* was similarly inserted into YCplac111 (pCB4342) and YCplac181 (pCB4356) after PCR with oligonucleotides WF1727 and UK0551.

The centromeric plasmid containing *HSE-eGFP* has previously been described [41] and was kindly provided by Martin Duennwald.

2.4. Western Blot Assay

Yeast extract prepared by grinding with glass beads [42] was separated by SDS-PAGE and transferred to PVDF membrane (Roche Applied Science, Penzberg, Germany). Anti-myc (9E10, Sigma-Aldrich, St. Louis, MO, USA) was used at a ratio of 1:5000. Secondary antibody (anti-Mouse IgG HRP, Promega, Madison, WI, USA) was used at a ratio of 1:10,000 and was detected using SuperSignal West Pico Chemiluminiscent Substrate (Thermo Scientific, Waltham, MA, USA).

2.5. Fluorescence Heat Shock Reporter

Yeast strains containing the heat shock response element (HSE)-*eGFP* reporter were grown to stationary phase in a medium lacking leucine and uracil, diluted 1:50 in the same medium, and grown for 6 h at 30 °C. Cell densities were normalized to OD_{600} before measuring fluorescence. Fluorescence was measured with a BioTek Synergy H1 microplate reader at an emission wavelength of 528 nm using Gen5 2.08 software. The mean relative fluorescence units were calculated across three technical and three biological replicates for each strain.

2.6. Mutual Information Analysis

Coevolving residues within SerRS were identified following the protocol outlined by Dickson and Gloor [43]. Briefly, crystal structures of SerRS from *Candida albicans* (3QNE; [44]), *Aquifex aeolicus* (2DQ3), *Pyrococcus horikoshii* (2ZR2; [45]), *Homo sapiens* (4L87; [46]), *Naegleria fowleri* (6BLJ), and *Thermus thermophilus* (1SET; [47]) from the Protein Data Bank were used to construct a structural alignment. ASN.1 files for each structure were obtained from the Molecular Modeling Database, and the structures were aligned in Cn3D [48] to obtain block structures for sequence alignment. Seryl-tRNA synthetase sequences were obtained from the UniProt Reference Clusters 50 (UniRef50) dataset as FASTA files (Supplemental File S1), imported into Cn3D and block aligned with the SerRS structures. Sequences that were not annotated as cytoplasmic SerRS, as well as hypothetical proteins, were removed. Coevolution scores were calculated with MIpToolset [49].

3. Results and Discussion

The processes involved in translation are conserved across all species. Not surprisingly, there is significant similarity in the key molecules involved [50,51]. In this work, our goal was to investigate the cross-species similarities and differences in tRNA function by analyzing the activity of human tRNA^{Ser} in *S. cerevisiae*. To begin the analysis, we aligned the tRNA^{Ser}$_{UGA}$ isodecoders from *S. cerevisiae* and human (Figure 1A). Of the 83 nucleotides, 60 are conserved in both sets of tRNAs. Fifteen positions are always different between the three identical yeast tRNAs and the four human tRNAs. The variable arms are conserved in length and G:C content.

3.1. Mistranslation Assay for tRNA^{Ser} Function

We previously identified a stress-sensitive allele (L187P) of the co-chaperone protein *TTI2* that is suppressed by mistranslation of L187P with serine [39]. A gene expressing yeast *SUP17* (*SctRNA^{Ser}*) with the UGA anticodon of the tRNA changed to UGG (to decode proline codons) cannot be introduced into yeast [39] due to high levels of mistranslation of serine at proline codons. Dampening the function of the tRNA with secondary mutations allows the introduction of tRNA^{Ser}$_{UGG}$ variants. If the tRNA$^{Ser}_{UGG}$ retains function, it is aminoacylated with serine and inserts (mistranslates) serine for proline at a level sufficient to suppress *tti2-L187P*. If the tRNA$^{Ser}_{UGG}$ variant is not functional, it can be introduced into yeast but does not mistranslate and thus does not suppress *tti2-L187P*. In our assay tRNA$^{Ser}_{UGG}$ competes with endogenous tRNA^{Pro} for the decoding of *tti2-L187P* mutation. We used this system to analyze the activity of human tRNA$^{Ser}_{UGA}$ in *S. cerevisiae* after converting the anticodon to UGG. A gene encoding human tRNA^{Ser} (chr2.trna21) with the UGA anticodon converted to UGG was inserted into the *SUP17* locus with ~300 bp of 5′ and 3′ flanking yeast sequence and cloned into the yeast *URA3* centromeric plasmid YCplac33 (see Figure S1 for the gene sequence). As shown in Figure 1B, the human gene expressing tRNA$^{Ser}_{UGG}$ (*Hst*RNA$^{Ser}_{UGG}$) could be transformed into the wild-type strain BY4742, in contrast to *Sct*RNA$^{Ser}_{UGG}$ [39]. The growth of this strain on a minimal plate was slightly slower than a strain containing YCplac33 alone and slightly faster than a strain containing *Sct*RNA$^{Ser}_{UGG}$ with a G26A mutation (Figure 1C; see Reference [39]). The reduced toxicity of the human tRNA$^{Ser}_{UGG}$ indicates that its activity in yeast is diminished relative to *S. cerevisiae* tRNA$^{Ser}_{UGG}$. To determine if the human tRNA^{Ser} functions in yeast, *Hst*RNA$^{Ser}_{UGG}$ was transformed into CY7020, which contains the crippled *tti2-L187P* allele. As shown by growth on medium containing 5% ethanol (Figure 1D), *Hst*RNA$^{Ser}_{UGG}$ suppressed the stress sensitive growth caused by *tti2-L187P*. The human tRNA thus has partial function in yeast.

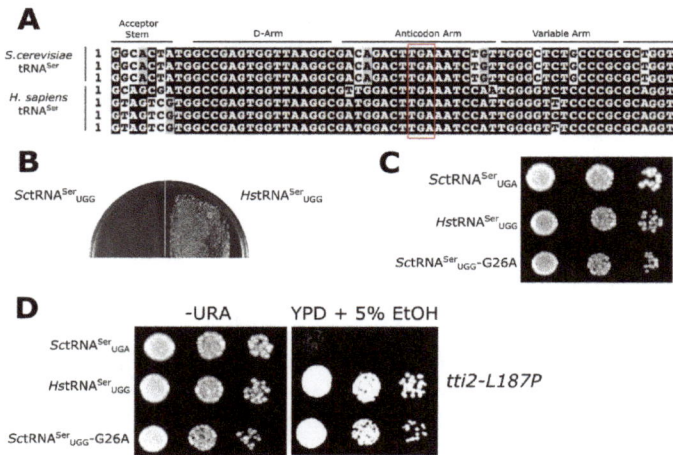

Figure 1. Human (*Homo sapiens*) tRNA$^{Ser}_{UGA}$ functions inefficiently in *Saccharomyces cerevisiae*. (**A**) Sequence alignment of genes encoding tRNA$^{Ser}_{UGA}$ isodecoders from *S. cerevisiae* and humans. The red box highlights the anticodon. Sequences were obtained from the GtRNAdb ([52]; http://gtrnadb.ucsc.edu/index.html) and aligned with MUSCLE (https://www.ebi.ac.uk/Tools/msa/muscle/). (**B**) *URA3* centromeric plasmid (YCplac33) expressing *S. cerevisiae* (*SUP17*) or human (tRNA-Ser-TGA-3-1) tRNA^{Ser} genes with their UGA anticodon converted to UGG for proline [*Sct*RNA$^{Ser}_{UGG}$ and *Hst*RNA$^{Ser}_{UGG}$] were transformed into the wild-type yeast strain BY4742 using a lithium acetate protocol and grown for 3 days at 30 °C. (**C**) The growth of BY4742 containing a centromeric plasmid expressing *Hst*RNA$^{Ser}_{UGG}$ was compared to strains containing wild-type *Sct*RNA$^{Ser}_{UGA}$ and the mistranslating *Sct*RNA$^{Ser}_{UGG}$-G26A. Strains were grown in medium lacking uracil for 40 h, then 10-fold serially diluted, spotted on minimal plates lacking uracil, and grown for 2 days at 30 °C. (**D**) Plasmids expressing *Sct*RNA$^{Ser}_{UGA}$, *Hst*RNA$^{Ser}_{UGG}$, or *Sct*RNA$^{Ser}_{UGG}$-G26A were transformed into the *tti2-L187P* strain CY7020, streaked for individual colonies, grown in minimal medium lacking uracil to stationary phase, then 10-fold serially diluted, spotted on minimal plates lacking uracil (-URA) or yeast peptone dextrose (YPD) containing 5% ethanol, and grown at 30 °C for 2 days (for -URA) or 3 days (for YPD + 5% ethanol).

Because 7 of the 15 differences always found between yeast and human tRNA^{Ser} are in the acceptor stem (Figure 1A), we engineered a chimeric tRNA with the human acceptor stem substituted on the *S. cerevisiae* tRNA and with a UGG anticodon (*ch*tRNA$^{Ser}_{UGG}$; Figure 2A). The chimeric tRNA$^{Ser}_{UGG}$ mimicked *Hs*tRNA$^{Ser}_{UGG}$ in its ability to be transformed into BY4742 (Figure 2B). As shown in Figure 2C, *ch*tRNA$^{Ser}_{UGG}$ was more toxic than *Hs*tRNA$^{Ser}_{UGG}$, suggesting that at least one other difference between *S. cerevisiae* and human tRNA^{Ser} exists outside the acceptor stem (discussed in Section 3.4). The chimeric tRNA was also functional as determined by its ability to suppress *tti2-L187P* when transformed into CY7020 (Figure 2D).

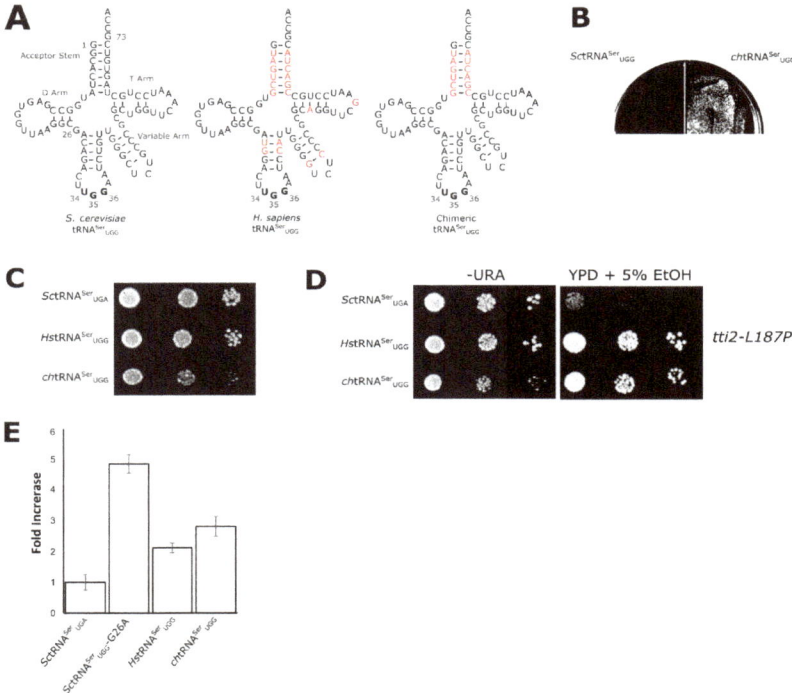

Figure 2. The human tRNA^{Ser} acceptor stem decreases the functionality of yeast tRNA^{Ser}. (**A**) Secondary structures of tRNA$^{Ser}_{UGG}$ derivatives used in these studies. The chimeric molecule contains the human acceptor stem on the yeast tRNA. Differences with the *S. cerevisiae* tRNA$^{Ser}_{UGG}$ (*SUP17*$_{UGG}$) are in red. (**B**) *URA3* centromeric plasmids expressing the chimeric tRNA (*ch*tRNA$^{Ser}_{UGG}$) with a UGG anticodon or *Sc*tRNA^{Ser} (*SUP17*) with a UGG anticodon were transformed into BY4742 and grown at 30 °C for 2 days on a minimal plate. (**C**) BY4742 expressing *Sc*tRNA$^{Ser}_{UGA}$, *Hs*tRNA$^{Ser}_{UGG}$, or *ch*tRNA$^{Ser}_{UGG}$ were grown overnight in minimal media, serially diluted, plated on media lacking uracil, and grown at 30 °C for 2 days. (**D**) Plasmids expressing *Sc*tRNA$^{Ser}_{UGA}$, *Hs*tRNA$^{Ser}_{UGG}$, or *ch*tRNA$^{Ser}_{UGG}$ were transformed into the *tti2-L187P* strain CY7020, streaked for individual colonies, grown in minimal medium for 40 h, 10-fold serially diluted, spotted on minimal plates lacking uracil (-URA) or yeast peptone dextrose (YPD) containing 5% ethanol, and grown at 30 °C for 2 days (for -URA) or 3 days (for YPD + 5% ethanol). (**E**) A reporter plasmid containing GFP expressed from a Hsf1-activated promoter was transformed into BY4742 also containing centromeric plasmids expressing wild-type *Sc*tRNA$^{Ser}_{UGA}$, *Hs*tRNA$^{Ser}_{UGG}$, or *ch*tRNA$^{Ser}_{UGG}$. Starter cultures were grown in minimal medium, diluted into fresh selective medium, and grown for eight hours. Induction of the Hsf1-GFP was measured by fluorescence at 528 nm. Numbers shown are the average of five biological replicates performed in duplicate with the standard deviation shown.

The mistranslation experiments we performed were a balance between improved growth due to suppression of *tti2-L187P* and toxicity due to proteome-wide mistranslation. To provide another indication of the toxicity arising from mistranslation due to each of the tRNAs, we measured heat shock induction in the strains using a reporter plasmid containing GFP expressed from an Hsf1-activated promoter (Figure 2E). *HstRNA*$^{Ser}_{UGG}$ and *chtRNA*$^{Ser}_{UGG}$ resulted in a heat shock response 2.1- and 2.8-fold, respectively, greater than the control. *SctRNA*$^{Ser}_{UGG}$-G26A resulted in a 4.8-fold increase in the heat shock response. These results suggest that the toxicity of human and chimeric tRNAs is the result of a loss of proteostasis due to mistranslation.

3.2. Evaluating Mechanisms for the Reduced Functionality of Human tRNA^{Ser}

Differences between yeast and human tRNA^{Ser} could be due to any step leading to the functionality of the tRNA. Included in this are factors that contribute to the cellular concentration of the tRNA. Although it is difficult to compare the levels of the mistranslating tRNAs because the genomically encoded copies of yeast tRNA$^{Ser}_{UGA}$ differ from *SctRNA*$^{Ser}_{UGG}$ by only one nucleotide, we addressed a possible role for the rapid tRNA decay (RTD) pathway in the function of *HstRNA*$^{Ser}_{UGG}$. *HstRNA*$^{Ser}_{UGG}$ was transformed into a *met22Δ* strain where the RTD pathway is inhibited [53,54]. If the mistranslating human tRNA was being turned over by the RTD pathway, we would expect it to be more toxic in a *met22Δ* strain, where it would accumulate at higher levels and result in increased levels of mistranslation. The toxicity of *HstRNA*$^{Ser}_{UGG}$ was not increased in the *met22Δ* strain (Figure 3A), suggesting the human tRNA is not being turned over by the RTD pathway.

If the functionality and resulting mistranslation of *chtRNA*$^{Ser}_{UGG}$ was reduced relative to *SctRNA*^{Ser} because of decreased aminoacylation, we would expect that toxicity would increase if human SerRS, SARS, was also introduced into yeast cells. We amplified the coding region of SARS by PCR and cloned it 3′ of the yeast *DED1* promoter and a myc tag. The myc-tagged SARS is expressed in *S. cerevisiae* strain BY4742 at a level comparable to *S. cerevisiae* Ses1 (Figure S2). Centromeric plasmids expressing *SctRNA*$^{Ser}_{UGA}$, *SctRNA*$^{Ser}_{UGG}$-G26A, *HstRNA*$^{Ser}_{UGG}$, and *chtRNA*$^{Ser}_{UGG}$ were then transformed into strains expressing SARS and Ses1 on 2μ plasmids. Their growth was compared on minimal media lacking uracil and leucine. As shown in Figure 3B, the presence of human SARS reduces the growth of the strain containing either *HstRNA*$^{Ser}_{UGG}$ or *chtRNA*$^{Ser}_{UGG}$ but not the strain containing *SctRNA*$^{Ser}_{UGG}$-G26A. The increased toxicity of *HstRNA*$^{Ser}_{UGG}$ and *chtRNA*$^{Ser}_{UGG}$ in the presence of SARS further indicates that the acceptor stem contributes to differences in the aminoacylation of human and yeast tRNA^{Ser}.

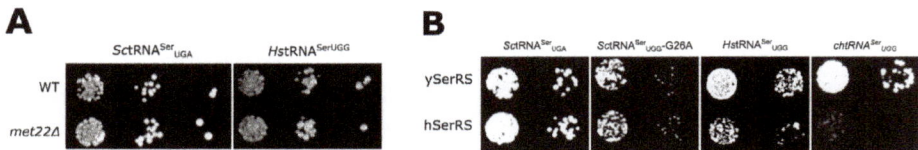

Figure 3. Evaluating mechanisms of decreased function of human tRNA$^{Ser}_{UGG}$ in yeast. (**A**) BY4742 (*MET22*) and CY7641 (*met22Δ*) containing either wild-type (WT) *SctRNA*$^{Ser}_{UGA}$ or *HstRNA*$^{Ser}_{UGG}$ were grown to stationary phase in media lacking uracil before cells were spotted in 10-fold serial dilutions onto a plate lacking uracil and grown at 30°. (**B**) Expression of human SerRS, SARS, increases the functionality and thus the toxicity of *HstRNA*$^{Ser}_{UGG}$ and *chtRNA*$^{Ser}_{UGG}$ in yeast. Yeast strain BY4742 containing a *URA3* centromeric plasmid expressing *SctRNA*$^{Ser}_{UGA}$, *SctRNA*$^{Ser}_{UGG}$-G26A. *HstRNA*$^{Ser}_{UGG}$ or *chtRNA*$^{Ser}_{UGG}$ and the *LEU2* multicopy plasmid YCplac181 containing myc-tagged yeast *SES1* (ySerRS) or human SARS (hSerRS) were grown to stationary phase in media lacking uracil and leucine. Cell densities were normalized, and cultures were spotted in 10-fold serial dilutions on the same media.

3.3. The tRNA^{Ser} 3:70 Base Pair Contributes to Species Specificity

To begin to map the key bases within the acceptor stem that could account for the functional difference of human tRNA^{Ser} in yeast, we first compared sequences of the *S. cerevisiae* tRNA^{Ser} isoacceptors. We hypothesized that key bases for yeast SerRS should be conserved. Five bases in the acceptor stem are conserved in all yeast serine tRNAs. These are the G1:C72 and C3:G70 bases pairs and U69 (Figure 4A; see Figure S3 for the sequence of the tRNAs). Figure 4B shows the bases that differ between yeast and human tRNA^{Ser}$_{UGA}$ isodecoders. Only G1:C72 is conserved for all these tRNAs (see sequences in Figure 1A). U69 and the 3:70 base pair are thus strong candidates for a difference determinant between *S. cerevisiae* and human tRNA^{Ser}. We evaluated the importance of U69 with a variant of *Sc*tRNA^{Ser}$_{UGG}$ containing the human G4:C69 base pair. A gene expressing this variant could not be transformed into yeast (Figure S4), indicating that it is fully functional and suggesting that U69 does not signal the specificity difference between yeast and human tRNA^{Ser}. To test the 3:70 base pair, we constructed a variant of *ch*tRNA^{Ser}$_{UGG}$ with the yeast C3:G70 base pair (*ch*tRNA^{Ser}$_{UGG}$-C3:G70). Unlike the unmodified chimeric gene, *ch*tRNA^{Ser}$_{UGG}$ = C3:G70 gave rise to extremely slow growing colonies when transformed into BY4742 (Figure 4C). The C3:G70 mutation thus increases the functionality and toxicity of the chimeric tRNA in *S. cerevisiae*.

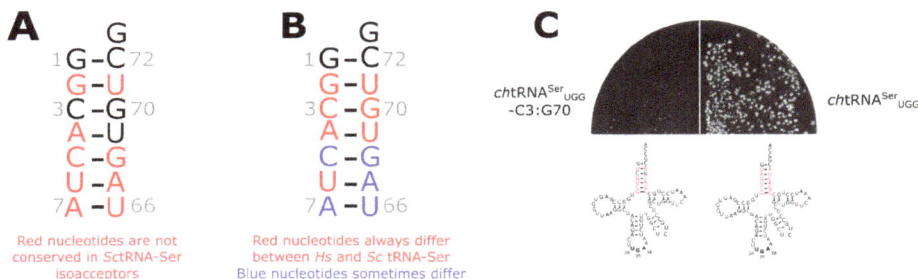

Figure 4. The acceptor stem 3:70 base pair is a determinant of yeast versus human tRNA^{Ser} specificity. (**A**) Variation in the acceptor stem of the *S. cerevisiae* tRNA^{Ser} isoacceptors. The acceptor stem of the tRNA encoded by *SUP17* is shown with the positions that vary in the different cytoplasmic tRNA^{Ser} isoacceptors colored in red. (**B**) Variation in the acceptor stems of tRNA^{Ser}$_{UGA}$ between *S. cerevisiae* and humans. The acceptor stem of the tRNA encoded by *SUP17* is shown with the bases always different between yeast and human colored in red and those sometimes-different colored in blue. (**C**) BY4742 expressing *ch*tRNA^{Ser}$_{UGG}$ or *ch*tRNA^{Ser}$_{UGG}$-C3:G70 on a *URA3* centromeric plasmid were transformed into BY4742 and grown at 30 °C for 2 days.

The aminoacylation domain of the class II aaRS enzymes contains antiparallel β sheets. The loop between strands 2 and 3 comprise the conserved motif 2, which interacts with the tRNA acceptor arm [34]. The crystal structures of SerRS from several organisms have been solved. Included in these are the human enzyme and a structure of the *T. thermophilus* enzyme in complex with tRNA^{Ser}. Unfortunately, in the latter, the acceptor stem is not fully resolved. Models for the interaction of SerRS with tRNA^{Ser} position motif 2 close to base pairs 3:70 and 4:69 of the acceptor stem (for example see Figure 5A and Reference [25]). To begin to identify what may account for the different acceptor stem preferences of the SerRS enzymes, we performed a sequence alignment of motif 2 from the human, *T. thermophilus*, *C. albicans*, and *S. cerevisiae* enzymes. The human and *T. thermophilus* enzymes prefer a A3:U70 base pair in the acceptor stem, whereas the two yeast enzymes prefer C3:G70. Two residues in motif 2 are conserved in the yeast enzymes that differ in the human and *T. thermophilus* enzymes (Figure 5B). Trp290 and Ala297 of the *S. cerevisiae* enzyme align with Arg and Gln, respectively, in *T. thermophilus* and human enzymes. Ala297 is not well conserved across yeast and fungal species containing C3:G70 tRNA^{Ser} (Figure S5); in contrast, the 290 position is conserved in yeast and is proximal to the base at position 70 (Figure 5A). We constructed a version of hSerRS, where residues T312 and R313 in motif 2 were converted to the corresponding yeast residues A and W. The mutated hSerRS was introduced into BY4742 containing wild-type $SctRNA^{Ser}_{UGA}$, $SctRNA^{Ser}_{UGG}$-G26A, $HstRNA^{Ser}_{UGG}$, or $chtRNA^{Ser}_{UGG}$ (Figure 5C). The wild-type human SerRS decreased the growth rate in strains containing the human or chimeric tRNA^{Ser} but did not affect the strains containing the yeast tRNAs. In contrast, the hSerRS TR to AW mutant did not alter the growth rate in these strains when compared to the yeast SerRS, suggesting that T312 and R313 are important for the function of the human SerRS on the human tRNA.

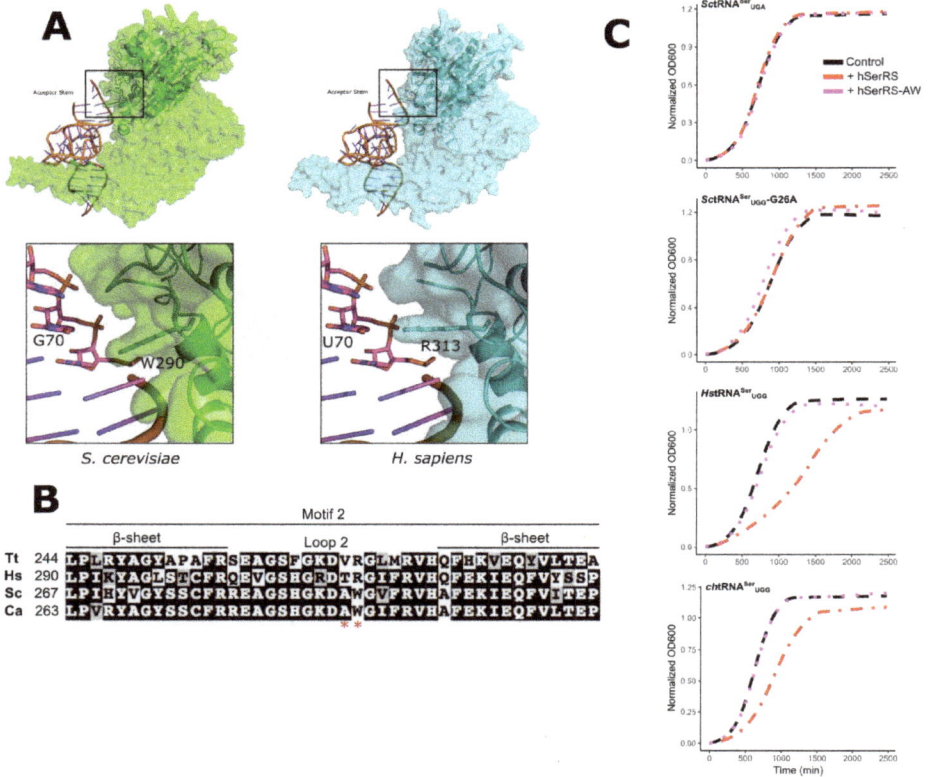

Figure 5. Sequence differences in Motif 2 of SerRS. (**A**) Model of motif 2 SerRS:tRNA^{Ser} interactions. The yeast (green) and human (cyan) SerRS were modeled using SWISS-MODEL [55] onto the *Thermus thermophilus* synthetase:tRNA co-crystal structure (PDB: 1SER; [12]). The region around motif 2 is boxed and expanded in the lower images. (**B**) Sequence alignment of SerRS motif 2 sequences from *T. thermophilus* (WP_024119136), *H. sapiens* (NP_006504), *S. cerevisiae* (NP_010306), and *Candida albicans* (XP_719967) using the default settings of MUSCLE (https://www.ebi.ac.uk/Tools/msa/muscle/). The red stars mark the residues mutated to the corresponding yeast residues. (**C**) Yeast strain BY4742 containing a *URA3* centromeric plasmid expressing $SctRNA^{Ser}_{UGA}$, $HstRNA^{Ser}_{UGG}$, $chtRNA^{Ser}_{UGG}$. or $SctRNA^{Ser}_{UGG}$-G26A and the *LEU2* multicopy plasmid YEplac181 containing myc-tagged ySerRS (black), hSerRS (red), or mutant hSerRS-AW (purple) were grown to stationary phase in media lacking uracil and leucine, diluted to an OD$_{600}$ ~ 0.1, and grown for 40 h at 30 °C. OD$_{600}$ was measured every 15 min to generate growth curves.

Though it is difficult to exclude contributions from all other mechanisms, such as the exact cellular levels of each tRNA and their interactions, we suggest that aminoacylation is a component of the different functionality of human tRNA^{Ser} in *S. cerevisiae*. For class II enzymes, motif 2 contacts the acceptor stem [23,56–58]. Interactions between SerRS and tRNA^{Ser} suggest that these interactions influence aminoacylation by *S. cerevisiae* SerRS [12,25]. In addition, the toxicity, and thus the inferred functionality, of both *Hst*RNA$^{Ser}_{UGG}$ and *cht*RNA$^{Ser}_{UGG}$ was enhanced when human SerRS was introduced into yeast. Furthermore, altering the motif 2 sequence of hSerRS to resemble the yeast sequence decreases the activity of the human enzyme with human tRNA^{Ser}. We do note that in our analysis, we compared one yeast serine isodecoders with one version of the same human isodecoders. Sequence differences in isodecoders families may result in different cross-species functionality. For tRNA^{Ser}, the major identity element is the variable arm, both its length and its sequence [6,13,59]. Our data support the argument that the acceptor stem of tRNA^{Ser} also has a role in its aminoacylation and that these changes are species-specific. The role of the acceptor stem of tRNA^{Ser} parallels that seen for many other aaRS enzymes [6]. Interestingly, Shiba et al. [31] observed that the positioning of identity elements for specific tRNA isoacceptors is generally common across species, but the exact nature of these sequences is often species-specific.

The possibility of a shared importance of the 3:70 base pair for aminoacylation by the class II enzymes AlaRS and SerRS is noteworthy. G3:U70 is unique for tRNA^{Ala} and highly conserved. The significance of the G3:U70 identity element for AlaRS has likely led to its parallel exclusion from other tRNAs. In fact, G3:C70 and A3:U70 show a somewhat limited distribution in non-alanine tRNAs in eukaryotic species (Figure S6), likely because of the mistranslation that would result from a single base mutation generating a G3:U70. These sequence limitations may have led to the 3:70 base pair's role in recognition arising in some other tRNAs in some species. For many yeast and fungal species, the C3:G70 base pair is conserved in tRNA^{Ser}-Ser. It is also conserved in *T. thermophilus* as A:U in all tRNA isoacceptors, which is interesting because of all the tRNAs, the four serine tRNAs are the only ones where A3:U70 is found.

3.4. Unique Aspects of the SerRS Motif 2

It is interesting that the loop sequence in motif 2 in SerRS is distinct from other aaRS (Figure S7). The placement of Trp in the motif 2 loop of yeast SerRS was previously observed by Lenhard et al. [60]. They demonstrated that mutations to the loop region affect functionality. Our model proposes that Trp290 of yeast SerRS is a discriminating factor in the sequence preferences of the enzyme for the tRNA. Arginine in motif 2 of SerRS is frequently found with tRNA^{Ser} containing the A3:U70 base pair and in species containing a broader range of acceptor stem sequences, including humans, whereas tryptophan is frequently found where C3:G70 is the sole or predominant sequence as seen in yeast and fungi. Eichert et al. [25] examined the structure of minihelices with the *T. thermophilus* SerRS enzyme. They found that the side chain of Arg267 of the *T. thermophilus* enzyme (the equivalent of Trp290 in *S. cerevisiae* Ses1) interacts with the phosphate backbone of bases C69 and U70 via a water network. The exact positioning of Trp290 in the *C. albicans* structure is difficult to predict because of low electron density in this region; however, the hydrophobic Trp290 side chain almost certainly will have different interactions than a polar arginine. Tryptophan is infrequently found in base interactions [61], but hydrophobic residues occur in nucleic acid interactions through base stacking. For example, a tryptophan in the Tn5 transpose base stacks with a thymine of its substrate DNA [62]. This interaction requires distortion of the DNA duplex. Similar hydrophobic interactions occurring after base flipping are observed for other enzymes, including methyltransferases and endonucleases [63]. Further experiments will be required to determine if correct positioning of the tRNA acceptor stem of the Trp290 enzymes involves distortion of the base pairing. Alternatively, the Trp290 containing enzymes may compensate for the lack of interaction through additional contacts. To address this possibility, we analyzed the mutual information between each pair of positions within the aminoacylation domain of the cytoplasmic SerRS family of proteins to determine residues coevolving with Trp290 (Figure S8; Supplemental File S2). No residue was found to specifically coevolve with Trp290. We do note that although the role of the distinct motif 2 sequence found in SerRS may be in tRNA recognition, motif 2 more generally plays a role in ATP binding. This function could be indirectly affected by our mutations.

The chimeric *S. cerevisiae* tRNA^{Ser} with the human acceptor stem was somewhat more toxic and induced a slightly greater heat shock response in yeast than the human tRNA^{Ser}. This suggests that some of the other base differences between yeast and human tRNA^{Ser} molecules contribute a partial effect to tRNA function. These may include the differences in the anticodon stem, specifically at position 29 (A in *S. cerevisiae*, G in humans); through random selection, we have identified a G mutation in the yeast tRNA as having reduced function [64]. It is unlikely that this mutation affects aminoacylation given the lack of involvement of the anticodon stem-loop in the reaction. There are also differences in the T-arms of yeast and human tRNAs, which may influence interactions of the tRNA with the elongation factor EF-1α.

3.5. Applications for Mistranslation

tRNAs that mistranslate have roles in synthetic biology and potentially therapeutically. The latter arises because most diseases are the result of missense mutations that could be corrected through mistranslation. The synthetic biology applications include expanding the diversity of expressed products (statistical proteins) and controlling expression while transferring DNA between species. All these cases require tRNAs that mistranslate at a level below a toxic threshold. Our results indicate that a simple approach to achieve this is to use the inherent differences in function seen in the tRNAs from different species.

Translation requires multiple independent yet connected processes. Overall, these are highly conserved across species. It is, however, the differences that allow the enzymes involved in translation to be the targets of antibiotic and antimicrobial agents. The subtle differences seen for SerRS and the species specificity of its interactions with tRNA^{Ser} suggest its potential as a therapeutic target.

Supplementary Materials: The following are available online at http://www.mdpi.com/2073-4425/9/12/612/s1. Figure S1: Human tRNA^{Ser} sequence, Figure S2: Expression of human SerRS in yeast, Figure S3: Gene sequence of *S. cerevisiae* tRNA^{Ser} isoacceptors for cytoplasmic translation, Figure S4: Human tRNA^{Ser}$_{UGG}$ with G4:C69 is toxic in *S. cerevisiae*, Figure S5: Alignment of cytoplasmic SerRS motif 2 sequences from yeast and fungal species, Figure S6: Percent distribution of the base pair at 3:70 in the tRNAs in *Saccharomyces cerevisiae*, *Homo sapiens*, *Candida albicans*, *Thermus thermophilus* and *Escherichia coli*, Figure S7: Motif 2 sequences from the yeast class II aaRS enzymes Hts1, Figure S8: Co-evolution of cytoplasmic SerRS, Table S1: Yeast strains used in this study, Table S2: Oligonucleotides used in this study, File S1: SerRS Coevolution Sequences, File S2: SerRS Coevolution Mutual Information Output.

Author Contributions: M.D.B. and C.J.B. conceived and designed the experiments; M.D.B., J.G. and Y.Z. performed the experiments; M.D.B. and S.M. analyzed the data and created the figures; G.B.G. guided the co-evolution analysis; M.D.B. and C.J.B wrote the paper.

Funding: This work was supported from the Natural Sciences and Engineering Research Council of Canada [RGPIN-2015-04394 to C.J.B.] and generous donations from Graham Wright and James Robertson to M.D.B. M.D.B. holds an NSERC Alexander Graham Bell Canada Graduate Scholarship (CGS-D).

Acknowledgments: We thank Martin Duennwald for providing the HSE-eGFP plasmid and Patrick Lajoie for his help with the growth curve analyses.

Conflicts of Interest: The authors declare no conflicts of interest.

References

1. Hoffman, K.S.; O'Donoghue, P.; Brandl, C.J. Mistranslation: From adaptations to applications. *Biochim. Biophys. Acta* **2017**. [CrossRef] [PubMed]
2. Giegé, R.; Jühling, F.; Pütz, J.; Stadler, P.; Sauter, C.; Florentz, C. Structure of transfer RNAs: Similarity and variability. *Wiley Interdiscip. Rev. RNA* **2012**, *3*, 37–61. [CrossRef] [PubMed]
3. Rich, A.; RajBhandary, U.L. Transfer RNA: Molecular structure, sequence, and properties. *Annu. Rev. Biochem.* **1976**, *45*, 805–860. [CrossRef] [PubMed]

4. De Duve, C. Transfer RNAs: The second genetic code. *Nature* **1988**, *333*, 117–118. [CrossRef] [PubMed]

5. Commans, S.; Lazard, M.; Delort, F.; Blanquet, S.; Plateau, P. tRNA anticodon recognition and specification within subclass IIb aminoacyl-tRNA synthetases. *J. Mol. Biol.* **1998**, *278*, 801–813. [CrossRef] [PubMed]

6. Giegé, R.; Sissler, M.; Florentz, C. Universal rules and idiosyncratic features in tRNA identity. *Nucleic Acids Res.* **1998**, *26*, 5017–5035. [CrossRef] [PubMed]

7. Imura, N.; Weiss, G.B.; Chambers, R.W. Reconstitution of alanine acceptor activity from fragments of yeast tRNA-Ala II. *Nature* **1969**, *222*, 1147–1148. [CrossRef] [PubMed]

8. Hou, Y.M.; Schimmel, P. Evidence that a major determinant for the identity of a transfer RNA is conserved in evolution. *Biochemistry* **1989**, *28*, 6800–6804. [CrossRef] [PubMed]

9. Hou, Y.M.; Schimmel, P. A Simple structural feature is a major determinant of the identity of a transfer RNA. *Nature* **1988**, *333*, 140–145. [CrossRef]

10. Hoffman, K.S.; Berg, M.D.; Shilton, B.H.; Brandl, C.J.; O'Donoghue, P. Genetic Selection for mistranslation rescues a defective co-chaperone in yeast. *Nucleic Acids Res.* **2017**, *45*, 3407–3421. [CrossRef]

11. Asahara, H.; Himeno, H.; Tamura, K.; Nameki, N.; Hasegawa, T.; Shimizu, M. *Escherichia coli* seryl-tRNA synthetase recognizes tRNA^{Ser} by its characteristics tertiary structure. *J. Mol. Biol.* **1994**, 738–748. [CrossRef]

12. Biou, V.; Yaremchuk, A.; Tukalo, M.; Cusack, S. The 2.9 Å crystal structure of *T. thermophilus* seryl-tRNA synthetase complexed with tRNA^{Ser}. *Science* **1994**, *263*, 1404–1410. [CrossRef] [PubMed]

13. Himeno, H.; Yoshida, S.; Soma, A.; Nishikawa, K. Only one nucleotide insertion to the long variable arm confers an efficient serine acceptor activity upon *Saccharomyces cerevisiae* tRNA^{Leu} in vitro. *J. Mol. Biol.* **1997**, *268*, 704–711. [CrossRef] [PubMed]

14. Pang, Y.L.J.; Poruri, K.; Martinis, S.A. tRNA synthetase: tRNA aminoacylation and beyond. *Wiley Interdiscip. Rev. RNA* **2014**, *5*, 461–480. [CrossRef] [PubMed]

15. Eriani, G.; Delarue, M.; Poch, O.; Gangloff, J.; Moras, D. Partition of tRNA synthetases into two classes based on mutually exclusive sets of sequence motifs. *Nature* **1990**, *347*, 203–206. [CrossRef] [PubMed]

16. Ibba, M.; Soll, D. Aminoacyl-tRNA synthesis. *Annu. Rev. Biochem.* **2000**, *69*, 617–650. [CrossRef] [PubMed]

17. Eriani, G.; Dirheimer, G.; Gangloff, J. Cysteinyl-tRNA synthetase: Determination of the last *E. Coli* aminoacyl-tRNA synthetase primary structure. *Nucleic Acids Res.* **1991**, *19*, 265–269. [CrossRef] [PubMed]

18. Sugiura, I.; Nureki, O.; Ugaji-Yoshikawa, Y.; Kuwabara, S.; Shimada, A.; Tateno, M.; Lorber, B.; Giegé, R.; Moras, D.; Yokoyama, S.; et al. The 2.0 Å crystal structure of *Thermus thermophilus* methionyl-tRNA synthetase reveals two RNA-binding modules. *Structure* **2000**, *8*, 197–208. [CrossRef]

19. Artymiuk, P.J.; Rice, D.W.; Poirrette, A.R.; Willet, P. A tale of two synthetases. *Nat. Struct. Biol.* **1994**, *1*, 758–760. [CrossRef] [PubMed]

20. Cusack, S.; Berthet-Colominas, C.; Härtlein, M.; Nassar, N.; Leberman, R. A Second class of synthetase structure revealed by X-ray analysis of *Escherichia coli* seryl-tRNA synthetase at 2.5 A. *Nature* **1990**, *347*, 249–255. [CrossRef]

21. Cusack, S.; Härtlein, M.; Leberman, R. Sequence, structural and evolutionary relationships between class 2 aminoacyl-tRNA synthetases. *Nucleic Acids Res.* **1991**, *19*, 3489–3498. [CrossRef] [PubMed]

22. Smith, T.F.; Hartman, H. The evolution of class II aminoacyl-tRNA synthetases and the first code. *FEBS Lett.* **2015**, *589*, 3499–3507. [CrossRef] [PubMed]

23. Ruff, M.; Krishnaswamy, S.; Boeglin, M.; Poterszman, A.; Mitschler, A.; Podjarny, A.; Rees, B.; Thierry, J.C.; Moras, D. Class II aminoacyl transfer RNA synthetases: Crystal structure of yeast aspartyl-tRNA synthetase complexed with tRNA^{Asp}. *Science* **1991**, *252*, 1682–1689. [CrossRef] [PubMed]

24. Cavarelli, J.; Moras, D. Recognition of tRNAs by aminoacyl-tRNA synthetases. *FASEB J.* **1993**, *7*, 79–86. [CrossRef] [PubMed]

25. Eichert, A.; Oberthuer, D.; Betzel, C.; Geßner, R.; Erdmann, V.A.; Fürste, J.P.; Förster, C. The seryl-tRNA synthetase/tRNA ser acceptor stem interface is mediated via a specific network of water molecules. *Biochem. Biophys. Res. Commun.* **2011**, *412*, 532–536. [CrossRef] [PubMed]

26. Bilokapic, S.; Maier, T.; Ahel, D.; Gruic-Sovulj, I.; Söll, D.; Weygand-Durasevic, I.; Ban, N. Structure of the unusual seryl-tRNA synthetase reveals a distinct zinc-dependent mode of substrate recognition. *EMBO J.* **2006**, *25*, 2498–2509. [CrossRef] [PubMed]

27. Vincent, C.; Borel, F.; Willison, J.C.; Laberman, R.; Härtlein, M. Seryl-tRNA synthetase from *Escherichia coli*: Functional evidence for cross-dimer tRNA binding during aminoacylation. *Nucleic Acids Res.* **1995**, *23*, 1113–1118. [CrossRef] [PubMed]

28. Wang, C.; Guo, Y.; Tian, Q.; Jia, Q.; Gao, Y.; Zhang, Q.; Zhou, C.; Xie, W. SerRS-tRNASec complex Structures reveal mechanism of the first step in selenocysteine biosynthesis. *Nucleic Acids Res.* **2015**, *43*, 10534–10545. [CrossRef]

29. Normanly, J.; Ogden, R.C.; Horvath, S.J.; Abelson, J. Changing the identity of a transfer RNA. *Nature* **1986**, *321*, 213–219. [CrossRef]

30. Normanly, J.; Ollick, T.; Abelson, J. Eight base changes are sufficient to convert a leucine-inserting tRNA into a serine-inserting tRNA. *Proc. Natl. Acad. Sci. USA* **1992**, *89*, 5680–5684. [CrossRef]

31. Shiba, K.; Stello, T.; Motegi, H.; Noda, T.; Musier-Forsyth, K.; Schimmel, P. Human lysyl-tRNA synthetase accepts nucleotide 73 variants and rescues *Escherichia coli* double-defective mutant. *J. Biol. Chem.* **1997**, *272*, 22809–22816. [CrossRef] [PubMed]

32. Ripmaster, T.L.; Shiba, K.; Schimmel, P. Wide cross-species aminoacyl-tRNA synthetase replacement in vivo: Yeast cytoplasmic alanine enzyme replaced by human polymyositis serum antigen. *Proc. Natl. Acad. Sci. USA* **1995**, *92*, 4932–4936. [CrossRef] [PubMed]

33. Rips, J.; Meyer-Schuman, R.; Breuer, O.; Tsabari, R.; Shaag, A.; Revel-Vilk, S.; Reif, S.; Elpeleg, O.; Antonellis, A.; Harel, T. MARS variant associated with both recessive interstitial lung and liver disease and dominant Charcot-Marie-Tooth disease. *Eur. J. Med. Genet.* **2018**. [CrossRef] [PubMed]

34. Burke, B.; Yang, F.; Chen, F.; Stehlin, C.; Chan, B.; Musier-Forsyth, K. Evolutionary coadaptation of the motif 2−acceptor stem interaction in the class II prolyl-tRNA synthetase system. *Biochemistry* **2000**, *39*, 15540–15547. [CrossRef] [PubMed]

35. Edwards, H.; Trézéguet, V.; Schimmel, P. An *Escherichia coli* tyrosine transfer RNA is a leucine-specific transfer RNA in the yeast *Saccharomyces cerevisiae*. *Proc. Natl. Acad. Sci. USA* **1991**, *88*, 1153–1156. [CrossRef] [PubMed]

36. Davis, L.; Chin, J.W. Designer proteins: Applications of genetic code expansion in cell biology. *Nat. Rev. Mol. Cell Biol.* **2012**, *13*, 168–182. [CrossRef] [PubMed]

37. Winzeler, E.A.; Davis, R.W. Functional analysis of the yeast genome. *Curr. Opin. Genet. Dev.* **1997**, *7*, 771–776. [CrossRef]

38. Tong, A.H.; Evangelista, M.; Parsons, A.B.; Xu, H.; Bader, G.D.; Pagé, N.; Robinson, M.; Raghibizadeh, S.; Hogue, C.W.; Bussey, H.; et al. Systematic genetic analysis with ordered arrays of yeast deletion mutants. *Science* **2001**, *294*, 2364–2368. [CrossRef]

39. Berg, M.D.; Hoffman, K.S.; Genereaux, J.; Mian, S.; Trussler, R.S.; Haniford, D.B.; O'Donoghue, P.; Brandl, C.J. Evolving mistranslating tRNAs through a phenotypically ambivalent intermediate in *Saccharomyces cerevisiae*. *Genetics* **2017**, *206*, 1865–1879. [CrossRef]

40. Hoffman, K.S.; Duennwald, M.L.; Karagiannis, J.; Genereaux, J.; McCarton, A.S.; Brandl, C.J. *Saccharomyces cerevisiae* Tti2 regulates PIKK proteins and stress response. *G3* **2016**, *6*, 1649–1659. [CrossRef]

41. Brandman, O.; Stewart-Ornstein, J.; Wong, D.; Larson, A.; Williams, C.C.; Li, G.-W.; Zhou, S.; King, D.; Shen, P.S.; Weibezahn, J.; et al. A Ribosome-bound quality control complex triggers degradation of nascent peptides and signals translation stress. *Cell* **2012**, *151*, 1042–1054. [CrossRef] [PubMed]

42. Saleh, A.; Lang, V.; Cook, R.; Brandl, C.J. Identification of native complexes containing the yeast coactivator/repressor proteins NGG1/ADA3 and ADA2. *J. Biol. Chem.* **1997**, *272*, 5571–5578. [CrossRef] [PubMed]

43. Dickson, R.J.; Gloor, G.B. Bioinformatics Identification of Coevolving Residues. In *Homing Endonucleases: Methods and Protocols*; Edgell, D.R., Ed.; Humana Press: Totowa, NJ, USA, 2014; pp. 223–243.

44. Rocha, R.; Pereira, P.J.B.; Santos, M.A.S.; Macedo-Ribeiro, S. Unveiling the structural basis for translational ambiguity tolerance in a human fungal pathogen. *Proc. Natl. Acad. Sci. USA* **2011**, *108*, 14091–14096. [CrossRef] [PubMed]

45. Itoh, Y.; Sekine, S. ichi; Kuroishi, C.; Terada, T.; Shirouzu, M.; Kuramitsu, S.; Yokoyama, S. Crystallographic and mutational studies of seryl-tRNA synthetase from the archaeon *Pyrococcus horikoshii*. *RNA Biol.* **2008**, *5*, 169–177. [CrossRef] [PubMed]

46. Xu, X.; Shi, Y.; Yang, X.-L. Crystal structure of human seryl-tRNA synthetase and ser-sa complex reveals a molecular lever specific to higher eukaryotes. *Structure* **2013**, *21*, 2078–2086. [CrossRef]

47. Belrhali, H.; Yaremchuk, A.; Tukalo, M.; Larsen, K.; Berthet-Colominas, C.; Leberman, R.; Beijer, B.; Sproat, B.; Als-Nielsen, J.; Grübel, G. Crystal structures at 2.5 angstrom resolution of seryl-tRNA synthetase complexed with two analogs of seryl adenylate. *Science* **1994**, *263*, 1432–1436. [CrossRef]

48. Wang, Y.; Geer, L.Y.; Chappey, C.; Kans, J.A.; Bryant, S.H. Cn3D: Sequence and Structure views for entrez. *Trends Biochem. Sci.* **2000**, *25*, 300–302. [CrossRef]

49. Dickson, R.J.; Gloor, G.B. The MIp toolset: An efficient algorithm for calculating mutual information in protein alignments. *arXiv*, 2013; arXiv:1304.4573.

50. Thompson, J.; Dahlberg, A.E. Testing the conservation of the translational machinery over evolution in diverse environments: Assaying *Thermus thermophilus* ribosomes and initiation factors in a coupled transcription—Translation system from *Escherichia coli*. *Nucleic Acids Res.* **2004**, *32*, 5954–5961. [CrossRef] [PubMed]

51. Ganoza, M.C.; Kiel, M.C.; Aoki, H. Evolutionary conservation of reactions in translation. *Microbiol. Mol. Biol. Rev.* **2002**, *66*, 460–485. [CrossRef]

52. Chan, P.P.; Lowe, T.M. GtRNAdb 2.0: An Expanded database of transfer RNA genes identified in complete and draft genomes. *Nucleic Acids Res.* **2016**, *44*, D184–D189. [CrossRef] [PubMed]

53. Dewe, J.M.; Whipple, J.M.; Chernyakov, I.; Jaramillo, L.N.; Phizicky, E.M. The yeast rapid tRNA decay pathway competes with elongation factor 1A for substrate tRNAs and acts on tRNAs lacking one or more of several modifications. *RNA* **2012**, *18*, 1886–1896. [CrossRef]

54. Chernyakov, I.; Whipple, J.M.; Kotelawala, L.; Grayhack, E.J.; Phizicky, E.M. Degradation of several hypomodified mature tRNA species in *Saccharomyces cerevisiae* is mediated by Met22 and the 5′-3′ exonucleases Rat1 and Xrn1. *Genes Dev.* **2008**, *22*, 1369–1380. [CrossRef] [PubMed]

55. Waterhouse, A.; Bertoni, M.; Bienert, S.; Studer, G.; Tauriello, G.; Gumienny, R.; Heer, F.T.; de Beer, T.A.P.; Rempfer, C.; Bordoli, L.; et al. SWISS-MODEL: Homology modelling of protein structures and complexes. *Nucleic Acids Res.* **2018**, *46*, W296–W303. [CrossRef] [PubMed]

56. Cavarelli, J.; Rees, B.; Ruff, M.; Thierry, J.C.; Moras, D. Yeast tRNA[Asp] recognition by its cognate class II aminoacyl-tRNA synthetase. *Nature* **1993**, *362*, 181–184. [CrossRef] [PubMed]

57. Goldgur, Y.; Mosyak, L.; Reshetnikova, L.; Ankilova, V.; Lavrik, O.; Khodyreva, S.; Safro, M. The Crystal structure of phenylalanyl-tRNA synthetase from *Thermus thermophilus* complexed with cognate tRNA (Phe). *Structure* **1997**, *5*, 59–68. [CrossRef]

58. Eiler, S.; Dock-Bregeon, A.-C.; Moulinier, L.; Thierry, J.C.; Moras, D. Synthesis of aspartyl-tRNA[Asp] in *Escherichia coli*–A snapshot of the second step. *EMBO J.* **1999**, *18*, 6532–6541. [CrossRef] [PubMed]

59. Achsel, T.; Gross, H.J. Identity determinants of human tRNA[Ser]: Sequence elements necessary for serylation and maturation of a tRNA with a long extra arm. *EMBO J.* **1993**, *12*, 3333–3338. [CrossRef]

60. Lenhard, B.; Filipic, S.; Landeka, I.; Ivan, S.; So, D. Defining the active site of yeast seryl-tRNA synthetase. *J. Biol. Chem.* **1997**, *272*, 1136–1141. [CrossRef]

61. Luscombe, N.M.; Laskowski, R.A.; Thornton, J.M. Amino acid—base interactions: A three-dimensional analysis of protein—DNA interactions at an atomic level. *Nucleic Acids Res.* **2001**, *29*, 2860–2874. [CrossRef]

62. Davies, D.R.; Goryshin, I.Y.; Reznikoff, W.S.; Rayment, I. Three-dimensional structure of the Tn5 synaptic complex transposition intermediate. *Science* **2000**, *289*, 77–85. [CrossRef] [PubMed]

63. Roberts, R.J.; Cheng, X. Base flipping. *Annu. Rev. Biochem.* **1998**, *67*, 181–198. [CrossRef] [PubMed]

64. Berg, M.D. Western University, London, Canada. in preparation.

GCAT
TACG
GCAT
genes

MDPI

Review

Alternative Biochemistries for Alien Life: Basic Concepts and Requirements for the Design of a Robust Biocontainment System in Genetic Isolation

Christian Diwo [1] and Nediljko Budisa [1,2,*]

[1] Institut für Chemie, Technische Universität Berlin Müller-Breslau-Straße 10, 10623 Berlin, Germany; christian.diwo@tu-berlin.de

[2] Department of Chemistry, University of Manitoba, 144 Dysart Rd, 360 Parker Building, Winnipeg, MB R3T 2N2, Canada

* Correspondence: nediljko.budisa@tu-berlin.de or nediljko.budisa@umanitoba.ca; Tel.: +49-30-314-28821 or +1-204-474-9178

Received: 27 November 2018; Accepted: 21 December 2018; Published: 28 December 2018

Abstract: The universal genetic code, which is the foundation of cellular organization for almost all organisms, has fostered the exchange of genetic information from very different paths of evolution. The result of this communication network of potentially beneficial traits can be observed as modern biodiversity. Today, the genetic modification techniques of synthetic biology allow for the design of specialized organisms and their employment as tools, creating an artificial biodiversity based on the same universal genetic code. As there is no natural barrier towards the proliferation of genetic information which confers an advantage for a certain species, the naturally evolved genetic pool could be irreversibly altered if modified genetic information is exchanged. We argue that an alien genetic code which is incompatible with nature is likely to assure the inhibition of all mechanisms of genetic information transfer in an open environment. The two conceivable routes to synthetic life are either de novo cellular design or the successive alienation of a complex biological organism through laboratory evolution. Here, we present the strategies that have been utilized to fundamentally alter the genetic code in its decoding rules or its molecular representation and anticipate future avenues in the pursuit of robust biocontainment.

Keywords: alternative amino acid and nucleotide repertoires; alternative core cellular chemistries; biocontainment; genetic firewall; genetic isolation; orthogonal central dogma of molecular biology; synthetic life; xenobiology

1. Introduction

The design and manufacturing of specialized biological systems promises to be the next giant leap in human technology to advance many aspects of our society in unimaginable ways. Technologies derived from this scientific groundwork are expected to be employed for fine chemical and pharmaceutical production [1,2], transform medicine and epidemic control [3,4], be employed for environmental remediation [5], mining [6] and crop fertilization [7], access novel renewable energy sources [8–10], and complement electronic circuits and computational devices [11,12].

The modification of metabolic pathways as well as the addition, deletion, minimization or integration of gene circuits (genes, gene clusters) transforms living organisms into autonomously acting tools [13] able to execute preprogrammed processes, generally termed genetically modified organisms (GMOs) [14]. The modern laboratory methods of synthetic biology (SB) take advantage of the modularity of living systems, mixing and matching traits from various species in order to create organisms with specific desired functionality [15]. The methods of SB aiming to transfer naturally evolved functions between organisms can be distinguished from the emerging field of xenobiology (XB), which aims to expand the framework of natural chemistries within living cells through the incorporation of non-natural building blocks [16,17].

Indeed, life on Earth is a reservoir of complex organized systems, displaying different levels of control over their respective dynamic environments and intricate physiological processes to achieve robust autonomy [18,19]. In this context, the genetic information encoding the organization of an organism can persist in two ways, either within a species through reproduction (vertical gene transfer, VGT), or between species through horizontal gene transfer (HGT). The pool of all possible manifestations of a certain mechanism of control that can be acquired by mutation and selection drastically expands with the complexification of biological systems. Thus, the exchange of components of the cellular organization from one path of evolution to another potentially can advance certain aspects of control which leads to a more robust population [20,21]. Exchanging information between species at the genetic level is only possible for individuals maintaining the same physicochemical character (DNA and RNA) and decoding logic (transcription and translation) in their genetic code, making the genetic code the lingua franca of life on Earth [17].

However, it is the universality of the genetic code which makes GMOs designed to be employed in open environments a potential threat, as leakage of synthetic genetic information polymers might contaminate the naturally evolved genetic pool and alter entire ecosystems [4]. There is an overwhelming consensus in both the public and scientific community that potentially harmful consequences of GMOs must be prevented by engineering appropriate safety measures before they can be safely employed [22,23]. Thus, the aim for a robust biocontainment system should be to achieve control over all possible mechanisms of proliferation of genetic information, which comprises preventing unintended VGT and HGT, as well as the possibility of circumvention of the biocontainment due to loss of genetically encoded safety mechanisms (genetic drift) [24]. Linking the expression of essential genes to the external supply of synthetic molecules has achieved a relatively high level of control over the reproduction of single-celled organisms [25]. The level of safety is usually measured in cells escaping the containment relative to the total cell count in a population. A biocontainment system is regarded as safe below 1 escapee in a population of 10^8 cells, which is the safety threshold proposed by the National Institutes of Health (NIH) [26]. Introducing dependency on synthetic molecules into the molecular complexes and processes along the flow of genetic information (central dogma) [27] seems to be an attractive target, directly impacting the universal characteristics of the genetic code. For example, the development of alternative genetic information storage molecules (xeno-nucleic acids = XNAs), or the alteration of the universal decoding logic through systematic introduction of non-canonical amino acids, should enable the development of biocontained synthetic organisms (Figure 1).

Tight containment of artificially altered genetic information is the primary concern of a biocontainment system. For such a system to be employed by industry, it must also enable standard engineering methodologies that do not impair the development of viable products [23]. Industrial needs are a legitimate aspect of a biocontainment system to be considered. However, this essay will be mainly concerned with the basic requirements and paths of implementation explored to create life in genetic isolation. Microorganisms are the preferred scaffold for developing biological tools (e.g., *E. coli*, *B. subtilis*, *S. cerevisiae*), where natural processes can be redesigned over relatively short timescales through the methods of SB and XB. As such, biocontainment concepts applied to single cellular organisms are presented. Approaches not concerned with altering the core chemistries of cellular organization are summarized elsewhere [28,29].

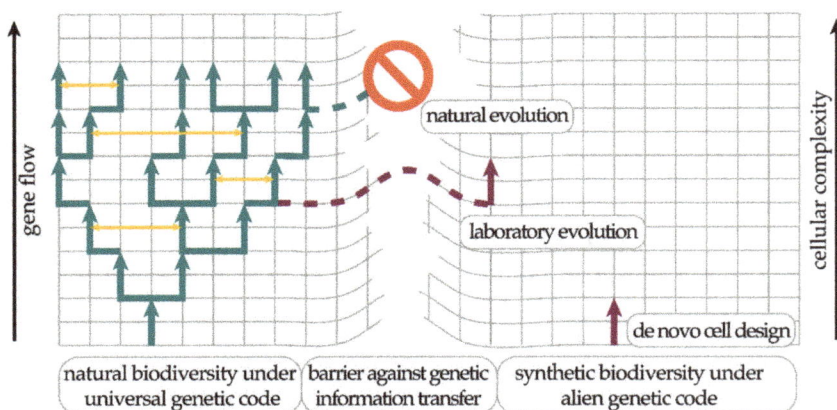

Figure 1. Biocontainment based on an alien genetic code. Life on Earth is a unity due to the existence of the universal genetic code. The exchange of genetic information from very different paths of evolution (horizontal gene transfer (HGT), yellow arrows) is facilitated by the universal genetic code fixing the basic core cellular chemistries (left side). Genetic information persists in time within a species through reproduction (vertical gene transfer (VGT), green arrows). A robust biocontainment system needs to restrict the flow of genetic information, which can be achieved through an alternative genetic code allowing for the exploration of drastically different cellular chemistries (right side). Currently, there are two promising experimental routes towards this goal: de novo (from scratch) cellular design (bottom-up) [30,31] or the successive alienation of a complex natural organism through laboratory evolution (top-down) [32,33].

2. Basic Considerations for an Alien Central Dogma

The cellular organization of even the simplest organisms is immensely complex and interconnected, to which the central dogma is an interpretation of the flow of information, illustrating a general theme conserved in every modern organism [27]. The molecular outputs (RNAs, proteins) along the ribosomal protein biosynthesis and their transforming processes (transcription, translation) possess a multitude of chemical and physical identities which are heavily interconnected [34], regulating the cellular processes over the complete cell cycle and in response to environmental cues. The topology of this network (who interacts with whom) is crucial for a cell to maintain robust autonomy [35] and manipulating these intricate associations for the purpose of biocontainment is a demanding task but thought to be very effective [28,36]. VGT is based on cellular reproduction, and as such depends on many synchronized processes in order to guarantee transmission of the complete genetic information to a robust next generation. Thus, intercepting any process along the central dogma will result in the cell losing its autonomy and prevents cell replication. For example, introducing dependencies to synthetic compounds into components of the central dogma results in trophic containment controlling cell survival through the external supplementation of the compound [24]. HGT, in turn, is a type of communication between cells and relies on the universality of the genetic code in its decoding logic and its physical representation. The main vectors of HGT between prokaryotes are transduction, conjugation or DNA uptake from the environment [37]. For successful HGT to take place, a cell is required to replicate or recombine, transcribe and translate received genetic information molecules into functional proteins and thus, HGT can be ruled out by altering the decoding rules (meaning), or the genetic information storage molecules (identity) of the genetic code [38].

3. Altering the Meaning of the Genetic Code

The idea of genetically encoding alternative cellular chemistries by incorporating non-canonical amino acids (ncAAs) into the proteome of microorganisms has spawned efforts to free codons from their canonical assignment and engineer the cellular translation machinery to accept synthetic compounds [39]. Genes harboring an alternative codon assignment will not be expressed in a cell with a standard codon assignment receiving the altered information, and thus the genetic information will be contained (Figure 2).

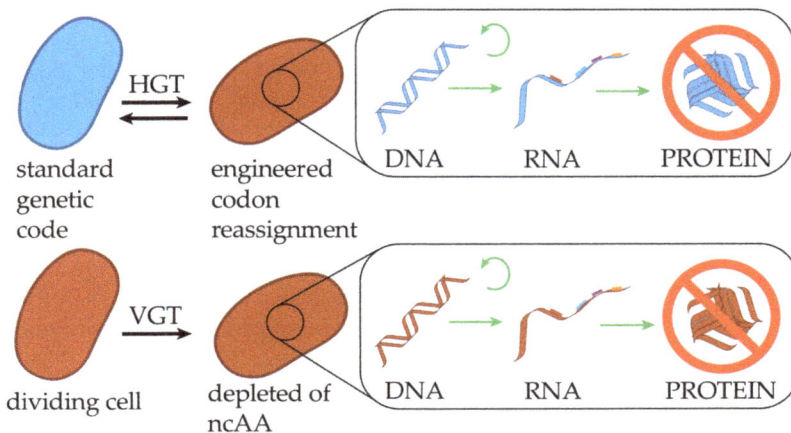

Figure 2. Restrictions of genetic information transfer in altering the meaning of the genetic code. In altering the codon assignment, cells receiving genetic information will be able to truthfully replicate, transcribe and translate the encoded information, however, the resulting polypeptides will not be functional. The same holds true for the genetic information transfer from genetically modified organisms (GMOs) to cells maintaining a standard genetic code. An engineered organism with a strict dependency for a non-canonical amino acid (ncAA) will not be capable of VGT in the absence of the ncAA.

DNA is encoding the cellular organization through several layers of its chemical identity. Codons are the logic units receiving their meaning through the coupling of a specific amino acid to a cognate tRNA adapter with the correct anticodon, a reaction catalyzed by a tRNA- and amino acid-specific aminoacyl-tRNA synthetase (aaRS). Eighteen out of 20 amino acids are encoded by multiple synonymous codons. Rather than being a redundancy, the degeneracy of the genetic code—61 codons encoding 20 + 2 amino acids and 3 codons encoding stop signals—allows for decision-making processes rooted at a basic level, where the codon distribution is directly informing translation processes at the ribosome, or gene regulation processes [40]. Direct reassignment of a codon will change all instances of the previously assigned amino acid in the proteome. The misincorporation of an amino acid at the reassigned codon will typically lead to misfolded and non-functional proteins [41], while the genome-wide substitution of a codon by a synonymously coding triplet is likely to drastically influence the cellular organization [40,42]. In recent decades, several schemes have been devised to overcome the challenges of cellular complexity and achieve partial genetic isolation through alteration of the canonical codon assignment.

3.1. Stop Codon Reassignment

Stop codon suppression (SCS) exploits the low abundance of the amber stop codon (UAG) in the *E. coli* genome for its reassignment into a sense codon. The introduction of an orthogonal tRNA:aaRS pair into the cell is used to facilitate the incorporation an ncAA at the command of the reassigned codon. However, with the functional canonical process in competition with the engineered assignment, a code ambiguity is created. The genome-wide substitution of the UAG codon with a synonymous codon allows for the deletion of the competing release factor (RF1), which mitigates the detrimental effects of the code ambiguity and enhances the ncAA incorporation efficiency [43]. Using this strain as a platform, highly contained organisms have been created through the design of essential proteins with functional dependencies on ncAA incorporation [24,41]. The design and selection of the proteins with the desired properties demands sophisticated methodologies, as incorporating UAG codons into genes might also influence mRNA secondary structures and ribosome-binding strength [44]. Once robust dependencies are found, combining several mutated sites into a few essential genes in one organism results in exceptionally low escape frequencies well below the suggested NIH threshold [41]. Essential for the robustness of the system are both the selectivity of the aaRS [45] and the tolerance of the protein towards regaining its function through misincorporation of canonical amino acids. Interestingly, Tack and colleagues developed a portable biocontainment system demonstrated on strains of *E. coli* and several other species of bacteria sensitive to ampicillin by modulating an ampicillin resistance gene to only be functional upon ncAA incorporation [46]. By supplying the antibiotic resistance gene and the tRNA:aaRS pair on a plasmid, they managed to contain the tested organisms over several hundred generations with escape rates one order of magnitude lower than the NIH threshold. This data suggests that genetic code ambiguity does not effectively reduce the containment if linked to a strong impact on cell survival.

While persistent in vivo incorporation of ncAAs into proteins has been achieved over many generations, the level of containment has yet to be improved [47]. The reassignment of a single non-sense codon can be readily overcome by evolution [24,41], or post-transcriptional modification (although demonstrated for a reassigned sense codon) [48] of a tRNA incorporating canonical amino acids at the command of the UAG codon. Such genes could in principle also be acquired by HGT. Although initially the UAG and RF1 deficient strain is less suitable as phage hosts [49], bacteriophage T7 has been able to adapt, overcoming the immunity of cells with an expanded genetic code [50]. This mechanism of immunity is traced back to a rescue pathway for stalled ribosomes and deleting the corresponding gene obliviates the immunity conferred by the UAG stop codon deletion [51]. The dependency of essential proteins to an additional ncAA through the reassignment of a second stop codon is conceivable, but potentially highly toxic for the cell [52].

3.2. Sense Codon Reassignment

A naturally evolved sense codon reassignment, exchanging the codons assigned to amino acids with entirely different chemical characters (Leu(CUG)Ser, non-polar to polar), does occur in a yeast fungus. Heterologous expression of genes containing the CUG codon in a similar organism with a standard genetic code leads to misfolded proteins, and thus the information transfer cannot be completed [53]. These findings suggest that the reassignment of a sense codon can be used to alter the cellular biochemistry on a proteome scale, ultimately interrupting HGT. The choice of the target codon, either degenerate or exclusively encoding an amino acid, will respectively expand the amino acid repertoire or substitute a canonical building block. However, there are currently no real examples of biocontainment achieved through a sense codon reassignment to an ncAA. The narrow choice of analogs which are tolerated for a proteome-wide incorporation limits the dependency of an organism to an ncAA.

It is assumed that robust reassignment of a sense codon can, in principle, be achieved by a small number of successive topology changes [35]. However, the exact mechanism of how variant genetic codes arise remains elusive and is subject to ongoing investigations [54]. Due to the lack of deep insights into mechanisms of cellular network plasticity [55], changes in the protein biochemistry on a proteome scale are usually invoked through laboratory evolution, forcing the cell to incorporate an increasing amount of an ncAA in order to survive [56]. Several long-term evolution experiments have yielded strains which are adapted to the complete replacement of tryptophan (encoded by a single codon) with a close chemical analog, whereas their ancestors are not able to show any growth under the same conditions [32,57]. However, more interesting for the purpose of biocontainment is to drive the cellular organizational topology beyond the ability to accept the canonical amino acid. A strain of *B. subtilis* has evolved to grow exclusively on several fluorinated tryptophan analogs, whereas it has lost the ability to accept the canonical amino acid tryptophan [58]. The adaptation is traced back to a mutated transporter and simply reverting a single mutation leads to a reconstitution of the ability to use the canonical amino acid [59]. Nevertheless, these findings indicate that obliviating the dependency of an organism for a canonical amino acid enables evolutionary processes to alter the cellular organization, which might lead to incompatibility for the canonical substrate.

In order to capture the assignment of a degenerate codon, highly codon-specific translation capabilities have to be added to an organism. Wobble pairing of canonical base pairs, as well as post-transcriptional modifications can alter the specificity of the codon–anticodon matching [60,61]. Several investigations have tackled the rare, degenerate, codon AGG to code for an ncAA by introducing an orthogonal tRNA:aaRS pair [62,63]. Bröcker and colleagues demonstrated that the specific requirements for selenocysteine translation (mRNA hair pin structure and special elongation factor [64]) can completely overwhelm the natural codon assignment when expressed in competition [65]. In order to completely eliminate ambiguous decoding of a captured codon, Bohlke and Budisa suggest exploiting the decoding mechanism of the AUA codon via a post-transcriptionally modified tRNA [61]. However, successful reassignment of the AUA codon has yet to be reported, whereas the AGG codon has been successfully reassigned to an ncAA.

These works demonstrate that, in principle, sense codon reassignments with close analogs are possible. However, the currently missing insight into which of the topology changes that have surfaced are necessary to accommodate alternative amino acids result in a lack of reproducibility and portability. Given enough time and data, these attempts may possibly lead to the discovery of fundamental engineering principles enabling the permanent alienation of the cellular biochemistry on a proteome scale [66].

3.3. Minimal Cell Design

Cellular organization has evolved employing the 20 canonical amino acids and their codons to confer maximal robustness [67]. However, under laboratory conditions many of the preprogrammed cellular responses might not be necessary, or be rather hindering for certain engineering approaches [68]. Thus, reducing the cellular complexity, by either reducing the size of the genome or by decreasing the degeneracy of the genetic code might help to uncover design principles and to simplify engineering approaches. Additionally, reducing the genetic code degeneracy may free multiple codons for the reassignment to ncAAs [47]. There has been noteworthy progress towards the construction of a so-called minimal cell (genome reduction and sense codon liberation), demonstrating the feasibility of computationally redesigning entire genomes.

In their latest iteration of a minimal genome, the Venter research group created a viable cell with a genome size of 531,000 base pairs named syn3.0, reducing the genome size of *M. mycoides* to approximately half [33]. Of the 473 genes encoded in the genome of syn3.0, 149 are of unknown function, but all seem essential in conditions considered to be optimal for cell growth (no competition for food, optimal temperature, supply of all small molecules in the medium).

Aiming to decrease the genetic code degeneracy, Lajolie and colleagues have devised an algorithm which replaces all instances of 13 rare codons in the 41 ribosomal protein-coding genes [40]. In 2016 they broke further ground by demonstrating the deletion of 7 rare codons, independently on each of 55 stretches of the *E. coli* genome (each ca. 1% of the genome) within certain constrains (for example, conserving relative codon usage, ribosomal binding sites and mRNA secondary structure) [42]. Due to the careful selection of codons, which might in any case be subject to low fidelity decoding due to wobble pairing, only limited growth defects and variations in transcription levels of affected genes can be observed [69].

Eliminating the ability of a cell to decode certain codons improves the genetic isolation, as foreign DNA from viruses, plasmids or other vectors can no longer be properly expressed, thus limiting the potential of HGT [42]. On the other hand, cells equipped with a minimal genome might be useful as a platform host to accept additional contained genetic information. An orthogonal central dogma (orthogonal replication, transcription and translation) could serve as a modular system to equip cells with some desired functionality [36]. A bottleneck to further investigation is the ability to incorporate large segments of synthetic DNA into the genome of living organisms or the creation of entire synthetic genomes. However, a more complete summary of the advances in the field of synthesizing and implementing whole genomes has been reviewed elsewhere [70,71].

4. Altering the Identity of the Genetic Code

Rather than trying to expand the cellular biochemistry through the replacement of preexisting coding events, researchers are exploring the feasibility of expanding the coding capacity of the genetic code. Codons are the naturally evolved coding units, with a coding capacity limited to 64 variations restricted by the four different nucleotides binned into units of three. Recent investigations have tested the ability to expand either the nucleotide base repertoire, or the codon unit to a higher binning. Any alteration of the fundamental coding units is likely to be poorly tolerated by the natural translation apparatus [72] or likely to fail the canonical replication and transcription machinery, thus leading towards the creation of a genetic firewall (Figure 3).

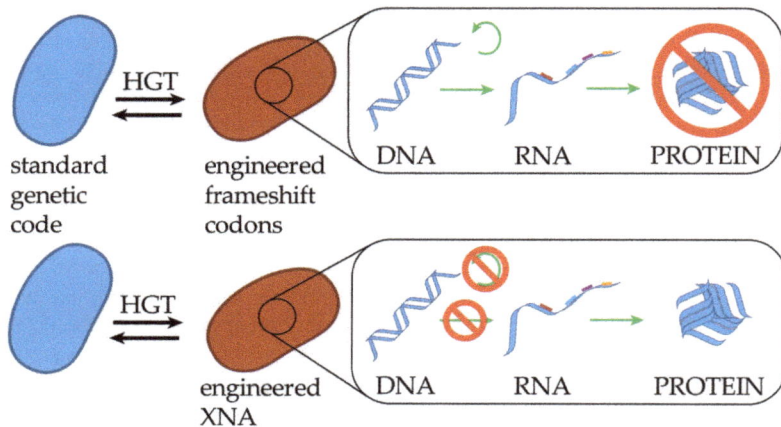

Figure 3. Restrictions of genetic information transfer in altering the identity of the genetic code. Similar to altering the canonical codon assignment, cells maintaining an expanded coding capacity, like quadruplet codons, will be able to replicate transcribe and translate genetic information received from a cell with a standard genetic code, where translation will not result in functional proteins. An organism strictly utilizing some kind of XNA backbone, while maintaining the canonical bases for genetic information storage, will not be able to replicate or transcribe genetic information in the form of DNA, but most likely be able to translate natural RNA into functional proteins.

4.1. Frameshift Codons

The advantages of a quadruplet code towards the creation of a genetic firewall seem obvious, in conferring a dramatic change to the fundamental coding identity of the genetic code, through a simple change in the unit convention brought about by only a few altered cellular components. Here, frameshift suppression is classified as a change in genetic code identity due to the expansion in coding capacity, although in order to convey these changes, the physical representation of the genetic information storing polymers does not need to be altered. In 2010 the group of Chin engineered a prokaryotic ribosome able to decode quadruplet codons, introducing an orthogonal in vivo translation system, which in theory would allow for 256 independent coding variants [73]. Linking the survival of a cell to the correct decoding of the AAGA codon, Neumann and colleagues discovered an evolved ribosome, with the ability to polymerize an ncAA in response to both the quadruplet codon and the UAG codon. Relying entirely on the promiscuity of canonical ribosomal protein biosynthesis, the group of Schulz found an orthogonal tRNA:aaRS pair able to efficiently decode a quadruplet codon without severe growth defects for the cell [74]. Expanding the codon capacity to four bases does not appear to be the limit. Hohsaka and colleagues demonstrated that a five-nucleotide based coding units can be considered plausible [75].

While in principle a quadruplet-based genetic code is possible, shortcomings in the understanding of genome engineering as explained in the Section 3.3 "Minimal Cell Design" hinder the progress towards compiling whole genomes. In order to enable the evolvability of ribosomes, they initially have to be freed from their natural duty by finding an orthogonal mRNA:ribosome pair [72]. The reversion of such a containment through HGT or evolution seems unlikely due to the highly toxic effects of a translation process that would be able to decode both canonical triplet and quadruplet codons simultaneously.

4.2. Synthetic Base Pairs

The creation of a synthetic base pair, expanding the canonical repertoire described by Watson and Crick [76], has been explored in various ways. Important for the in vivo incorporation of such base pairs into the DNA molecule is the transport into the cell, the enzymatic synthesis in the complex intracellular medium, and the tolerance towards the natural error elimination mechanisms. Retaining the information over replication cycles as well as retrieving the information in transcription and translation with high fidelity are fundamental requirements for a synthetic genetic information polymer [66]. This makes the modification of the components at the basis of the central dogma an extensive and challenging endeavor, which so far has not been developed successfully.

The four canonical base pairs rely on size and hydrogen-bonding complementation for their correct matching. Conserving these pairing rules may conserve the physicochemical character of the molecule [77], possibly allowing for an easier in vivo implementation. Chemical alteration of a canonical base pair is naturally only observed in bacteriophages, probably counteracting host restriction enzymes [78]. The research group of Benner pioneered the efforts to find additional base pairs relying on the canonical matching principles. While proving successful in vitro enzymatic polymerization [79] and replication [80] as well as stable transcription into RNA [81], the implementation into a living cell is so far difficult to achieve [82]. The directed evolution experiment of Marlière and colleagues aiming to exchange a canonical base in the entire genome of *E. coli* for a halogenated analog found that it is indeed possible to eliminate all of the natural thymidine down to the detection limit of 1.5% [83]. In their study, chlorouracil is carefully selected as precursor for intracellular conversion to a nucleotide, preserving the ultrastructure and hydrogen bonding characteristics of natural DNA. Although the resulting strain does not have expanded coding capabilities, DNA fragments containing the non-natural nucleotide will most likely be excluded from decoding by the canonical replication or transcription machinery of non-adapted cells.

Instead of relying on hydrogen bonding for base pairing, hydrophobic base pairs have been found useful for the in vivo expansion of the genetic alphabet. The Romesberg group successfully demonstrated the importance of exogenously phosphorylated unnatural nucleotides and their faithful in vivo replication in *E. coli* [84]. However, the plasmid-based replication of a single nucleotide at a specific position does not allow an evaluation for its usefulness in biocontainment. In later experiments, they demonstrated chromosomal incorporation and replication as well as assignment of the unnatural codon to an ncAA [85,86]. Thus far, this represents the most sophisticated platform of an in vivo genetic alphabet extension that has been developed. However, there have been legitimate concerns regarding the choice of the hydrophobic base pair which relies on stabilization of the surrounding canonical nucleotides for matching, possibly restricting the applicability for its extended use throughout the genome [87].

4.3. Alternative DNA Backbone Motifs

Synthetic DNA-like polymers (XNAs) have been engineered using non-natural sugar moieties or phosphate linkers of the DNA backbone, while conserving the four canonical bases. XNAs have been constructed with the ability to encode information, as well as bearing the potential for evolution [88]. XNAs are not readily processed by the natural replication and transcription enzymes [23], which makes it necessary to develop the according molecular machinery facilitating the intracellular processing of the genetic information. The same considerations as for the in vivo implementation of synthetic bases mentioned above are applicable for XNA molecules.

Threose nucleic acid (TNA) is a sugar-modified type of XNA that is able to bind to the reverse complement of RNA and DNA. Furthermore, it shows good stability in the intracellular milieu and a high resistance against natural nuclease enzymes [89]. The complementation of DNA and TNA single strands allows for the transcription of information from one molecule to the other, demonstrated through enzymatic transcription of DNA templates into TNA [90] and enzymatic reverse transcription of TNA templates into DNA [91]. In recent advances, the Chaput research group demonstrates a powerful approach computational approach for enzyme engineering, which was used to drastically improve TNA polymerase speed and accuracy. They used computational design together with sampling and pooling of beneficial mutations in a single enzyme, which leads to a an improved polymerase [92] able to polymerize TNA with an unnatural base [93]. However, the enzymes developed thus far lack orthogonality, and thus cannot easily be implemented in vivo, an issue concerning most of the developed XNAs [94].

Liu and colleagues recently published their progress towards the in vivo implementation of XNAs. They developed an XNA backbone modified in its sugar-phosphate moiety together with a cognate synthase with low affinity to natural nucleotides [95]. Although not demonstrating in vivo replication, their experiments show the poor acceptance of their XNA-DNA chimeras as a substrate for the endogenous *E. coli* replication machinery. However, it remains unclear if the proposed XNA chassis can be further developed to accommodate all necessary requirements to sustain life. The implementation of XNA molecules as genetic information storage in a cell-free system may certainly result in a robust biocontained system suitable for certain physically contained applications [28,77].

Complementing a minimal genome cell with an XNA based orthogonal central dogma, carrying contained information could be a fruitful strategy to achieve biocontainment in a precursor to true alien life [36].

5. The Farther the Safer

The premise "the farther the safer" [22], referring to the alienation of the core cellular chemistries which have naturally evolved under the universal genetic code, is still a legitimate claim if the aim is to eliminate the eventuality of genetic information transfer between naturally evolved organisms and GMOs.

The historical landscape of evolution with its specific dynamic environmental boundary conditions [96–98] and advantageous mechanisms of information transfer [20] has formed a narrow trajectory which seems to restrict the possibilities of cellular biochemistries to be explored [69]. Overcoming these restrictions by altering the meaning and identity of the genetic code will allow for drastically divergent evolutionary paths unable to exchange genetic information with, or revert back to, the naturally evolved counterpart (Figure 1).

The route to this new synthetic world will lead through vastly uncharted territory. There are two conceivable routes. Synthetic life can either be reached through a bottom up de novo cell design [30], or through the top down introduction of multiple ncAAs and the successive exchange of components of complex cellular biochemistry [31]. Promising progress has been reported towards de novo cell synthesis [99,100]. In the scope of this review we will however only discuss the top-down approach.

Although nucleotide and amino acid stereochemistry is decoupled in ribosomal protein biosynthesis through the use of tRNAs, it is the amino acid repertoire which shaped the chemical identity of the genetic code [34]. Organisms with an artificial amino acid repertoire thus far only maintain their installed biocontainment if essential proteins are dependent on the physicochemical character and related functional features of these synthetic building blocks. The stabilization of the topology changes necessary to accommodate the synthetic building blocks on a proteome scale remains the limiting factor, demonstrating the robustness of the naturally evolved cellular organization.

Yu and colleagues [59] suspect that the safeguards against the proteome-wide substitution of a canonical amino acid with a close analog are encrypted in only a few cellular components and depend on the amino acid and the analog. Their study shows that the correct selective pressure does allow for the propagation of mutations strengthening an alternative genetic code. The organism which is auxotroph for tryptophan abolishes the uptake mechanism of the amino acid, a mutation that would otherwise be fatal. This somehow fragile containment of VGT could be strengthened by restricting the promiscuity of an engineered aaRS to only incorporate the ncAA, a strategy which has been proposed earlier by Bohlke and Budisa [61]. The topology changes surfacing in such a strain during a laboratory evolution experiment would certainly be interesting to observe. Until now, a number of organisms have been adapted to a variety of environmental conditions or equipped with desired metabolic traits through directed evolution approaches, all within moderate timeframes [101]. However, the insights necessary to establish such deep-rooted chemical changes like a sense codon reassignment are lacking, which impedes the design of appropriate starting or selection conditions for such evolutionary experiments [102]. More information on the topic can be retrieved from a recent review, which accumulates the extensive knowledge concerned with the knobs and dials of modern laboratory evolution [101].

Beyond facilitating HGT, the universal genetic code is thought to limit the adverse effects of information misinterpretation which may occur during the processes along the central dogma [103]. This feature usually complicates the stabilization of alternative genetic codes. However, there do exist some noteworthy exceptions of organisms or organelles having evolved deviating codon assignments [104–106]. For instance, *Micrococcus luteus* has been identified as an organism lacking six codon assignments, which could be exploited for the incorporation of ncAAs into proteins, or as a starting point for laboratory evolutions [22,107]. Similarly, the evolutionary path of mitochondria, which have lost a large amount of their cellular complexity while also maintaining an alternative codon assignment, could inform engineering approaches aiming to establish alternative genetic codes [108]. To date, the exact mechanism of how the assignment of a codon can change remains elusive. Recent studies predict the loss of a tRNA [54], or its deactivation due to a post-transcriptional modification, as the main drivers [109]. Detailed investigations into the mechanisms influencing these naturally occurring deviations from the standard genetic code might surface effective protocols with which to sustainably gain control over the process in laboratory evolutions.

6. Alienation Far beyond the Canonical Chemistries of Life

Life on earth is limited to the repertoire of 20 amino acids, which ensures protein function through reoccurring structural motifs based on hydrogen bonding. Thereby it becomes apparent that proline and glycine usually not participate in the β-sheet and α-helix secondary structures, but break these ordered formations to form loops [110]. This architecture has been extensively explored by nature through combinations of ordered formations as well as through post-translational modifications. Therefore, a long-term perspective would be to seek for other scaffolds that do allow a functional proteome to be based on different chemical architectures and alternative principles of protein folding. The creation of a 'fundamentally new' alien life would therefore require the use of radically new building blocks (and not the mere modifications of existing ones). Attributes and perspectives of alternative genetic codes whose repertoires are based on derivatives of proline [111], sarcosine, ornithine and other 'alien' building blocks are most recently elaborated by Kubyshkin and Budisa (Trends in Biotechnology, 2019, under revision).

If codon diversity and availability can be identified as a major hurdle to establish new chemistries into complex organisms, multiple-sense codon reassignments, quadruplet codons or XNAs with alternative base pairs might be the entry point for the sustainable incorporation of multiple synthetic building blocks into complex organisms. However, as explained above, the degeneracy of the genetic code mitigates potential adverse effects of genetic code misinterpretation. How technologies like quadruplet codons or sense codon reassignments affect the viability of cells in this regard remains to be determined. A synthetic organism based on an alien genetic code with a decoding logic incompatible to naturally evolved organisms, will be excluded from any HGT (receiving and transmitting). As XNA molecules have been shown to be capable of heredity and evolution [88], XNA-based organisms seem like a plausible vehicle for implementing proteome wide chemical changes. However, there is no experimental evidence that these organisms would absolutely depend on xeno-nutrients to survive, and thus VGT may not automatically be tightly controlled [22]. In the future, stronger, standardized tests have to be applied to newly created xeno-organisms with estranged genetic codes in order to quantify the robustness of the engineered containment against VGT, HGT and genetic drift in complex, diverse environments [13]. Only a synthetic organism that has been thoroughly tested for its absolute dependency on an otherwise inaccessible synthetic compound will be safe to employ in any complex open environment as long as no parts are toxic for any biological system in contact. However, depending on the time of interaction between the synthetic and biological world, an evolutionary drift towards preserving the genetic material and its contained information might be possible. Thus, for now, the attribute robust biocontainment cannot be assigned to any of the here presented technologies.

Author Contributions: Conceptualization, N.B. and C.D.; writing—original draft preparation, C.D.; writing—review and editing, C.D. and N.B.; visualization, C.D.; supervision, N.B.

Funding: This research received no external funding.

Acknowledgments: We acknowledge support by the German Research Foundation and the Open Access Publication Funds of TU Berlin. We thank Sean McKenna for critically reading the manuscript. NB thanks the 'Fachgruppe' Organic Chemistry of Institute of Chemistry/Technical University of Berlin for an extremely dynamic environment that has promoted his intellectual development and inspired him to make important decisions. Some of the ideas and thoughts elaborated here are the result of our intellectual interaction with Philippe Marliere (Xenobiology), Dirk Schulze-Makuch (Astrobiology) and Marcello Barbieri (Code Biology) over the past eight years. Special thanks go to Vladimir Kubyshkin for continuous, insightful and inspiring critical discussions about the chemistry of life.

Conflicts of Interest: The authors declare no conflict of interest.

References

1. Keasling, J.D. Manufacturing Molecules Through Metabolic Engineering. *Science* **2010**, *330*, 1355–1358. [CrossRef] [PubMed]

2. Agostini, F.; Völler, J.S.; Koksch, B.; Acevedo-Rocha, C.G.; Kubyshkin, V.; Budisa, N. Biocatalysis with Unnatural Amino Acids: Enzymology Meets Xenobiology. *Angew. Chem. Int. Ed.* **2017**, *56*, 9680–9703. [CrossRef] [PubMed]

3. Gao, G.-P.; Alvira, M.R.; Wang, L.; Calcedo, R.; Johnston, J.; Wilson, J.M. Novel adeno-associated viruses from rhesus monkeys as vectors for human gene therapy. *Proc. Natl. Acad. Sci. USA* **2002**, *99*, 11854–11859. [CrossRef]

4. Noble, C.; Adlam, B.; Church, G.M.; Esvelt, K.M.; Nowak, M.A. Current CRISPR gene drive systems are likely to be highly invasive in wild populations. *eLife* **2018**, *7*, 1–30. [CrossRef] [PubMed]

5. Hlihor, R.M.; Gavrilescu, M.; Tavares, T.; Favier, L.; Olivieri, G. Bioremediation: An Overview on Current Practices, Advances, and New Perspectives in Environmental Pollution Treatment. *Biomed. Res. Int.* **2017**, *2017*, 3–5. [CrossRef] [PubMed]

6. Dunbar, W.S. Biotechnology and the Mine of Tomorrow. *Trends Biotechnol.* **2017**, *35*, 79–89. [CrossRef] [PubMed]

7. Li, H.; Rasheed, A.; Hickey, L.T.; He, Z. Fast-Forwarding Genetic Gain. *Trends Plant Sci.* **2018**, *23*, 184–186. [CrossRef]

8. Gust, D.; Moore, T.A.; Moore, A.L. Mimicking photosynthetic solar energy transduction. *Acc. Chem. Res.* **2001**, *34*, 40–48. [CrossRef]

9. Lynd, L.R.; Van Zyl, W.H.; McBride, J.E.; Laser, M. Consolidated bioprocessing of cellulosic biomass: An update. *Curr. Opin. Biotechnol.* **2005**, *16*, 577–583. [CrossRef]

10. He, L.; Du, P.; Chen, Y.; Lu, H.; Cheng, X.; Chang, B.; Wang, Z. Advances in microbial fuel cells for wastewater treatment. *Renew. Sustain. Energy Rev.* **2017**, *71*, 388–403. [CrossRef]

11. Rivnay, J.; Owens, R.M.; Malliaras, G.G. The rise of organic bioelectronics. *Chem. Mater.* **2014**, *26*, 679–685. [CrossRef]

12. Shipman, S.L.; Nivala, J.; Macklis, J.D.; Church, G.M. CRISPR-Cas encoding of a digital movie into the genomes of a population of living bacteria. *Nature* **2017**, *547*, 345–349. [CrossRef]

13. Schmidt, M.; de Lorenzo, V. Synthetic bugs on the loose: Containment options for deeply engineered (micro)organisms. *Curr. Opin. Biotechnol.* **2016**, *38*, 90–96. [CrossRef] [PubMed]

14. Cohen, S.N.; Chang, A.C.Y.; Boyer, H.W.; Helling, R.B. Construction of Biologically Functional Bacterial Plasmids In Vitro. *Proc. Natl. Acad. Sci. USA* **1973**, *70*, 3240–3244. [CrossRef] [PubMed]

15. Agapakis, C.M.; Silver, P.A. Synthetic biology: Exploring and exploiting genetic modularity through the design of novel biological networks. *Mol. Biosyst.* **2009**, *5*, 704–713. [CrossRef] [PubMed]

16. Endy, D. Foundations for engineering biology. *Nature* **2005**, *438*, 449–453. [CrossRef] [PubMed]

17. Kubyshkin, V.; Budisa, N. Synthetic alienation of microbial organisms by using genetic code engineering: Why and how? *Biotechnol. J.* **2017**, *12*, 1–8. [CrossRef]

18. Kitano, H. Towards a theory of biological robustness. *Mol. Syst. Biol.* **2007**, *3*. [CrossRef]

19. Ruiz-Mirazo, K.; Briones, C.; De La Escosura, A. Chemical roots of biological evolution: The origins of life as a process of development of autonomous functional systems. *Open Biol.* **2017**, *7*. [CrossRef]

20. Syvanen, M. Cross-species gene transfer; implications for a new theory of evolution. *J. Theor. Biol.* **1985**, *112*, 333–343. [CrossRef]

21. Vetsigian, K.; Woese, C.; Goldenfeld, N. Collective evolution and the genetic code. *Proc. Natl. Acad. Sci. USA* **2006**, *103*, 10696–10701. [CrossRef] [PubMed]

22. Marliere, P. The farther, the safer: A manifesto for securely navigating synthetic species away from the old living world. *Syst. Synth. Biol.* **2009**, *3*, 77–84. [CrossRef] [PubMed]

23. Schmidt, M. Diffusion of synthetic biology: A challenge to biosafety. *Syst. Synth. Biol.* **2008**, *2*, 1–6. [CrossRef] [PubMed]

24. Mandell, D.J.; Lajoie, M.J.; Mee, M.T.; Takeuchi, R.; Kuznetsov, G.; Norville, J.E.; Gregg, C.J.; Stoddard, B.L.; Church, G.M. Biocontainment of genetically modified organisms by synthetic protein design. *Nature* **2015**, *518*, 55–60. [CrossRef] [PubMed]

25. Agmon, N.; Tang, Z.; Yang, K.; Sutter, B.; Ikushima, S.; Cai, Y.; Caravelli, K.; Martin, J.A.; Sun, X.; Choi, W.J.; et al. Low escape-rate genome safeguards with minimal molecular perturbation of *Saccharomyces cerevisiae*. *Proc. Natl. Acad. Sci. USA* **2017**, *114*, E1470–E1479. [CrossRef] [PubMed]

26. Wilson, D.J. Nih guidelines for research involving recombinant dna molecules. *Account. Res.* **1993**, *3*, 177–185. [CrossRef] [PubMed]

27. Crick, F. Central dogma of molecular biology. *Nature* **1970**, *227*, 561. [CrossRef]

28. Lee, J.W.; Chan, C.T.Y.; Slomovic, S.; Collins, J.J. Next-generation biocontainment systems for engineered organisms. *Nat. Chem. Biol.* **2018**, *14*, 530–537. [CrossRef]

29. Whitford, C.M.; Dymek, S.; Kerkhoff, D.; März, C.; Schmidt, O.; Edich, M.; Droste, J.; Pucker, B.; Rückert, C.; Kalinowski, J. Auxotrophy to Xeno-DNA: An exploration of combinatorial mechanisms for a high-fidelity biosafety system for synthetic biology applications. *J. Biol. Eng.* **2018**, *12*. [CrossRef]

30. Schwille, P. Bottom-Up Synthetic Biology: Engineering in a Tinkerer's World. *Science* **2011**, *333*, 1252–1254. [CrossRef]

31. Blain, J.C.; Szostak, J.W. Progress Toward Synthetic Cells. *Annu. Rev. Biochem.* **2014**, *83*, 615–640. [CrossRef] [PubMed]

32. Hoesl, M.G.; Oehm, S.; Durkin, P.; Darmon, E.; Peil, L.; Aerni, H.R.; Rappsilber, J.; Rinehart, J.; Leach, D.; Söll, D.; et al. Chemical Evolution of a Bacterial Proteome. *Angew. Chem. Int. Ed.* **2015**, *54*, 10030–10034. [CrossRef] [PubMed]

33. Hutchison, C.A.; Chuang, R.Y.; Noskov, V.N.; Assad-Garcia, N.; Deerinck, T.J.; Ellisman, M.H.; Gill, J.; Kannan, K.; Karas, B.J.; Ma, L.; et al. Design and synthesis of a minimal bacterial genome. *Science* **2016**, *351*. [CrossRef] [PubMed]

34. Kubyshkin, V.; Acevedo-Rocha, C.G.; Budisa, N. On universal coding events in protein biogenesis. *BioSystems* **2018**, *164*, 16–25. [CrossRef] [PubMed]

35. Ciliberti, S.; Martin, O.C.; Wagner, A. Robustness can evolve gradually in complex regulatory gene networks with varying topology. *PLoS Comput. Biol.* **2007**, *3*, e15. [CrossRef] [PubMed]

36. Liu, C.C.; Jewett, M.C.; Chin, J.W.; Voigt, C.A. Toward an orthogonal central dogma. *Nat. Chem. Biol.* **2018**, *14*, 103–106. [CrossRef] [PubMed]

37. Thomas, C.M.; Nielsen, K.M. Mechanisms of, and barriers to, horizontal gene transfer between bacteria. *Nat. Rev. Microbiol.* **2005**, *3*, 711–721. [CrossRef]

38. Herdewijn, P.; Marlière, P. Toward safe genetically modified organisms through the chemical diversification of nucleic acids. *Chem. Biodivers.* **2009**, *6*, 791–808. [CrossRef]

39. Vargas-Rodriguez, O.; Sevostyanova, A.; Söll, D.; Crnković, A. Upgrading aminoacyl-tRNA synthetases for genetic code expansion. *Curr. Opin. Chem. Biol.* **2018**, *46*, 115–122. [CrossRef]

40. Lajoie, M.J.; Kosuri, S.; Mosberg, J.A.; Gregg, C.J.; Zhang, D.; Church, G.M. Probing the limits of genetic recoding in essential genes. *Science* **2013**, *342*, 361–363. [CrossRef]

41. Rovner, A.J.; Haimovich, A.D.; Katz, S.R.; Li, Z.; Grome, M.W.; Gassaway, B.M.; Amiram, M.; Patel, J.R.; Gallagher, R.R.; Rinehart, J.; et al. Recoded organisms engineered to depend on synthetic amino acids. *Nature* **2015**, *518*, 89–93. [CrossRef] [PubMed]

42. Ostrov, N.; Landon, M.; Guell, M.; Kuznetsov, G.; Teramoto, J.; Cervantes, N.; Zhou, M.; Singh, K.; Napolitano, M.G.; Moosburner, M.; et al. Design, synthesis, and testing toward a 57-codon genome. *Science* **2016**, *353*, 819–822. [CrossRef] [PubMed]

43. Lajoie, M.J.; Rovner, A.J.; Goodman, D.B.; Aerni, H.-R.; Haimovich, A.D.; Kuznetsov, G.; Mercer, J.A.; Wang, H.H.; Carr, P.A.; Mosberg, J.A.; et al. Genomically Recoded Organisms Expand Biological Functions. *Science* **2013**, *342*, 357–360. [CrossRef] [PubMed]

44. Acevedo-Rocha, C.G.; Budisa, N. Xenomicrobiology: a roadmap for genetic code engineering. *Microb. Biotechnol.* **2016**, *9*, 666–676. [CrossRef] [PubMed]

45. Kunjapur, A.M.; Stork, D.A.; Kuru, E.; Vargas-Rodriguez, O.; Landon, M.; Söll, D.; Church, G.M. Engineering posttranslational proofreading to discriminate nonstandard amino acids. *Proc. Natl. Acad. Sci. USA* **2018**, *115*, 619–624. [CrossRef] [PubMed]

46. Tack, D.S.; Ellefson, J.W.; Thyer, R.; Wang, B.; Gollihar, J.; Forster, M.T.; Ellington, A.D. Addicting diverse bacteria to a noncanonical amino acid. *Nat. Chem. Biol.* **2016**, *12*, 138–140. [CrossRef] [PubMed]

47. Lajoie, M.J.; Söll, D.; Church, G.M. Overcoming Challenges in Engineering the Genetic Code. *J. Mol. Biol.* **2016**, *428*, 1004–1021. [CrossRef]

48. Biddle, W.; Schmitt, M.A.; Fisk, J.D. Modification of orthogonal tRNAs: Unexpected consequences for sense codon reassignment. *Nucleic Acids Res.* **2016**, *44*, 10042–10050. [CrossRef]

49. Ma, N.J.; Isaacs, F.J. Genomic Recoding Broadly Obstructs the Propagation of Horizontally Transferred Genetic Elements. *Cell Syst.* **2016**, *3*, 199–207. [CrossRef]

50. Hammerling, M.J.; Ellefson, J.W.; Boutz, D.R.; Marcotte, E.M.; Ellington, A.D.; Barrick, J.E. Bacteriophages use an expanded genetic code on evolutionary paths to higher fitness. *Nat. Chem. Biol.* **2014**, *10*, 178–180. [CrossRef]

51. Ma, N.J.; Hemez, C.F.; Barber, K.W.; Rinehart, J.; Isaacs, F.J. Organisms with alternative genetic codes resolve unassigned codons via mistranslation and ribosomal rescue. *eLife* **2018**, *7*, 1–23. [CrossRef] [PubMed]

52. Wan, W.; Huang, Y.; Wang, Z.; Russell, W.K.; Pai, P.J.; Russell, D.H.; Liu, W.R. A facile system for genetic incorporation of two different noncanonical amino acids into one protein in *Escherichia coli*. *Angew. Chem. Int. Ed.* **2010**, *49*, 3211–3214. [CrossRef] [PubMed]

53. Santos, M.A.S.; Cheesman, C.; Costa, V.; Moradas-Ferreira, P.; Tuite, M.F. Selective advantages created by codon ambiguity allowed for the evolution of an alternative genetic code in Candida spp. *Mol. Microbiol.* **1999**, *31*, 937–947. [CrossRef] [PubMed]

54. Kollmar, M.; Mühlhausen, S. How tRNAs dictate nuclear codon reassignments: Only a few can capture non-cognate codons. *RNA Biol.* **2017**, *14*, 293–299. [CrossRef] [PubMed]

55. Schmitt, M.A.; Biddle, W.; Fisk, J.D. Mapping the Plasticity of the *Escherichia coli* Genetic Code with Orthogonal Pair-Directed Sense Codon Reassignment. *Biochemistry* **2018**, *57*, 2762–2774. [CrossRef] [PubMed]

56. Hoesl, M.G.; Budisa, N. Expanding and Engineering the Genetic Code in a Single Expression Experiment. *ChemBioChem* **2011**, *12*, 552–555. [CrossRef] [PubMed]

57. Wong, J.T. Membership mutation of the genetic code: loss of fitness by tryptophan. *Proc. Natl. Acad. Sci. USA* **1983**, *80*, 6303–6306. [CrossRef] [PubMed]

58. Mat, W.K.; Xue, H.; Wong, J.T.F. Genetic code mutations: The breaking of a three billion year invariance. *PLoS ONE* **2010**, *5*, e12206. [CrossRef]

59. Yu, A.C.S.; Yim, A.K.Y.; Mat, W.K.; Tong, A.H.Y.; Lok, S.; Xue, H.; Tsui, S.K.W.; Wong, J.T.F.; Chan, T.F. Mutations enabling displacement of tryptophan by 4-fluorotryptophan as a canonical amino acid of the genetic code. *Genome Biol. Evol.* **2014**, *6*, 629–641. [CrossRef]

60. Crick, F.H.C. Codon—anticodon pairing: The wobble hypothesis. *J. Mol. Biol.* **1966**, *19*, 548–555. [CrossRef]

61. Bohlke, N.; Budisa, N. Sense codon emancipation for proteome-wide incorporation of noncanonical amino acids: Rare isoleucine codon AUA as a target for genetic code expansion. *FEMS Microbiol. Lett.* **2014**, *351*, 133–144. [CrossRef] [PubMed]

62. Mukai, T.; Yamaguchi, A.; Ohtake, K.; Takahashi, M.; Hayashi, A.; Iraha, F.; Kira, S.; Yanagisawa, T.; Yokoyama, S.; Hoshi, H.; et al. Reassignment of a rare sense codon to a non-canonical amino acid in *Escherichia coli*. *Nucleic Acids Res.* **2015**, *43*, 8111–8122. [CrossRef] [PubMed]

63. Lee, B.S.; Shin, S.; Jeon, J.Y.; Jang, K.S.; Lee, B.Y.; Choi, S.; Yoo, T.H. Incorporation of Unnatural Amino Acids in Response to the AGG Codon. *ACS Chem. Biol.* **2015**, *10*, 1648–1653. [CrossRef] [PubMed]

64. Fu, X.; Söll, D.; Sevostyanova, A. Challenges of site-specific selenocysteine incorporation into proteins by *Escherichia coli*. *RNA Biol.* **2018**, *15*, 461–470. [CrossRef] [PubMed]

65. Bröcker, M.J.; Ho, J.M.L.; Church, G.M.; Söll, D.; O'Donoghue, P. Recoding the genetic code with selenocysteine. *Angew. Chem. Int. Ed.* **2014**, *53*, 319–323. [CrossRef] [PubMed]

66. Torres, L.; Krüger, A.; Csibra, E.; Gianni, E.; Pinheiro, V.B. Synthetic biology approaches to biological containment: pre-emptively tackling potential risks. *Essays Biochem.* **2016**, *60*, 393–410. [CrossRef] [PubMed]

67. Grosjean, H.; Westhof, E. An integrated, structure- and energy-based view of the genetic code. *Nucleic Acids Res.* **2016**, *44*, 8020–8040. [CrossRef]

68. Posfai, G. Emergent Properties of Reduced-Genome *Escherichia coli*. *Science* **2006**, *312*, 1044–1046. [CrossRef]

69. Mukai, T.; Lajoie, M.J.; Englert, M.; Söll, D. Rewriting the Genetic Code. *Annu. Rev. Microbiol.* **2017**, *71*, 557–577. [CrossRef]

70. Chari, R.; Church, G.M. Beyond editing to writing large genomes. *Nat. Rev. Genet.* **2017**, *18*, 749–760. [CrossRef]

71. Kuo, J.; Stirling, F.; Lau, Y.H.; Shulgina, Y.; Way, J.C.; Silver, P.A. Synthetic genome recoding: new genetic codes for new features. *Curr. Genet.* **2018**, *64*, 327–333. [CrossRef] [PubMed]

72. Rackham, O.; Chin, J.W. A network of orthogonal ribosome·mrna pairs. *Nat. Chem. Biol.* **2005**, *1*, 159–166. [CrossRef] [PubMed]

73. Neumann, H.; Wang, K.; Davis, L.; Garcia-Alai, M.; Chin, J.W. Encoding multiple unnatural amino acids via evolution of a quadruplet-decoding ribosome. *Nature* **2010**, *464*, 441–444. [CrossRef] [PubMed]

74. Anderson, J.C.; Wu, N.; Santoro, S.W.; Lakshman, V.; King, D.S.; Schultz, P.G. An expanded genetic code with a functional quadruplet codon. *Proc. Natl. Acad. Sci. USA* **2004**, *101*, 7566–7571. [CrossRef] [PubMed]

75. Hohsaka, T.; Ashizuka, Y.; Murakami, H.; Sisido, M. Five-base codons for incorporation of nonnatural amino acids into proteins. *Nucleic Acids Res.* **2001**, *29*, 3646–3651. [CrossRef] [PubMed]

76. Watson, J.D.; Crick, F.H.C. Molecular Structure of Nucleic Acids: A Structure for Deoxyribose Nucleic Acid. *Nature* **1953**, *171*, 737–738. [CrossRef] [PubMed]

77. Pinheiro, V.B.; Holliger, P. The XNA world: Progress towards replication and evolution of synthetic genetic polymers. *Curr. Opin. Chem. Biol.* **2012**, *16*, 245–252. [CrossRef]

78. Warren, R.A.J. Modified Bases in Bacteriophage DNAs. *Annu. Rev. Microbiol.* **1980**, *34*, 137–158. [CrossRef]

79. Yang, Z.; Sismour, A.M.; Sheng, P.; Puskar, N.L.; Benner, S.A. Enzymatic incorporation of a third nucleobase pair. *Nucleic Acids Res.* **2007**, *35*, 4238–4249. [CrossRef]

80. Sismour, A.M. PCR amplification of DNA containing non-standard base pairs by variants of reverse transcriptase from Human Immunodeficiency Virus-1. *Nucleic Acids Res.* **2004**, *32*, 728–735. [CrossRef]

81. Leal, N.A.; Kim, H.J.; Hoshika, S.; Kim, M.J.; Carrigan, M.A.; Benner, S.A. Transcription, reverse transcription, and analysis of RNA containing artificial genetic components. *ACS Synth. Biol.* **2015**, *4*, 407–413. [CrossRef] [PubMed]

82. Chen, F.; Zhang, Y.; Daugherty, A.B.; Yang, Z.; Shaw, R.; Dong, M.; Lutz, S.; Benner, S.A. Biological phosphorylation of an Unnatural Base Pair (UBP) using a Drosophila melanogaster deoxynucleoside kinase (DmdNK) mutant. *PLoS ONE* **2017**, *12*, e0174163. [CrossRef] [PubMed]

83. Marlière, P.; Patrouix, J.; Döring, V.; Herdewijn, P.; Tricot, S.; Cruveiller, S.; Bouzon, M.; Mutzel, R. Chemical evolution of a bacterium's genome. *Angew. Chem. Int. Ed.* **2011**, *50*, 7109–7114. [CrossRef] [PubMed]

84. Malyshev, D.A.; Dhami, K.; Lavergne, T.; Chen, T.; Dai, N.; Foster, J.M.; Corrêa, I.R.; Romesberg, F.E. A semi-synthetic organism with an expanded genetic alphabet. *Nature* **2014**, *509*, 385–388. [CrossRef] [PubMed]

85. Zhang, Y.; Romesberg, F.E. Semisynthetic Organisms with Expanded Genetic Codes. *Biochemistry* **2018**, *57*, 2177–2178. [CrossRef]

86. Zhang, Y.; Ptacin, J.L.; Fischer, E.C.; Aerni, H.R.; Caffaro, C.E.; San Jose, K.; Feldman, A.W.; Turner, C.R.; Romesberg, F.E. A semi-synthetic organism that stores and retrieves increased genetic information. *Nature* **2017**, *551*, 644–647. [CrossRef]

87. Hettinger, T.P. Helix instability and self-pairing prevent unnatural base pairs from expanding the genetic alphabet. *Proc. Natl. Acad. Sci. USA* **2017**, *114*, E6476–E6477. [CrossRef]

88. Pinheiro, V.B.; Taylor, A.I.; Cozens, C.; Abramov, M.; Renders, M.; Zhang, S.; Chaput, J.C.; Wengel, J.; Peak-Chew, S.Y.; McLaughlin, S.H.; et al. Synthetic genetic polymers capable of heredity and evolution. *Science* **2012**, *336*, 341–344. [CrossRef]

89. Culbertson, M.C.; Temburnikar, K.W.; Sau, S.P.; Liao, J.; Bala, S.; Chaput, J.C. Bioorganic & Medicinal Chemistry Letters Evaluating TNA stability under simulated physiological conditions. *Bioorg. Med. Chem. Lett.* **2016**, *26*, 2418–2421. [CrossRef]

90. Ichida, J.K.; Horhota, A.; Zou, K.; Mclaughlin, L.W.; Szostak, J.W. High fidelity TNA synthesis by Therminator polymerase. *Nucleic Acids Res.* **2005**, *33*, 5219–5225. [CrossRef]

91. Dunn, M.R.; Chaput, J.C. Reverse Transcription of Threose Nucleic Acid by a Naturally Occurring DNA Polymerase. *Chembiochem* **2016**, 1804–1808. [CrossRef] [PubMed]

92. Dunn, M.R.; Otto, C.; Fenton, K.E.; Chaput, J.C. Improving Polymerase Activity with Unnatural Substrates by Sampling Mutations in Homologous Protein Architectures. *ACS Chem. Biol.* **2016**. [CrossRef] [PubMed]

93. Mei, H.; Shi, C.; Jimenez, R.M.; Wang, Y.; Kardouh, M.; Chaput, C. Synthesis and polymerase activity of a fluorescent cytidine TNA triphosphate analogue. *Nucleic Acids Res.* **2017**, *45*, 5629–5638. [CrossRef] [PubMed]

94. Schmidt, M. Xenobiology: A new form of life as the ultimate biosafety tool. *BioEssays* **2010**, *32*, 322–331. [CrossRef] [PubMed]

95. Liu, C.; Cozens, C.; Jaziri, F.; Rozenski, J.; Maréchal, A.; Dumbre, S.; Pezo, V.; Marlière, P.; Pinheiro, V.B.; Groaz, E.; et al. Phosphonomethyl Oligonucleotides as Backbone-Modified Artificial Genetic Polymers. *J. Am. Chem. Soc.* **2018**, *140*, 6690–6699. [CrossRef] [PubMed]

96. Hartman, H.; Smith, T. The Evolution of the Ribosome and the Genetic Code. *Life* **2014**, *4*, 227–249. [CrossRef] [PubMed]

97. Smith, T.F.; Hartman, H. The evolution of Class II Aminoacyl-tRNA synthetases and the first code. *FEBS Lett.* **2015**, *589*, 3499–3507. [CrossRef]

98. Granold, M.; Hajieva, P.; Toşa, M.I.; Irimie, F.-D.; Moosmann, B. Modern diversification of the amino acid repertoire driven by oxygen. *Proc. Natl. Acad. Sci. USA* **2017**, *115*, 201717100. [CrossRef]

99. Forster, A.C.; Church, G.M. Synthetic biology projects in vitro. *Genome Res.* **2007**, 1–6. [CrossRef]

100. Adamala, K.P.; Martin-Alarcon, D.A.; Guthrie-Honea, K.R.; Boyden, E.S. Engineering genetic circuit interactions within and between synthetic minimal cells. *Nat. Chem.* **2017**, *9*, 431–439. [CrossRef]

101. Van den Bergh, B.; Swings, T.; Fauvart, M.; Michiels, J. Experimental Design, Population Dynamics, and Diversity in Microbial Experimental Evolution. *Microbiol. Mol. Biol. Rev.* **2018**, *82*, e00008-18. [CrossRef] [PubMed]

102. Tack, D.S.; Cole, A.C.; Shroff, R.; Morrow, B.R.; Ellington, A.D. Evolving Bacterial Fitness with an Expanded Genetic Code. *Sci. Rep.* **2018**, *8*, 3288. [CrossRef] [PubMed]

103. Freeland, S.J.; Hurst, L.D. The genetic code is one in a million. *J. Mol. Evol.* **1998**, *47*, 238–248. [CrossRef] [PubMed]

104. Knight, R.; Freeland, S.; Landweber, L. Rewiring the keyboard: evolvability ogf the genetic code. *Nat. Rev. Genet.* **2001**, *2*. [CrossRef] [PubMed]

105. Palmer, J.D.; Adams, K.L.; Cho, Y.; Parkinson, C.L.; Qiu, Y.L.; Song, K. Dynamic evolution of plant mitochondrial genomes: mobile genes and introns and highly variable mutation rates. *Proc. Natl. Acad. Sci. USA* **2000**, *97*, 6960–6966. [CrossRef] [PubMed]

106. Baranov, P.V.; Atkins, J.F.; Yordanova, M.M. Augmented genetic decoding: Global, local and temporal alterations of decoding processes and codon meaning. *Nat. Rev. Genet.* **2015**, *16*, 517–529. [CrossRef] [PubMed]

107. Kowal, A.K.; Oliver, J.S. Exploiting unassigned codons in Micrococcus luteus for tRNA-based amino acid mutagenesis. *Nucleic Acids Res.* **1997**, *25*, 4685–4689. [CrossRef]

108. Chandel, N.S. Evolution of Mitochondria as Signaling Organelles. *Cell Metab.* **2015**, *22*, 204–206. [CrossRef]

109. Nagao, A.; Ohara, M.; Miyauchi, K.; Yokobori, S.I.; Yamagishi, A.; Watanabe, K.; Suzuki, T. Hydroxylation of a conserved tRNA modification establishes non-universal genetic code in echinoderm mitochondria. *Nat. Struct. Mol. Biol.* **2017**, *24*, 778–782. [CrossRef]

110. Krieger, F.; Mo, A.; Kiefhaber, T. Effect of Proline and Glycine Residues on Dynamics and Barriers of Loop Formation in Polypeptide Chains. *J. Am. Chem. Soc.* **2005**, 3346–3352. [CrossRef]

111. Kubyshkin, V.; Grage, S.L.; Bürck, J.; Ulrich, A.S.; Budisa, N. Transmembrane Polyproline Helix. *J. Phys. Chem. Lett.* **2018**, *9*, 2170–2174. [CrossRef] [PubMed]

genes

MDPI

Article

Development of a Transformation Method for Metschnikowia borealis and other CUG-Serine Yeasts

Zachary B. Gordon [1,2], Maximillian P.M. Soltysiak [3], Christopher Leichthammer [1], Jasmine A. Therrien [1], Rebecca S. Meaney [1], Carolyn Lauzon [1], Matthew Adams [1], Dong Kyung Lee [3], Preetam Janakirama [1], Marc-André Lachance [3] and Bogumil J. Karas [1,2,*]

[1] Designer Microbes Inc., London, ON N6G 4X8, Canada; zgordon2@uwo.ca (Z.B.G.); cleichth@uwo.ca (C.L.); jasmine.alyssa.therrien@gmail.com (J.A.T.); rmeaney2@uwo.ca (R.S.M.); carolyn.lauzon@gmail.com (C.L.); adams.mil@hotmail.com (M.A.); preetam.janakirama@gmail.com (P.J.)
[2] Department of Biochemistry, Schulich School of Medicine and Dentistry, University of Western Ontario, London, ON N6A 5C1, Canada
[3] Department of Biology, University of Western Ontario, London, ON N6A 5B7, Canada; msoltys4@uwo.ca (M.P.M.S); dlee335@uwo.ca (D.K.L.); lachance@uwo.ca (M.-A.L.)
[*] Correspondence: bkaras@uwo.ca

Received: 10 November 2018; Accepted: 18 January 2019; Published: 23 January 2019

Abstract: Yeasts belonging to the *Metschnikowia* genus are particularly interesting for the unusual formation of only two needle-shaped ascospores during their mating cycle. Presently, the meiotic process that can lead to only two spores from a diploid zygote is poorly understood. The expression of fluorescent nuclear proteins should allow the meiotic process to be visualized in vivo; however, no large-spored species of *Metschnikowia* has ever been transformed. Accordingly, we aimed to develop a transformation method for *Metschnikowia borealis*, a particularly large-spored species of *Metschnikowia*, with the goal of enabling the genetic manipulations required to study biological processes in detail. Genetic analyses confirmed that *M. borealis*, and many other *Metschnikowia* species, are CUG-Ser yeasts. Codon-optimized selectable markers lacking CUG codons were used to successfully transform *M. borealis* by electroporation and lithium acetate, and transformants appeared to be the result of random integration. Mating experiments confirmed that transformed-strains were capable of generating large asci and undergoing recombination. Finally, random integration was used to transform an additional 21 yeast strains, and all attempts successfully generated transformants. The results provide a simple method to transform many yeasts from an array of different clades and can be used to study or develop many species for various applications.

Keywords: genome engineering; synthetic biology; yeasts; *Metschnikowia*; genetic tools; DNA delivery; CUG-Ser

1. Introduction

For decades, yeasts have proven to be tremendously useful in the study of life's fundamental processes. As single-celled models for eukaryotic life, they have short generation times, can be easily manipulated, and contain thousands of conserved genes—many of which are vitally important across nearly all living organisms [1,2]. While the discovery of new species provides us with new opportunities, genetic tools previously developed for *Saccharomyces cerevisiae* and other model organisms must be re-optimized for use in newly discovered species [3,4]. Altered membrane composition, modified gene expression, and alternative codon usage can prevent the seamless application of established transformation methods onto new species. Such transformation methods are crucial in the development of targeted knockouts, changes in gene expression, and fusion proteins that allow the study of biological processes in detail.

Of particular interest among yeasts are the 81 or more species assigned to the genus *Metschnikowia*, some of which have become popular for their varied applications [5]. For example, *M. pulcherrima* is of importance to winemakers for its generation of pleasant aromas in high-quality wines, and its anti-microbial properties further its potential as an industrial microorganism [6,7]. Similarly, *M. bicuspidata* has been studied extensively for its role as a pathogen, and other species (including *M. agaves* and *M. hawaiiensis*) are being studied as potential anti-aging agents in cosmetics [5].

Metschnikowia species are particularly intriguing, due to the unusual formation of only two needle-shaped ascospores during their sexual cycle. A two-spored meiotic product stands in sharp contrast to the four spores arising from meiosis in many other yeast species (like *S. cerevisiae*) [5]. Although the occurrence of meiotic recombination has been demonstrated in some *Metschnikowia* species [8] the fate of the nuclei during meiosis remains unclear, and is yet to be elucidated. The expression of fluorescent nuclear fusion proteins should allow the meiotic process to be visualized in vivo, but the generation of such fusions is greatly hindered by the lack of genetic tools available for *Metschnikowia* species. *M. pulcherrima* is the only species ever to have been transformed [9]. Unfortunately, the small size of its ascospores and the difficulty of generating abundant asci make *M. pulcherrima* less than ideal for the study of meiotic nuclei.

In contrast, *M. borealis* is one of many *Metschnikowia* species that form ascospores that can reach 20–50 times the size of normal budding cells [10]. The giant, elongated spores would greatly facilitate the task of elucidating the meiotic process in *Metschnikowia* species. The initial selection of *M. borealis* for study is further justified by the fact that the species has the highest maximum growth temperature among the large-spored *Metschnikowia* species (37 °C), making it more likely to survive the high-temperature incubations required by many of the conventional yeast transformation methods [11,12]. Therefore, we aimed to develop and optimize a transformation method for *M. borealis* to enable the study of the species in more detail.

Our approach builds on the conventional methods used to transform other yeast species, such as *S. cerevisiae* and *M. pulcherrima*. After identifying antibiotics that can be used as selectable markers, we developed various genetic constructs carrying the appropriate genes to allow growth on selective plates. We then tested conventional transformation techniques, such as lithium acetate transformation and electroporation to introduce the various genetic cassettes [11]. It was originally anticipated that differences in membrane composition, antibiotic sensitivity, and origins of replication may pose significant obstacles in our ability to seamlessly apply classical transformation methods from *S. cerevisiae* onto *M. borealis*. Accordingly, we developed a systematic approach entailing a variety of different transformation methods, using a variety of different antibiotic resistance markers to develop a successful transformation method for *M. borealis* and other CUG-Ser yeasts.

2. Materials and Methods

2.1. Microbial Strains and Growth Conditions

Metschnikowia borealis strains UWOPS 96-101.1 (*MATα*) and SUB 99-207.1 (*MATa*), and other strains that were used in transformation experiments were obtained from the yeast collection of the Department of Biology, University of Western Ontario, where they are kept frozen in liquid nitrogen. *Saccharomyces cerevisiae* VL6-48 (ATCC MYA-3666: *MATα his3-Δ200 trp1-Δ1 ura3-52 lys2 ade2-1 met14 cir⁰*). All strains were grown in YPAD broth (1% yeast extract, 2% peptone, 0.01% adenine hemisulfate, 2% D-glucose) at 30 °C with shaking at 225 rpm, with the exception of *C.* aff. *bentonensis*, *M. bicuspidata* and *M. orientalis*, which were grown at 27 °C. Stationary cultures were grown on the same media containing 1% agar at the same growth temperatures.

2.2. Codon-Optimization of KanMX and Sh ble

To codon-optimize the KanMX and Sh *ble* selectable markers, each CTG codon was changed to a TTG, and the resulting gene sequence was uploaded to the OPTIMIZER tool (http://genomes.urv.es/OPTIMIZER/) [13]. The codon frequencies from *M. borealis* in three highly expressed yeast genes—*ADH1*, *TDH3*, and *ENO1*—were used to guide optimization. A table was generated using the Codon Usage Database (https://www.kazusa.or.jp/codon/) and entered into OPTIMIZER to guide codon-optimization. The method used for optimization was one amino acid to one codon, with a maximum of 25 allowed codon changes. Optimized genes were called *MbKanMX* and *MbShBle*.

2.3. Yeast DNA Isolation

Genomic DNA was isolated from *M. borealis* using a modified alkaline lysis method. Individual colonies were suspended in 1.0 mL of sterile double-deionized water (sddH$_2$O) and pelleted at 3000× g for 5 min. Cells were resuspended in 250 μL P1 Buffer (Qiagen, Valencia, CA, USA), 0.25 μL of 14 M β-mercaptoethanol, and 12.5 μL zymolyase solution (10 mg mL^{-1} zymolyase 20T in 25% glycerol), and incubated at 37 °C for 1 h. Following incubation, 250 μL of P2 Buffer (Qiagen) was added, gently mixed, and the suspension was incubated at room temperature for 5 min. Subsequently, 250 μL P3 Buffer (Qiagen) was added, gently mixed, and the suspension was incubated on ice for 10 min. Cellular debris was pelleted at max speed for 10 min at 4 °C, and the supernatant was transferred to a separate tube. Next, 750 μL ice-cold isopropanol was added to the supernatant, and the mixture was frozen at −80 °C for 15 min. The DNA was pelleted at top speed for 10 min at 4 °C, washed with 750 μL ice-cold 70% ethanol, and dissolved in 50 μL Tris-EDTA (TE) pH 8.0.

2.4. Lithium Acetate Transformations

The lithium acetate transformations were conducted as described previously, with some modifications [14]. Briefly, a 50 mL cell culture was grown to an OD$_{600}$ of 1.0 and pelleted at 6000× g for 5 min. The pellet was washed 3 times with and resuspended in 1 mL sterile water. Next, 100 μL of cells were transferred to a sterile tube and pelleted to remove the supernatant. The pellet was resuspended in 360 μL of transformation mix (33% PEG 3350 with 0.1 M LiOAc, 100 μg salmon sperm carrier DNA, and 1 μg DNA), and vortexed briefly. The mixture was incubated with shaking at 39 °C and 300 rpm for 30 min, and then pelleted to remove the transformation mix. The pellet was resuspended in 1 mL of YPAD in a 1.5 mL Eppendorf tube, recovered for 2 h at 30 °C and 225 rpm, plated on the appropriate selective media, and incubated at 30 °C for 2 days.

2.5. Electroporation Conditions

Electroporation experiments were conducted as previously described [15], with slight modifications. Briefly, cells were grown to an OD$_{600}$ of 1.5 in 50 mL YPAD and pelleted at 5000× g for 5 min. The pellet was resuspended in 10 mL 0.1 M LiOAc in 1×TE, pH 7.5, and incubated with shaking at 150 rpm and 30 °C for 1 h. Next, 250 μL 1 M DTT was added, and cells were returned to the incubator for 30 min. After adding 40 mL ice-cold sterile water, cells were pelleted, washed with 25 mL ice-cold sterile water, washed again with 5 mL ice-cold 1 M sorbitol, and resuspended in 250 μL 1 M sorbitol. Electro-competent cells were placed at −80 °C in 100 μL aliquots. When needed, cells were quickly thawed in a 37 °C water bath and immediately placed on ice. Electroporation reactions were prepared in 0.2 cm cuvettes with 40 μL electro-competent cells, 5 μL DNA (1 μg), and 20 μg single-stranded carrier DNA, and conditions were 1.8 kV, 200 Ω, and 25 μF. Cells were collected from each cuvette in 1 mL of YPAD and allowed to recover in Eppendorf tubes for 2 h at 30 °C and 225 rpm. For each transformation, 20 μL were plated on the appropriate selective media and incubated at 30 °C for 2 days.

2.6. Locating the Insertion Site

Metschnikowia borealis cells were transformed with PCR-linearized *CaNAT1* flanked by 60 base-pair sequences of homology to the *HIS3* promoter and terminator, and DNA was isolated from 5 colonies by Alkaline Lysis as described above. DNA samples were digested with CfoI

(Promega, Madison, WI, USA), with expected cut sites every ~600 base-pairs in the *M. borealis* genome and no cut sites within *CaNAT1*. Digested samples were then ligated using T4 DNA Ligase (New England BioLabsWhitby, ON, Canada). Qiagen Multiplex (Cat No. 206143) was used to PCR-amplify the unknown sequences surrounding the insertion site, using primers that bind within *CaNAT1* and that were oriented to amplify the surrounding sequence. PCR conditions were 95 °C melting for 30 s, 60 °C annealing for 90 s, and 72 °C extension for 90 s, cycled 35 times. PCR products were sent for sequencing at the DNA Sequencing Facility and Robarts Research Institute (London, ON, Canada). To identify insertion sites, sequences adjacent to the inserted marker were analyzed using BLASTn with Geneious 11.1.5 to search the *M. borealis* genome [16]. For secondary verification, primers were designed to anneal ~150 base-pairs upstream and downstream of the insertion site, and PCR was used to observe the insertion of the 693-base-pair cassette.

2.7. Mating Experiments

M. borealis MATa was transformed with *MbShBle*, and used in mating experiments with a *MATα* strain that had been transformed with *CaNAT1*. *MbShBle*-transformed *MATa* was thinly spread on 2% agar plates containing ½ strength glucose medium lacking histidine and uracil (Teknova, catalogue number C7221). *CaNAT1*-transformed *MATα* was then spread perpendicular to *MATa*, creating a grid with many sections where both strains were mixed. The cells were incubated for 2 days at room temperature. Approximately 20% of the cells were subsequently collected and resuspended in 25 mL of YPAD, and then grown at 30 °C for 6 h with shaking at 225 rpm. Next, 1 mL aliquots of culture were plated on YPAD containing 200 mg L^{-1} zeocin and 75 mg L^{-1} nourseothricin and incubated at 30 °C. After two days, resulting colonies were collected, serially diluted, and plated on YPAD with 200 mg L^{-1} zeocin and 75 mg L^{-1} nourseothricin. Purified recombinant colonies were grown overnight in YPAD, serially diluted, and plated alongside wildtype *MATa*, wildtype *MATα*, *MbShBle*-transformed *MATa*, and *CaNAT1*-transformed *MATα*.

3. Results

3.1. Identification of Antibiotics that Inhibit Growth of M. borealis

Before designing genetic cassettes (containing selection marker flanked by homology/promoter/terminator sequences) for transformation, it was necessary to identify antibiotics that inhibit the growth of *M. borealis* to enable us to choose appropriate antibiotic resistance markers. Cultures of *M. borealis MATa* and *MATα* were grown to OD$_{600}$ of 1.5, pelleted, and resuspended to an OD$_{600}$ of 3.0. Cultures were spot-diluted in 5 μL aliquots on YPAD agar plates with various concentrations of geneticin, zeocin, or nourseothricin (Supplementary Table S1); and it was determined that each antibiotic was effective at inhibiting growth at 400, 125, and 50 mg L^{-1}, respectively. To further prevent the occurrence of spontaneous colonies, higher concentrations of nourseothricin (75 mg L^{-1} or 100 mg L^{-1}) and zeocin (200 mg L^{-1}) were used for selection.

3.2. Growth Rate of M. borealis

Many conventional yeast transformation methods require that a culture to be grown to a specific optical density at 600 nm; accordingly, it was necessary to determine the growth rate of *M. borealis*, to enable us to accurately calculate the growth time to the required optical density. Three cultures of each mating type of *M. borealis* (MATa and MATα) were grown to mid-log phase (OD$_{600}$ = 1.0), and diluted in 50 mL of YPAD to an optical density of 0.065. Each culture was grown at 30 °C with shaking at 225 rpm, and OD$_{600}$ was recorded at 1 h time points for 8 h. The average OD$_{600}$ of the three cultures for each mating type was recorded, and the doubling time was calculated as 76 min for mating type a and 74 min for mating type α (Supplementary Figure S1).

3.3. M. borealis Codon Usage

A series of transformation attempts were made with *M. borealis* using genetic cassettes containing the *NAT*, *KanMX*, and *Sh ble* genes; however, no colonies were ever obtained for any of the transformation attempts. While investigating factors that would prevent us from obtaining colonies of *M. borealis*, the question of genetic code became significant. Some species (including *M. bicuspidata* and *M. fructicola*) have been reported to recognize a CUG codon as a serine (yeast alternative nuclear code) [17,18], which fits within the current view that the Metschnikowiaceae belong to the CUG-Ser 1 clade of budding yeasts [19]. However, the only published literature regarding the genetic code of *M. pulcherrima* indicated that the species translates the CUG codon to leucine (standard code) [20]. The *NAT*, *KanMX*, and *Sh ble* cassettes used in the transformation attempts contain 8, 4, and 5 CUG codons in their open reading frames, respectively.

In an attempt to determine the genetic code of *M. borealis*, the coding sequence of *URA3* from *M. borealis* was translated with the ExPASy translate tool, using both the standard code and yeast alternative nuclear code, and aligned with the 30 top hits from a *URA3*-BLASTx search using MUSCLE (Figure 1A). A leucine aligned at a highly conserved serine or threonine position when *URA3* was translated with the standard genetic code; however, a serine aligned at that position when translated with the yeast alternative nuclear code.

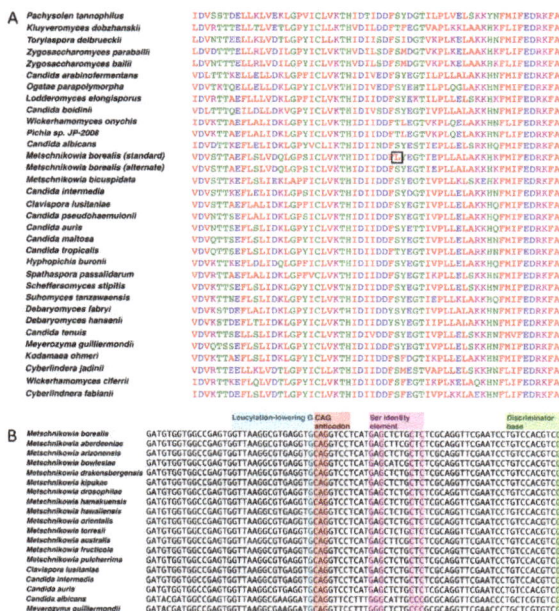

Figure 1. Bioinformatic analyses of the genetic code of *M. borealis*. (**A**) Sequence alignment of *URA3* (encodes Orotidine 5′-phosphate decarboxylase). The coding sequence of *URA3* from *M. borealis* was submitted as a query sequence to BLASTx, using the standard code for the predicted protein sequence. The protein sequences of the top 30 hits were aligned in MUSCLE with the *M. borealis URA3* gene, translated in both the standard code and yeast alternative code using the ExPASy translate tool. A CUG codon aligned at a highly conserved serine or threonine amino acid position, producing a leucine when *URA3* was translated using the standard code (black square). (**B**) tRNA$_{CAG}$^{Ser} sequences aligned for many species of *Metschnikowia* and various CUG-Ser yeasts. Sequences were aligned in MEGA6 and annotated as described previously [18].

An examination of available Metschnikowiaceae genome sequences by tRNAscan-SE v2.0 [21] revealed that all species encoded at least one tRNA with a CAG anticodon. The retrieved tRNA sequences were analyzed through the tRNAscan-SE on-line web server, aimed at estimating the most probable translated amino acid. All *Metschnikowia* species that were examined had serine as the most probable translation with scores of 85 or more. Genome sequences for *Candida* aff. *bentonensis* or relatives are not available. Further evidence arose from aligning the sequence for $tRNA_{CAG}$ from *M. borealis* with the $tRNA_{CAG}^{Ser}$ of related CUG-Ser clade yeast species (Figure 1B). The $tRNA_{CAG}$ from *M. borealis* and other Metschnikowiaceae contain the serine discriminator base and identity elements, indicating that the tRNAs identified are in fact $tRNA_{CAG}^{Ser}$. No $tRNA_{CAG}^{Leu}$ could be identified in any of the *Metschnikowia* species that were analyzed, including *M. pulcherrima*.

3.4. Transformation of M. borealis with Codon-Optimized Cassettes

Optimizer was used to codon-optimize the *NAT*, *KanMX*, and Sh *ble* genes for expression in *M. borealis* based on the codon frequencies in three highly expressed yeast genes: *ADH1*, *TDH3*, and *ENO1*. The optimized genes were translated with the ExPASy translate tool, using both the standard code and yeast alternative nuclear code, and aligned with BLAST to ensure that the same protein product would be produced. No differences were found between the wildtype gene products and the optimized gene products, regardless of whether the optimized genes were translated with the standard code or yeast alternative nuclear code. The optimized *NAT* gene was substantially similar to *CaNAT1*, a gene optimized for expression in *Candida albicans* [22]. Due to its previous success in transforming *C. albicans*, another CUG-Ser species, *CaNAT1* was used in this experiment. Codon-optimized KanMX and Sh *ble* genes were called *MbKanMX* and *MbShble*.

Optimizer was used to codon-optimize the *NAT*, *KanMX*, and Sh *ble* genes for expression in *M. borealis* based on the codon frequencies in three highly expressed yeast genes: *ADH1*, *TDH3*, and *ENO1*. The optimized genes were translated with the ExPASy translate tool, using both the standard code and yeast alternative nuclear code, and aligned with BLAST to ensure that the same protein product would be produced. No differences were found between the wildtype gene products and the optimized gene products, regardless of whether the optimized genes were translated with the standard code or yeast alternative nuclear code. The optimized *NAT* gene was substantially similar to *CaNAT1*, a gene optimized for expression in *Candida albicans* [22]. Due to its previous success in transforming *C. albicans*, another CUG-Ser species, *CaNAT1* was used in this experiment. Codon-optimized KanMX and Sh *ble* genes were called *MbKanMX* and *MbShble*.

The second set of antibiotic resistance genes were created by reintroducing the same 8, 4, and 5 CUG codons found in the wildtype *NAT*, *KanMX*, and Sh *ble* genes, respectively, as a control for codon usage. All six codon-optimized genes (three in standard code and three in yeast alternative nuclear code) were ordered as separate G-block fragments from Integrated DNA Technologies. All genes were amplified with primers containing homologous hooks to the *M. borealis ADH1* promoter and terminator and were used to transform the species by electroporation. Transformants were plated on YPAD with appropriate concentrations of each antibiotic (75 mg L^{-1} nourseothricin for *CaNAT1*, 200 mg L^{-1} zeocin for *MbShBle*, and 400 mg L^{-1} geneticin for *MbKanMX*) alongside negative controls. Several colonies appeared for all samples that were transformed using the yeast alternative nuclear code, but no colonies appeared on any of the other plates (Figure 2). Similar results were seen using our best lithium acetate protocol. Selected transformants from the electroporation experiment were verified by genotyping (Supplementary Figure S2). Transformation efficiencies were determined for each method by transforming *MATa* and *MATα* by each method in triplicate, and were calculated to be approximately 1600 and 817 CFUs per µg of DNA for electroporation, and 9 and 4 CFUs per µg of DNA for lithium acetate transformation, respectively (Supplementary Table S2).

The relatively high efficiency of electroporation was believed to be attributable to homologous recombination, because each cassette contained by 60 base-pairs of homology to the *ADH1* promoter and terminator. Based on this assumption, we attempted to generate histidine auxotrophs by transforming

M. borealis with *CaNAT1* flanked by 60 base-pair sequences of the *HIS3* promoter and terminator. After transformation, ten colonies of each making type were spot-diluted and re-isolated, but all grew on media lacking histidine. Subsequent genotyping by PCR identified the presence of both *HIS3* and *CaNAT1*, but failed to obtain bands across the *HIS3* promoter and terminator junctions with *CaNAT1*: Bands that would be expected in the case of a successfully targeted knockout (Supplementary Figure S3). Despite the random integration of our cassettes, double marker transformation experiments revealed that two cassettes were inserted in approximately 2% of transformants (Supplementary Table S3).

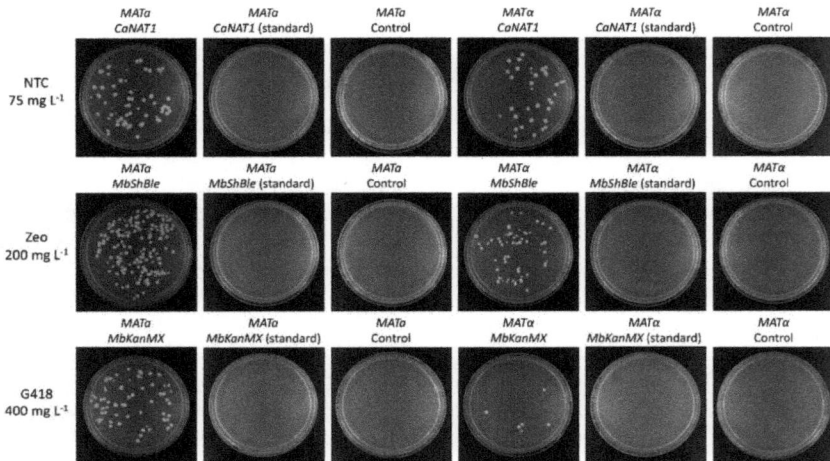

Figure 2. Electroporation of *M. borealis* with *CaNAT1*, *MbKanMX*, and *MbShBle* in standard and yeast alternative nuclear code. Electro-competent cells of *M. borealis* (*MATa* and *MATα*) were transformed with 1 µg of PCR-amplified *CaNAT1*, *MbShBle*, or *MbKanMx*, in standard or alternative code, flanked by 60 base-pair sequences of the *M. borealis ADH1* promoter and terminator, using the electroporation protocol outlined in this study. After recovery, 20 µg of each transformation was plated on YPAD with the appropriate antibiotic and incubated at 30 °C for two days. NTC = nourseothricin; Zeo = zeocin; G418 = geneticin.

3.5. Location of Insertion Site

Subsequent screening of more than 30 *MATa* and 30 *MATα* transformants for *HIS3* knockouts failed to identify any successful insertions in the *HIS3* gene. To identify the correct insertion sites, DNA was isolated from 6 *MATα* transformants, digested with CfoI, circularized with T4 ligase, and the surrounding genomic DNA was PCR-amplified with primers binding within the inserted *CaNAT1* cassette (oriented to amplify the surrounding sequence) (Supplementary Figure S3). We were successful in obtaining a PCR-fragment from the third *MATα* transformant (clone 3), and the amplified gDNA was sequenced. Analysis with BLASTn indicated that the *CaNAT1* cassette was inserted in scaffold bor-_s886 of *M. borealis* UWOPS 96-101.1, as annotated in GenBank (PRJNA312754), just upstream of the gene for peroxin-14. PCR was used to amplify this site in wildtype *M. borealis*, as well as all six isolated transformants, and the insertion was only observed at this site in clone 3.

3.6. Mating Experiments

To confirm that transformed-strains of *M. borealis MATa* and *MATα* were capable of successfully mating, *MbShBle*-transformed *MATa* and *CaNAT1*-transformed *MATα* were mated to develop ascospores (Figure 3A). Recombinant cells that contain both markers were isolated by serial dilution and genotyped to confirm that they contain both *MbShBle* and *CaNAT1* (Figure 3B).

Mated cells were then plated alongside wildtype *MATa* and *MATα*, *MbShBle*-transformed *MATa*, and *CaNAT1*-transformed *MATα* on YPAD, YPAD with zeocin, YPAD with nourseothricin, and YPAD with nourseothricin and zeocin for comparison (Figure 3C).

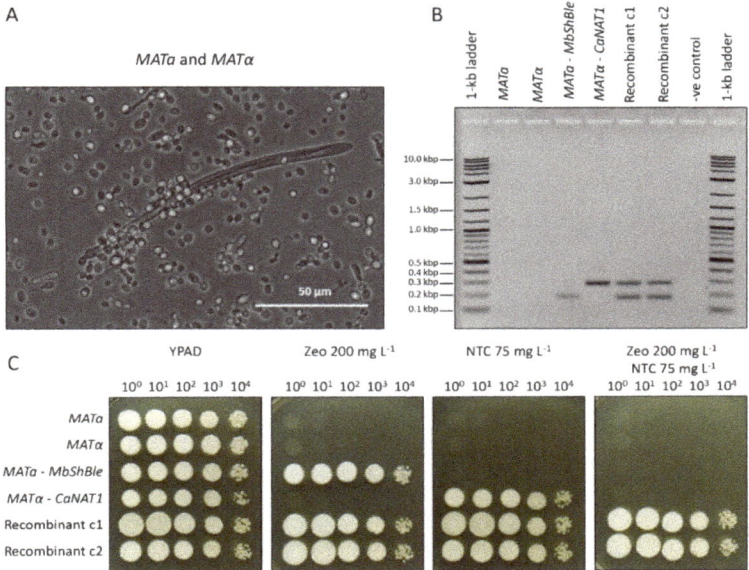

Figure 3. Mating experiments with marker-transformed *M. borealis*. (**A**) Photograph of a large mating ascospore. *MbShBle*-transformed *MATa* and *CaNAT1*-transformed *MATα* were mated for two days on ½-strength glucose broth lacking histidine and uracil (Teknova), collected in YPAD, and photographed under a light microscope. (**B**) Genotyping results of marker-transformed diploids. Recombinant colonies harbouring both selectable markers were purified by spot dilution on YPAD with 200 mg L^{-1} zeocin and 75 mg L^{-1} nourseothricin. Recombinant colonies were genotyped by Qiagen multiplex with primers that amplify the *CaNAT1* and *MbShBle* genes, alongside marker-transformed haploid strains and untransformed controls. *MbShBle* product size (**C**) Spot-dilutions of mated-colonies alongside transformed haploid strains and untransformed controls on YPAD, YPAD with 200 mg L^{-1} zeocin, YPAD with 75 mg L^{-1} nourseothricin, and YPAD with 200 mg L^{-1} zeocin and 75 mg L^{-1} nourseothricin. Plates were incubated at 30 °C for two days. NTC = nourseothricin; Zeo = zeocin; G418 = geneticin.

3.7. Transformation of other CUG-Ser Yeasts

To determine if our electroporation method would yield transformants of other yeasts by random integration, we transformed strains of 21 yeast species with *CaNAT1* cassettes (in standard and yeast alternative nuclear code) flanked by the same 60-base-pair minimal *ADH1* promoter and terminator from *M. borealis* (Figure 4; Supplementary Table S4). Transformants were plated on YPAD with 75–200 mg L^{-1} nourseothricin and incubated at 27–30 °C for 24–72 h until colonies appeared. All strains yielded colonies for the alternative-coded *CaNAT1*, but only *S. cerevisiae* and *Candida* aff. *bentonensis* yielded colonies when transformed with standard-coded *CaNAT1* (Figure 4; Supplementary Table S4). It's important to note that the alternative-coded *CaNAT1* lacks CUG codons and is therefore degenerate between the standard code and yeast alternative code; however, the standard-coded *CaNAT1* uses CUG codons to translate leucine and is only translatable to CUG-Leu yeasts.

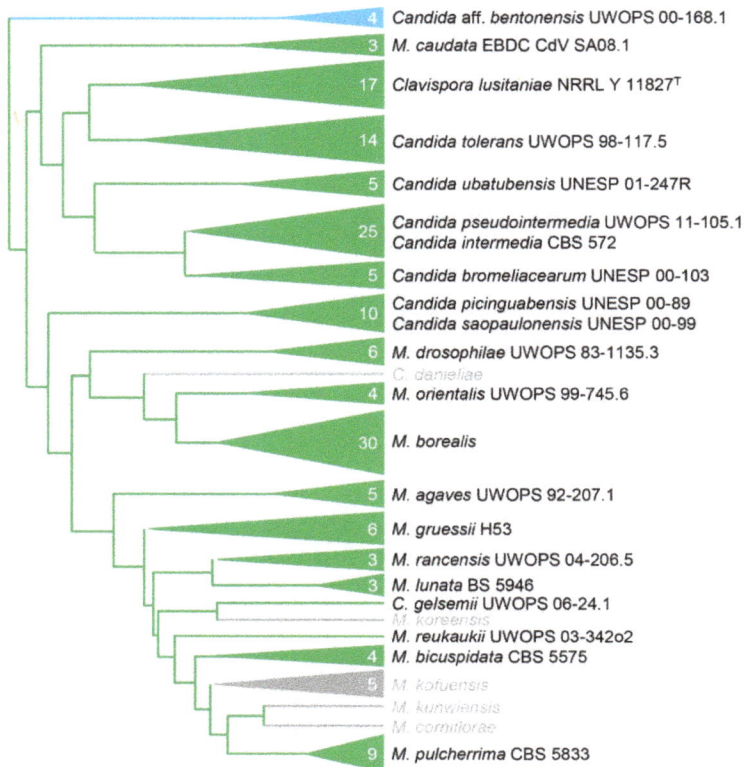

Figure 4. A phylogeny of the Metschnikowiaceae, modified from Lachance (2016) [5], showing the taxonomic range of successful transformations in the family. Representative strains were successfully transformed with only the alternative-coded *CaNAT1* construct (green) or with both the alternative-coded and the standard-coded *CaNAT1* (blue) construct. Species in grey were not examined.

4. Discussion

Through pioneering experiments, we were successful in determining the necessary conditions to transform *Metschnikowia borealis*, and the method was effective in many other yeast species. We were able to identify the $tRNA_{CAG}^{Ser}$ in *M. borealis*, as well as many other *Metschnikowia* species, and transformation experiments confirmed that many other—if not all—*Metschnikowia* species are also CUG-Ser 1 clade yeasts (Figures 2B and 4). It is unclear why *M. pulcherrima* was previously reported to be a CUG-Leu yeast [9].

Once we determined that *M. borealis* utilizes the yeast alternative nuclear code, the species was successfully transformed with selectable markers that contain homologous hooks to the *HIS3* promoter and terminator, with the high efficiency. Transformants were first believed to have been obtained through a targeted knockout of *HIS3* by homologous recombination; however, all colonies grew on medium lacking histidine, and genotyping confirmed the presence of both the selectable marker and the wildtype *HIS3* gene. Additionally, we failed to observe any of the expected genotyping bands associated with the knockout junctions. This indicated that the cassettes were not incorporated into the genome by targeted integration, but by another mechanism with high efficiency.

Although we only identified and confirm one insertion site (in clone 3), the *HIS3*-targeted cassette was found to insert in the site just upstream of the peroxin-14 gene in clone 3, and not in any of the other transformants tested (Supplementary Figure S2C). Therefore, the cassette must have inserted in different locations in the other transformants, indicating that *M. borealis* appears to integrate DNA cassettes randomly into its genome. Even though the species is not efficient at targeted integration, its high efficiency at random integration makes it simple to transform and incorporate exogenous sequences into the genome.

Finally, our electroporation transformation method was successful across all CUG-Ser yeasts that were attempted, as well as *S. cerevisiae* and *Candida* aff. *bentonensis*. This confirms the suspicion [5] that *C. bentonensis* is not a member of the Metschnikowiaceae. The results of this study indicate that the species is in fact a CUG-Leu yeast. Fortunately, the *CaNAT1*, *MbShBle*, and *MbKanMX* genes used in our experiments lack the use of a CUG codon and are degenerate to the standard and yeast alternative nuclear code. Therefore, the markers would likely be universal to all CUG-Leu, CUG-Ser, and CUG-Ala yeasts [19].

Through the use of random integration, the methods outlined in this study make it possible to transform a large range of yeasts and would likely be successful in numerous additional species in a wide array of clades. Random integration can be used to introduce elements to many understudied yeast species to enable their study in more detail. In particular, it is now possible to design codon-optimized fluorescent nuclear-fusion proteins that can be used to elucidate the unusual meiotic process in *Metschnikowia* species that leads to the production of only two ascospores from a diploid zygote.

Supplementary Materials: The following are available online at http://www.mdpi.com/2073-4425/10/2/78/s1, Figure S1: Growth rate of M. borealis MATa and MATα, Figure S2: Genotyping transformants, Figure S3: Identification of insertion sites; Table S1: Identification of antibiotic sensitivity, Table S2: Transformation efficiencies for M. borealis, Table S3: Double marker transformation of M. borealis, Table S4: Transformation results for additional yeast strains.

Author Contributions: Conceptualization, Z.B.G., M.P.M.S., P.J., M.-A.L., B.J.K.; methodology, Z.B.G., M.P.M.S., C.L., J.A.T., R.S.M., C.L., M.A., D.K.L., P.J., M.-A.L., B.J.K; validation, Z.B.G., M.P.M.S., C.L., J.A.T., R.S.M., C.L., M.A., D.K.L., P.J., M.-A.L., B.J.K; formal analysis, Z.B.G., M.P.M.S., C.L., J.A.T., R.S.M., C.L., M.A., D.K.L., P.J., M.-A.L., B.J.K; investigation, Z.B.G., M.P.M.S., C.L., J.A.T., R.S.M., C.L., M.A., D.K.L., P.J., M.-A.L., B.J.K; resources, M.-A.L., B.J.K; data curation, Z.B.G., M.P.M.S., C.L., J.A.T., R.S.M., C.L., M.A., D.K.L., P.J., M.-A.L., B.J.K; writing—original draft preparation, Z.B.G., M.P.M.S., R.S.M., M.A., M.-A.L., B.J.K; writing—review and editing, Z.B.G., M.P.M.S., R.S.M., M.A., M.-A.L., B.J.K; visualization, Z.B.G., M.P.M.S., C.L., J.A.T., R.S.M., C.L., M.A., D.K.L., P.J., M.-A.L., B.J.K; supervision, Z.B.G., P.J., M.-A.L., B.J.K; project administration M-A.L., B.J.K; funding acquisition, M.-A.L., B.J.K.

Funding: This research was funded by: Designer Microbes Inc.; NSERC Experience grants to Z.B.G., C.L., C.L., R.S.M. The B.J.K. lab is also supported by Natural Sciences and Engineering Research Council of Canada (NSERC), grant number: RGPIN-2018-06172.

Conflicts of Interest: B.J.K. is Chief Executive Officer of Designer Microbes Inc. B.J.K., P.J. and M-A.L. hold Designer Microbes Inc. stock.

References

1. Deng, J.; Deng, L.; Su, S.; Zhang, M.; Lin, X.; Wei, L.; Minai, A.A.; Hassett, D.J.; Lu, L.J. Investigating the predictability of essential genes across distantly related organisms using an integrative approach. *Nucl. Acids Res.* **2011**, *39*, 795–807. [CrossRef] [PubMed]

2. Zhang, R.; Ou, H.-Y.; Zhang, C.-T. DEG: A database of essential genes. *Nucl. Acids Res.* **2004**, *32*, D271–D272. [CrossRef] [PubMed]

3. De Backer, M.D.; Maes, D.; Vandoninck, S.; Logghe, M.; Contreras, R.; Luyten, W.H. Transformation of *Candida albicans* by electroporation. *Yeast Chichester Engl.* **1999**, *15*, 1609–1618. [CrossRef]

4. Yildirim, S.; Thompson, M.G.; Jacobs, A.C.; Zurawski, D. V.; Kirkup, B.C. Evaluation of parameters for high efficiency transformation of *Acinetobacter baumannii. Sci. Rep.* **2016**, *6*, 22110. [CrossRef] [PubMed]

5. Lachance, M.-A. *Metschnikowia*: Half tetrads, a regicide and the fountain of youth. *Yeast Chichester Engl.* **2016**. [CrossRef] [PubMed]

6. González-Pombo, P.; Pérez, G.; Carrau, F.; Guisán, J.M.; Batista-Viera, F.; Brena, B.M. One-step purification and characterization of an intracellular beta-glucosidase from *Metschnikowia pulcherrima*. *Biotechnol. Lett.* **2008**, *30*, 1469–1475. [CrossRef] [PubMed]

7. Macwilliam, I.C. A survey of the antibiotic powers of yeasts. *J. Gen. Microbiol.* **1959**, *21*, 410–414. [CrossRef]

8. Marinoni, G.; Piškur, J.; Research, M.L.-F. Ascospores of large-spored *Metschnikowia* species are genuine meiotic products of these yeasts. *FEMS Yeast Res.* **2003**, *3*, 85–90. [CrossRef] [PubMed]

9. Nigro, F.; Sialer, M.M.F.; Gallitelli, D. Transformation of *Metschnikowia pulcherrima* 320, biocontrol agent of storage rot, with the Green Fluorescent Protein gene. *J. Plant Pathol.* **1999**, *81*, 205–208.

10. Lachance, M.-A.; Hurtado, E.; Hsiang, T. A stable phylogeny of the large-spored *Metschnikowia* clade. *Yeast Chichester Engl.* **2016**, *33*, 261–275. [CrossRef] [PubMed]

11. Kawai, S.; Hashimoto, W.; Murata, K. Transformation of *Saccharomyces cerevisiae* and other fungi. *Bioeng. Bugs* **2010**, *1*, 395–403. [CrossRef] [PubMed]

12. Lachance, M.-A.; Rosa, C.A.; Starmer, W.T.; Schlag-Edler, B.; Baker, J.S.F.; Bowles, J.M. *Metschnikowia continentalis* var. *borealis*, *Metschnikowia continentalis* var. *continentalis*, and *Metschnikowia hibisci*, new heterothallic haploid yeasts from ephemeral flowers and associated insects. *Can. J. Microbiol.* **1998**, *44*, 279–288. [CrossRef]

13. Puigbò, P.; Guzmán, E.; Romeu, A.; Garcia-Vallvé, S. OPTIMIZER: A web server for optimizing the codon usage of DNA sequences. *Nucl. Acids Res.* **2007**, *35*, W126–W131. [CrossRef] [PubMed]

14. Gietz, R.D. Yeast transformation by the LiAc/SS carrier DNA/PEG method. *Methods Mol. Biol. Clifton* **2014**, *1205*, 1–12. [CrossRef]

15. Froyd, C.A.; Kapoor, S.; Dietrich, F.; Rusche, L.N. The deacetylase Sir2 from the yeast *Clavispora lusitaniae* lacks the evolutionarily conserved capacity to generate subtelomeric heterochromatin. *PLoS Genet.* **2013**, *9*, e1003935. [CrossRef] [PubMed]

16. Kearse, M.; Moir, R.; Wilson, A.; Stones-Havas, S.; Cheung, M.; Sturrock, S.; Buxton, S.; Cooper, A.; Markowitz, S.; Duran, C.; et al. Geneious Basic: An integrated and extendable desktop software platform for the organization and analysis of sequence data. *Bioinform. Oxf. Engl.* **2012**, *28*, 1647–1649. [CrossRef] [PubMed]

17. Mühlhausen, S.; Kollmar, M. Molecular phylogeny of sequenced saccharomycetes reveals polyphyly of the alternative yeast codon usage. *Genome Biol. Evol.* **2014**, *6*, 3222–3237. [CrossRef] [PubMed]

18. Riley, R.; Haridas, S.; Wolfe, K.H.; Lopes, M.R.; Hittinger, C.T.; Göker, M.; Salamov, A.A.; Wisecaver, J.H.; Long, T.M.; Calvey, C.H.; et al. Comparative genomics of biotechnologically important yeasts. *Proc. Natl. Acad. Sci. USA* **2016**, *113*, 9882–9887. [CrossRef] [PubMed]

19. Krassowski, T.; Coughlan, A.Y.; Shen, X.-X.; Zhou, X.; Kominek, J.; Opulente, D.A.; Riley, R.; Grigoriev, I. V.; Maheshwari, N.; Shields, D.C.; et al. Evolutionary instability of CUG-Leu in the genetic code of budding yeasts. *Nat. Commun.* **2018**, *9*, 1887. [CrossRef] [PubMed]

20. Sugita, T.; Nakase, T. Non-universal usage of the leucine CUG codon and the molecular phylogeny of the genus *Candida*. *Syst. Appl. Microbiol.* **1999**, *22*, 79–86. [CrossRef]

21. Lowe, T.M.; Chan, P.P. tRNAscan-SE On-line: Integrating search and context for analysis of transfer RNA genes. *Nucl. Acids Res.* **2016**, *44*, W54–W57. [CrossRef] [PubMed]

22. Shen, J.; Guo, W.; Köhler, J.R. CaNAT1, a heterologous dominant selectable marker for transformation of *Candida albicans* and other pathogenic *Candida* species. *Infect. Immunity* **2005**, *73*, 1239–1242. [CrossRef] [PubMed]

![genes logo] GCAT TACG GCAT *genes*

MDPI

Review

The Role of 3′ to 5′ Reverse RNA Polymerization in tRNA Fidelity and Repair

Allan W. Chen [1], Malithi I. Jayasinghe [2], Christina Z. Chung [1], Bhalchandra S. Rao [2], Rosan Kenana [1], Ilka U. Heinemann [1,*] and Jane E. Jackman [2,*]

[1] Department of Biochemistry, The University of Western Ontario, 1151 Richmond Street, London, ON N6A 5C1, Canada; achen325@uwo.ca (A.W.C.); cchung88@uwo.ca (C.Z.C.); rkenana@uwo.ca (R.K.)
[2] Department of Chemistry and Biochemistry, Center for RNA Biology, Ohio State Biochemistry Program, The Ohio State University, Columbus, OH 43210, USA; jayasinghearachchige.1@osu.edu (M.I.J.); brao@arrakistx.com (B.S.R.)
*   Correspondence: ilka.heinemann@uwo.ca (I.U.H.); Jackman.14@osu.edu (J.E.J.)

Received: 28 February 2019; Accepted: 21 March 2019; Published: 26 March 2019

Abstract: The tRNA[His] guanylyltransferase (Thg1) superfamily includes enzymes that are found in all three domains of life that all share the common ability to catalyze the 3′ to 5′ synthesis of nucleic acids. This catalytic activity, which is the reverse of all other known DNA and RNA polymerases, makes this enzyme family a subject of biological and mechanistic interest. Previous biochemical, structural, and genetic investigations of multiple members of this family have revealed that Thg1 enzymes use the 3′ to 5′ chemistry for multiple reactions in biology. Here, we describe the current state of knowledge regarding the catalytic features and biological functions that have been so far associated with Thg1 and its homologs. Progress toward the exciting possibility of utilizing this unusual protein activity for applications in biotechnology is also discussed.

Keywords: reverse polymerization; tRNA editing; tRNA repair; protein engineering; synthetic biology

---

1. Introduction

During DNA strand synthesis, 5′ to 3′ polymerization conveys significant advantages, particularly in the removal of mismatched nucleotides by the exonuclease activity of DNA polymerases [1,2]. This 3′ to 5′ nucleotide removal regenerates a polymerizable 3′-OH end, whereas the removal of a 5′ nucleotide generates a 5′-monophosphate, which is thought to be incapable of immediate replacement with another nucleotide (Figure 1) [2]. However, 3′ to 5′ or reverse polymerization is possible, despite the restrictions of extending a 5′-monophosphate end [3]. For such a 3′-5′ polymerase to function, the 5′-monophosphate end could be activated by ATP, creating an intermediate that is competent for nucleophilic attack by the 3′-OH of an incoming nucleotide, which is then a mechanistically similar process to forward polymerization (Figure 1) [4,5]. Indeed, enzymes that are capable of catalyzing this reverse polymerase reaction, which are members of the tRNA[His] guanylyltransferase (Thg1) superfamily, have been recently discovered [6–10]. Despite the possible advantages of reverse polymerization, such as coupled leading strands in DNA replication as opposed to the generation of Okazaki fragments, 3′ to 5′ polymerization appeared to be more limited in biology, and was initially only observed to function in several reactions associated with tRNA repair and processing. However, newer data suggests broader substrate specificity exhibited by some members of the Thg1 family known as Thg1-like proteins (TLPs), thus raising the potential for additional roles for 3′ to 5′ polymerization yet to be discovered.

Reverse 3' to 5' polymerization   **Forward 5' to 3' polymerization**

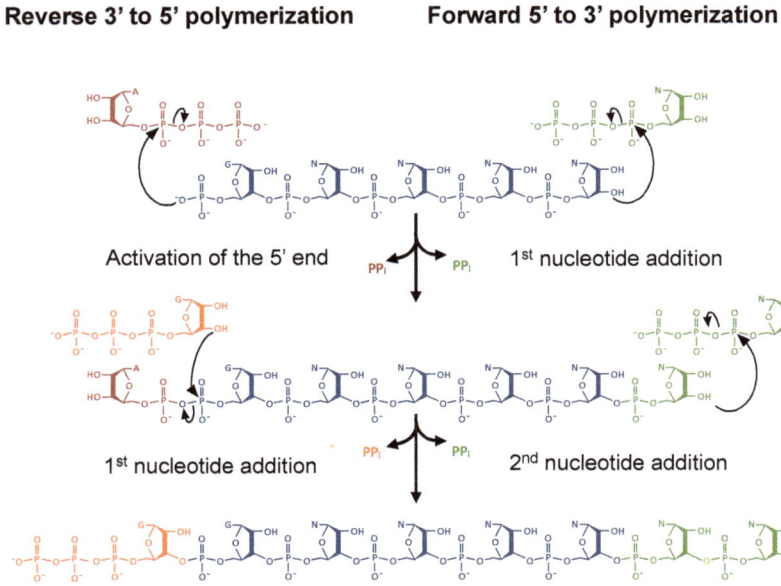

Figure 1. Forward 5′ to 3′ and reverse 3′ to 5′ polymerization are mechanistically similar. Ligation of a nucleotide to an RNA strand utilizes chemical energy that has been stored in the high-energy phosphoanhydride bonds in the nucleotide triphosphate. During forward polymerization, the triphosphate of the incoming nucleotide is hydrolyzed and provides energy to form a phosphodiester bond. For reverse polymerization, an initial activation step of a monophosphorylated RNA 5′ end (such as by adenylylation, which is shown here) is required; subsequent nucleotide additions utilize energy derived from hydrolysis of the phosphodiester bond of the RNA 5′-end triphosphate.

2. Thg1 Maintains tRNA^{His} Aminoacylation Fidelity

2.1. G$_{-1}$ is a tRNA^{His} Identity Element

Translational accuracy is dependent upon both tRNA selection by the ribosome and prior tRNA recognition and charging by a specific cognate aminoacyl-tRNA synthetase (aaRS) [11,12]. The side chain of the histidine residue, with its associated acid–base properties that contribute to both the protein structure and catalytic mechanisms, must be reliably and appropriately incorporated during translation [11,13]. Therefore, the aminoacylation of an amino acid to its respective tRNA must be specific, especially in the case of tRNA^{His} due to its particular properties that are difficult to chemically mimic with another natural amino acid, possibly explaining why tRNA^{His} relies on several key identity elements for proper aminoacylation.

For tRNA^{His} in all domains of life, one of these identity elements is a distinctive extra guanine nucleotide at the 5′-end (the −1 position) [14] (Figure 2). The 5′ guanylate residue at the −1 position (G$_{-1}$) serves as a key identity element for histidyl-tRNA synthetase (HisRS), which charges tRNA^{His} with its respective histidine [15–17]. In bacteria and many archaea, the G$_{-1}$ residue is genomically encoded and transcribed in the precursor tRNA transcript, and subsequent cleavage of the 5′ leader sequence by ribonuclease P (RNase P) yields a mature tRNA^{His} with its identity-establishing G$_{-1}$ element (tRNA^{His}G$_{-1}$) (Figure 2) [14,18,19].

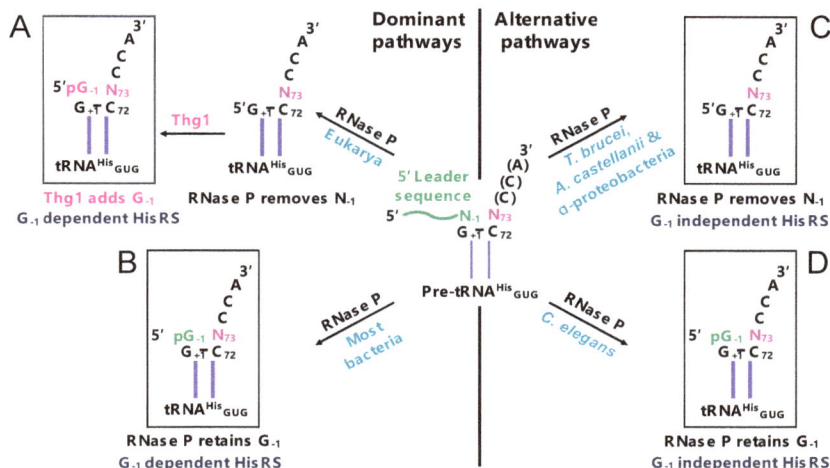

Figure 2. Different pathways to establish tRNA^{His} identity. (**A**) tRNA^{His} identity in many eukaryotes. RNase P removes N_{-1} from pre-tRNA^{His} during the removal of 5′ leader sequence (shown in green). Then, tRNA^{His} guanylyltransferase (Thg1) post-transcriptionally adds G_{-1} (shown in pink). Histidyl-tRNA synthetase (HisRS) recognizes the Thg1-incorporated G_{-1} for the accurate histidylation of tRNA^{His} in eukaryotes. (**B**) tRNA^{His} identity in bacteria. G_{-1} is encoded in tRNA^{His} genes in most bacteria, and RNase P retains the genomically encoded G_{-1} during the removal of the 5′ leader sequence. HisRS recognizes the RNase P-retained G_{-1} for the accurate histidylation of tRNA^{His} in bacteria. (**C**) tRNA^{His} identity in several groups of α-proteobacteria, *Trypanosoma brucei* and *Acanthamoeba castellanii*. RNase P removes N_{-1} from precursor tRNA^{His} during the removal of the 5′ leader sequence. An atypical HisRS in these organisms is capable of aminoacylating G_{-1}-lacking tRNA^{His} during histidylation. (**D**) tRNA^{His} identity in *Caenorhabditis elegans* (an atypical eukaryote). G_{-1} is encoded in tRNA^{His} genes, and RNase P retains the genomically encoded G_{-1}, which is similar to bacteria. HisRS in *C. elegans* is capable of aminoacylating both G_{-1}-containing and G_{-1}-lacking tRNA^{His}.

In many eukaryotes, the establishment of tRNA^{His} identity follows a different pathway. The G_{-1} residue is not encoded in the tRNA^{His} gene, but rather is added post-transcriptionally by Thg1 following RNase P cleavage (Figure 2A) [20,21]. Consistent with the necessity for the G_{-1} determinant of tRNA^{His} identity in protein fidelity, deletion or silencing of the gene encoding Thg1 results in severe growth defects in yeast, humans, and *Dictyostelium discoideum*, which is consistent with the wide conservation of Thg1 in Eukarya [21–23]. The addition of G_{-1} by Thg1 in eukaryotes occurs opposite of the conserved terminal A_{73} and is the result of a non-templated 3′-5′ addition [20,21]. A few exceptions to the necessity of G_{-1} have already been identified; a group of α-proteobacteria and several protozoan eukaryotes (*Acanthamoeba castellanii* and *Trypanosoma brucei)* do not genomically encode or post-transcriptionally add the G_{-1} identity element; the absence of the otherwise highly conserved extra nucleotide is accommodated by an atypical HisRS recognition of tRNA^{His} (Figure 2, right panel) [24–27].

2.2. Thg1 Structurally Resembles Canonical 5′ to 3′ Polymerases

Human Thg1 (HsThg1) is encoded by 269 amino acids with a calculated molecular weight of 32 kDa, but purification by gel exclusion chromatography eluted a protein of a molecular weight of ~165 kDa. This suggests the formation of a higher order multimer in solution, and is consistent with existing determinations of dimer-of-dimer tetrameric forms of active Thg1 enzymes [4,5,28–30]. Despite Thg1 enzymes sharing no obvious sequence similarity to known polymerases, a surprising structural homology exists between Thg1 and T7 DNA polymerase and DNA polymerase II (family pol I and pol α, respectively), as the enzymes share the same catalytic palm domain (Figure 3) [4,5]. The superposition of Thg1 with the aforementioned polymerases displays similar positioning of the three conserved carboxylate residues in the polymerase active site to the pol I family; HsThg1 carboxylates D29, D76, and E77 correspond to T7 DNA polymerases D475, D654, and E655. This similar carboxylate positioning suggests that the Thg1 mechanism of reverse polymerization may share features with the forward polymerization of the pol I family. Moreover, in the context of the overall structure, each monomeric subunit resembles a polymerase "hand" shape with the palm and fingers domains. This discovery revealed that forward and reverse polymerization can be accommodated by the same catalytic palm domain [4,5].

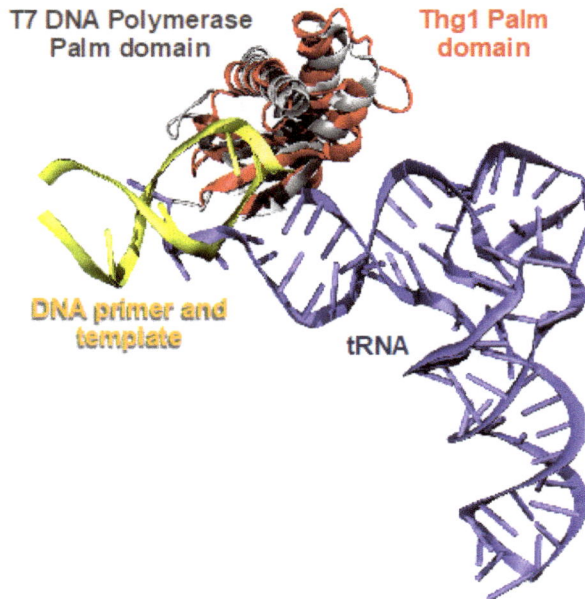

Figure 3. Forward and reverse polymerization is catalyzed by structurally similar enzymes but require opposing substrate orientation. The tRNA substrate (blue) of Thg1 (red) is in an opposing orientation compared to the DNA substrate and template (yellow) approaching the T7 DNA polymerase palm domain (grey).

Crystal structures of eukaryotic Thg1 enzymes showed that monomeric Thg1 is composed of a six-strand antiparallel β-sheet flanked by three or four α-helices on each side, along with a protruding long arm composed of two antiparallel β-strands (Figure 3). Each monomer forms a dimer, mediated by salt bridge formation and hydrogen bonding between the β-sheet and an α-helix [4,5]. Alanine scanning mutagenesis in *Saccharomyces cerevisiae* Thg1 (ScThg1) of conserved residues responsible for mediating dimer formation has shown strongly diminished G_{-1} addition activity [4,31]. Dimer-of-dimer formation stems from initial dimer formation; the α-helix of the N-terminus in monomer A interacts with the nucleotide binding site of monomer B and ultimately forms an intertwined N-terminus segment, which again interacts with a symmetrical dimer to form its active tetrameric form [4,5]. The crystal structures of Thg1 from *Candida albicans* and *S. cerevisiae* corroborate similar findings in protein folding and quaternary structure, which is suggestive that Thg1 variants fold similarly, and are active in tetrameric form [4,5,28,30]

The cocrystal structure of *C. albicans* Thg1 (CaThg1) in complex with tRNA subsequently provided insight into the mechanistic details of reverse polymerization. Compared to forward polymerases, such as the T7 DNA polymerase, the RNA substrate of Thg1 approaches the catalytic palm domain from the opposing site (Figure 3) [5]. Thus, the direction of polymerization is dependent on the direction of substrate approach to the catalytic core; Thg1 forces incoming nucleotides to approach from the opposite orientation of canonical forward polymerases, which is reflective of the overall domain organization of the enzyme (Figure 3). The fingers domain, which in part forms the nucleotide-binding site, is situated on opposite sides of the palm domain in Thg1, forcing a reversed substrate approach. Thus, the domain organization of Thg1 cannot accommodate forward polymerization, just as forward polymerases do not accommodate reverse polymerization [5]. Overall, these data revealed the molecular basis of reverse nucleotide addition, and showed that while reverse and forward polymerization can be catalyzed by the same catalytic core, substrate orientation may be the deciding factor in determining the directionality of nucleotide addition.

2.3. The Molecular Basis for tRNA Recognition

Thg1 recognizes $tRNA^{His}$ through its GUG anticodon, as demonstrated by the ability of Thg1 to add G_{-1} to a mutagenized $tRNA^{Phe}$ that has been altered to contain the His anticodon ($tRNA^{Phe}_{GUG}$) and subsequently validated through a cocrystal structure of *C. albicans* Thg1 in complex with $tRNA^{Phe}_{GUG}$. The coordination of tRNA molecules occurs in a molar ratio of 4:2, where two tRNA molecules bind a Thg1 tetramer in parallel orientation, and each tRNA is coordinated by three subunits of the tetramer [5]. Common identity elements of the coordinated tRNA interact with the Thg1 tetramer; the acceptor stem is coordinated by the intertwined N-terminus of a Thg1 dimer, and the T arm is situated near polar residues on the rear surface of the catalytic core of the same dimer. This surface interaction in tRNA stabilization is analogous to the thumb domain in canonical forward polymerases. However, the anticodon, which is another distinctive identity element, is coordinated by the opposite dimer, and the G_{-1} addition of this tRNA occurs solely in one dimer of the tetramer, specifically, the dimer that coordinates the acceptor stem and T arm positioning. The structural superposition of a crystal structure of ScThg1 and $CaThg1-tRNA^{Phe}_{GUG}$ discovered a conserved secondary structure in the fingers domain, and suggested conserved dual RNA-binding surfaces that were originally elucidated in CaThg1. Both the acceptor stem's sugar phosphate backbone and the anticodon loop are coordinated by the fingers domain; the latter fingers–anticodon interaction is base-specific [5].

Recognition of the $tRNA^{His}$ anticodon has been rationalized structurally through the $CaThg1:tRNA^{Phe}_{GUG}$ complex [5]. The G_{34}, U_{35}, and G_{36} nucleotides that make up the $tRNA^{His}$ anticodon were all observed to be coordinated by specific residues of CaThg1. All three anticodon bases are tightly coordinated, and mutations of the coordinating amino acids lead to a disruption or reduction in enzyme activity [4,5,31]. The anticodon loop structure itself is stabilized by interactions with the sugar phosphate backbone of U_{35}. Anticodon base G_{36} is coordinated in a groove formed by two α-helices flanking the central β-sheet and stabilized by a stacking interaction with H154, which is an essential residue in the eukaryotic-specific sequence motif HINNLYN [5,32]. The structure of Thg1 in complex with $tRNA^{Phe}_{GUG}$ suggests the molecular basis for the fairly stringent dependence of Thg1 on anticodon recognition, which is in agreement with its biological function of establishing $tRNA^{His}$ identity.

2.4. The Molecular Basis for Non-Watson–Crick G_{-1} Addition: tRNA Activation

The resemblance between canonical forward polymerase structures and Thg1 in the overall structure raises questions about how exactly Thg1 performs reverse polymerization. In forward polymerization, the catalytic core coordinates the two catalytic metal ions with the first Mg^{2+} (Mg^{2+} A) promoting deprotonation of the 3′-OH in the polymerizing DNA strand, and the second Mg^{2+} (Mg^{2+} B) coordinating the triphosphate moiety of the incoming nucleotide. By analogy, Mg^{2+} A and Mg^{2+} B in Thg1 family polymerases use similar chemical features to catalyze the first step of G_{-1} addition by activating the 5′-end of $tRNA^{His}$ with ATP (Figures 1 and 4). In this case, Mg^{2+} A would promote the nucleophilic attack of the tRNA 5′-phosphate on the α-phosphate of the activating ATP nucleotide, which is also bound by Mg^{2+} B. The exact mechanism of the activation step is not entirely known, but the GTP-bound crystal structures of HsThg1, ScThg1, and CaThg1 have elucidated binding interactions in the activation of the nucleotide-binding site and revealed a possible mechanism of how Thg1 differentiates between ATP and GTP in the initial activation of tRNA^{His} [4,5]. Nucleotidyl transfer in HsThg1 and ScThg1 is mediated largely by guanine base stacking against A37 and F42. The interaction is furthered by hydrogen bonding via amide, carbonyl, and the side chain moieties of D47, A43, and H34, respectively [4,5]. The superimposition of CaThg1–ATP and CaThg1–GTP cocrystal structures revealed that the nucleotide responsible for initial activation, ATP, resides deeper in the nucleotide-binding pocket than GTP, and interacts directly with D47 and K44. Mutational analysis of a ScThg1 D44A mutant greatly decreased the catalytic efficiency, suggesting that the D44 interaction is involved in G_{-1} addition activation [33].

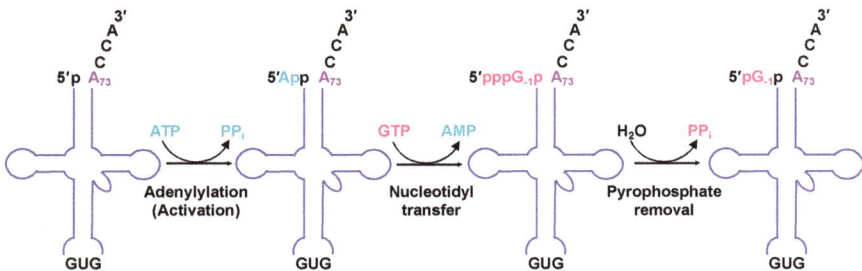

Figure 4. Eukaryotic Thg1 catalyzes a three-step reaction. ScThg1 adds G_{-1} in three consecutive steps. First, the 5′-end of a monophosphorylated tRNA^{His} (p-tRNA^{His}) is activated by an ATP to generate a 5′-adenylylated tRNA^{His} (App–tRNA^{His}) intermediate (adenylation activation). Second, the 3′-hydroxyl of an incoming GTP nucleotide attacks the 5′-end of the App–tRNA^{His}, which then releases AMP and adds the GTP to yield a triphosphorylated tRNA^{His} (pppG$_{-1}$p–tRNA^{His}) intermediate (nucleotidyl transfer). Third, pyrophosphate (PP_i) from the pppG$_{-1}$p–tRNA^{His} is released, creating a mature pG$_{-1}$p–tRNA^{His} (pyrophosphate removal).

2.5. The Molecular Basis for Non-Watson–Crick G_{-1} Addition: Nucleotidyl Transfer

Thg1 family enzymes have a distinct site (at least partially separable from the activation site) for the binding of the incoming nucleotide used for the nucleotidyl transfer step (Figures 1 and 4). This can be clearly seen in HsThg1–dGTP and CaThg1–GTP crystal structures [4,5]. This observation is further supported by the kinetic data showing distinct functions for highly conserved residues in ScThg1. ScThg1 residues R27, K96, and R133 play a more significant role during the nucleotidyl transfer step based on more significant defects in the observed rate of this reaction with alanine variants compared to the other steps in G_{-1} addition [34]. However, these three residues only interact with the triphosphate moiety of the partially visualized incoming dGTP nucleotide in the HsThg1–dGTP crystal structure and the completely visualized GTP in the CaThg1–GTP structure. Due to the lack of direct contacts with the base or ribose of the GTP bound to the nucleotidyl transfer site in the CaThg1–GTP structure, neither of these structures explain how Thg1 specifically positions or recognizes its highly preferred nucleotide, GTP, over any other NTP, as observed repeatedly in many in vitro assays. Thus, the precise mechanism of Thg1-catalyzed nucleotidyl transfer to create the non-Watson–Crick G_{-1}–A_{73} base pair step is yet to be determined.

2.6. The Molecular Basis for Non-Watson–Crick G_{-1} Addition: Pyrophosphate Removal

In the final step of the reaction catalyzed during tRNA^{His} maturation by Thg1, the enzyme removes the 5′-pyrophosphate from the added G_{-1} nucleotide, yielding the 5′-monophosphate that is required for recognition by HisRS (Figure 4) [17,21,33]. This reaction requires the same two metal ion active sites as used for the previous two steps of the reaction, since alanine mutations to the metal coordinating carboxylate residues completely eliminate the ability of the enzyme to catalyze this reaction in vitro [33]. However, additional residues that participate directly in the chemistry of this reaction have not yet been identified [33,35]. Interestingly, the removal of this pyrophosphate also removes an activated 5'-triphosphorylated end that could be extended further by the enzyme in the reverse polymerase reaction. Consistent with this idea, removal of the 5'-pyrophosphate by ScThg1 occurs much more efficiently in the context of the non-Watson–Crick base paired 5'-ends and 3'-ends of the tRNA than with a Watson–Crick base-paired end, thus preventing the further extension of this tRNA and limiting addition to the single essential G_{-1} nucleotide [36].

2.7. Maintenance of tRNA^{His} Fidelity

Maintaining tRNA identity is essential for translational fidelity, although a certain amount of mistranslation can be tolerated by the cell [12,37]. In eukaryotes, G_{-1} addition is essential to establish tRNA^{His} identity, and consequently, Thg1 is required for cell survival. In *S. cerevisiae*, the conditional depletion of Thg1 leads to the accumulation of unguanylated and uncharged tRNA^{His} and growth arrest [8]. Interestingly, tRNA^{His} in Thg1-depleted cells contains elevated levels of a 5-methylcytidine (m^5C) modification at residues C_{48} and C_{50} [38]. While the change in nucleoside methylation is specific to tRNA^{His}, it is most likely the result of the growth arrest rather than a direct consequence of G_{-1} depletion. Upon conditional Thg1 depletion, these cells are arrested in prophase or G2. Thg1 was also shown to interact with the Orc2 component of the origin recognition complex, and loss of this interaction impairs DNA replication and nuclear division, although the precise molecular basis for this apparent connection of Thg1 to DNA metabolism remains unknown [39]. Thus, not surprisingly, Thg1 activity is essential in eukaryotes such as *S. cerevisiae*, which depend on Thg1 to maintain tRNA^{His} identity.

3. Thg1-Like Proteins Function in tRNA Repair

3.1. Thg1-Like Proteins Are Functionally and Phylogenetically Distinct from Thg1

The biological function of Thg1 has been investigated in detail, but Thg1-like proteins (TLPs, which are also referred to as archaeal-type Thg1) only more recently gained attention. First identified in archaea, homologs have now been identified and characterized from several bacteria and eukaryotes. While related in sequence and structure, Thg1 and TLPs are phylogenetically and functionally distinct [32,40]. Where Thg1 is essential in many eukaryotes to establish tRNA^{His} identity through the post-transcriptional addition of G_{-1}, most bacteria and most archaea instead genomically encode the G_{-1} identity element (Figure 2) [14,18–20,41]. Interestingly, in vitro analysis of bacterial and archaeal Thg1 homologs demonstrated that they are capable of catalyzing an analogous addition of G_{-1} to bacterial/archaeal tRNA^{His} transcripts that lack the G_{-1} nucleotide, which would necessarily require the templated polymerization of G_{-1} opposite to the C_{73} that is universally found in these tRNA^{His} [23,40]. Using these different biochemical capabilities and distinct sequence features as a functional classification system, Thg1 superfamily enzymes can thus be classified into two groups. The first group entails the bona fide Thg1 homologs, which are found only in Eukarya, and post-transcriptionally establishes tRNA^{His} identity via G_{-1}. The second group includes TLP homologs that are found in all three domains of life, and have not yet been implicated biologically in tRNA^{His} identity, but for which this function would be in many cases redundant, if at all observed in vivo [23,32,40,42–44].

A phylogenetic analysis of candidate Thg1/TLP genes from all domains of life have grouped nearly all eukaryotic Thg1 variants into a single group; a second clade containing eukaryotic TLPs also contains archaeal and bacterial TLPs, which further supports the classification of Thg1 and TLPs as distinct enzymes [32,43,45]. A secondary trend further groups bacterial and archaeal Thg1 into two separate phylogenetic groups, and suggests that bacterial Thg1 was not descended from earlier bacterial ancestors [32,43,45].

3.2. The Discovery of TLPs in Bacteria and Archaea

The first characterization of a non-eukaryotic Thg1 homolog was carried out in *Methanosarcina acetivorans* [43]. Initially, the enzyme was annotated as two open reading frames split by a UAG Stop codon. While the enzyme activity of the split enzyme could be reconstituted in vitro, the protein is translated into a single protein in vivo, linking the two frames by a genetically encoded pyrrolysine at the UAG codon [43]. While in *M. acetivorans* the UAG codon could still be signaling for translation termination, it is unlikely that Thg1 is translated as two halves in vivo [46]. To date, Thg1 homologs belonging to the TLP clade have now been characterized biochemically from a diverse range of eight different bacterial and archaeal species. The common ability of all enzymes to catalyze strictly templated 3′ to 5′ addition reactions has been universally observed in assays of these enzymes with both tRNA^{His} and non-tRNA^{His} substrates [23,29,35,40,45,47]. The question of the precise nature of the biological substrates of these enzymes in their relevant organisms has not been entirely addressed, but some likely possibilities based on in vitro characterization so far include analogous tRNA^{His} maturation reactions to those employed by eukaryotic Thg1 (albeit in many cases redundant with co-transcriptional mechanisms of obtaining the G$_{-1}$ nucleotide) and/or alternative RNA repair reactions catalyzed on substrates such as 5′-truncated tRNAs that remain to be identified.

The idea that bacterial and archaeal homologs of Thg1 might participate in G$_{-1}$ addition to tRNA^{His} was an obvious extension of the known function of these enzymes in eukaryotes, and the initial in vitro characterization of several enzymes demonstrated activities that were consistent with this type of role [40,45,47]. It is notable that endogenous tRNA^{His} genes in these species universally encoding a C$_{73}$ nucleotide in place of the A$_{73}$ found in eukaryotes was also consistent with the known preference of these enzymes to catalyze Watson–Crick-dependent 3′ to 5′ addition reactions. Thus, it was not surprising to see that TLPs were not able to efficiently add G$_{-1}$ to wild-type A$_{73}$-containing eukaryotic tRNA^{His} substrates, either in vitro or in vivo [23,40,45,47]. Interestingly, the *Bacillus thuringiensis* TLP homolog (BtTLP) is unusual compared to several other tested enzymes of this family, in that it is able to support the growth of a yeast Δ*thg1* strain, presumably by acting in place of ScThg1 on tRNA^{His} [23,40,47]. Indeed, sequences of tRNA^{His} derived from the BtTLP-complemented *S. cerevisiae* Δ*thg1* strain reveal that the tRNA contains almost exclusively U$_{-1}$, instead of the canonical G$_{-1}$ residue [48]. In this case, the BtTLP evidently maintains its preference for incorporating Watson–Crick base-paired nucleotides by incorporating U$_{-1}$ across from the A$_{73}$ discriminator nucleotide. Although the minor growth defect of the BtTLP-expressing strains may then be attributed to the presence of the non-canonical U$_{-1}$ nucleotide on tRNA^{His}, the ability of *S. cerevisiae* HisRS to accept other N$_{-1}$-containing tRNA^{His} substrates in vitro with only a fivefold to sixfold loss in efficiency is consistent with the observed viability of the BtTLP-expressing strain.

3.3. TLPs Catalyze Multiple Nucleotide Additions

Interestingly, TLP homologs differ in terms of the number of nucleotide additions that are observed in vitro with tRNA transcripts mimicking the mature tRNA^{His} that could be a substrate for this kind of activity in vivo. The ability to add multiple 5′-nucleotides to these types of tRNA^{His} is a direct consequence of the presence of a C$_{73}$ discriminator, which results in the formation of three consecutive C-nucleotides in the 3′-C$_{73}$CCA end. These types of substrates were initially shown to cause multiple G-addition reactions even in the context of eukaryotic Thg1, and several bacterial/archaeal homologs (the TLPs from *Methanopyrus kandleri*, *Pyrobaculum aerophilum*, and *B. thuringiensis*) behave similarly, catalyzing multiple G-additions [9,23,35,40,47]. However, other homologs (such as from *Myxococcus xanthus*, *Methanothermobacter thermoautotrophicus*, *Methanosarcina barkeri*, and *M. acetivorans*) appear to limit nucleotide addition to only a single guanine addition, despite the presence of an extended C-template at the tRNA 3′-end [40]. The molecular basis for these distinct behaviors is not yet known, but suggests that not all homologs in Bacteria and Archaea would react similarly in terms of any possible role in the maturation or maintenance of tRNA^{His} [9,23,35,40,47].

The observation of alternative RNA substrate specificity exhibited by many enzymes in the TLP clade also suggested the possibility of non-tRNA^{His} related functions for these enzymes. The identification of certain biochemical properties that distinguish TLPs from bona fide Thg1 counterparts lent support to these ideas. In addition to the previously described ability of TLPs to efficiently catalyze Watson–Crick base pair-dependent addition reactions, all of the bacterial and archaeal homologs investigated to date exhibit the preferential repair of 5′-truncated tRNAs instead of 5′ nucleotide addition to full-length tRNA^{His} [23,29,35]. This tRNA repair activity is not species-specific, as opposed to G$_{-1}$ addition in tRNA^{His} identity establishment by bona fide Thg1 [23,26,31,42]. *M. acetivorans* TLP (MaTLP) has also demonstrated similar 5′-truncated tRNA repair as previously described with BtTLP; with preference to additions forming Watson–Crick base pairs [23,29,40]. Similarly, *Ignicoccus hospitalis* catalyzes an extended tRNA repair on truncated RNA substrates in vitro, adding up to 13 nucleotides in a templated reaction to restore a full tRNA^{His} [35]. These data indicate that TLP function may indeed be found in tRNA repair rather than G$_{-1}$ addition.

3.4. TLPs in Eukaryotes: Multiple TLPs Encoded by Dictyostelium Discoideum

Analysis of the eukaryotic slime mold, *Dictyostelium discoideum*, revealed that at least four Thg1 family enzymes appear to be present in the proteome (Figure 5). One of these genes is a bona fide Thg1 (DdiThg1), catalyzing the G$_{-1}$ addition to tRNA^{His}, and thus establishing tRNA^{His} identity [42,44]. The remaining three genes are characterized by sequence similarity and phylogeny as TLPs. Two of these TLPs (DdiTLP2 and DdiTLP3) are mitochondrial enzymes that catalyze distinct and non-redundant functions to add G$_{-1}$ to mitochondrial tRNA^{His} (DdiTLP2) or to repair the 5′-end of mitochondrial tRNA (mt-tRNA) during a process known as tRNA 5′-editing (DdiTLP3) [44]. Both of these reactions take advantage of the 3′ to 5′ polymerase function that is the biochemically-preferred activity of TLP enzymes.

Although DdiTLP2 catalyzes a reaction that is on the surface very similar to that of a bona fide Thg1, there are several critical features that distinguish these two activities, which is consistent with the evolutionary distinct nature of these two enzymes. First, DdiTLP2 does not use the tRNA^{His} GUG anticodon to recognize the tRNA for the addition of the G$_{-1}$ nucleotide, and its specific mechanism of tRNA^{His} recognition remains unknown [44]. Second, the reaction catalyzed by DdiTLP2 is not essential for specifying tRNA^{His} identity, since *D. discoideum* Δ*dditlp2* deletion strains are viable, but completely lack the G$_{-1}$ nucleotide on the mt-tRNA^{His} [44]. DdiTLP3's role in tRNA 5′-editing also utilizes 3′ to 5′ polymerase function, but to repair mt-tRNA that have been truncated at their 5′-ends due to the removal of one or more incorrectly base-paired nucleotides encoded in the precursor tRNA [49]. This 5′-end repair step is essential for *D. discoideum*, and likely for many other single-celled eukaryotes that similarly encode mt-tRNA with 5′-mismatches [34,47,49–54]. Presumably, the TLPs encoded by these species are capable of participating in 5′-editing, although the identity of specific enzymes that participate in this process has so far only been demonstrated in *D. discoideum* [32,33,41–44]. Interestingly, the third TLP encoded in *D. discoideum* (DdiTLP4) catalyzes an essential function that remains unknown (Figure 5), although its ability to catalyze 3′ to 5′ polymerase activity on non-tRNA RNA substrates broadens the potential impact of these enzymes in terms of RNA processing and/or repair [44]. The continued functional characterization of TLP enzymes from diverse domains of life will be needed to provide a comprehensive picture of all of the biological reactions associated with this unusual family of enzymes (Figure 5).

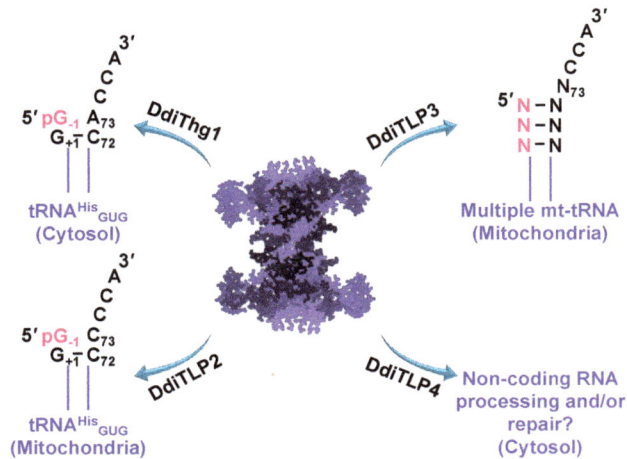

Figure 5. Non-redundant physiological roles for Thg1 and Thg1-like proteins (TLPs) in *Dictyostelium discoideum* tRNA 5′-editing. DdiThg1 catalyzes cytosolic tRNA^{His} maturation by incorporating G_{-1} across from A_{73}. DdiTLP2 catalyzes mitochondrial tRNA^{His} maturation by incorporating G_{-1} across from C_{73}. DdiTLP3 catalyzes mitochondrial tRNA 5′-editing by repairing 5′-truncated tRNAs resulting from the removal of one or more mismatched nucleotides encoded in the precursor tRNAs. DdiTLP4 function remains unknown to date, but potentially this essential cytosolic enzyme is involved in non-coding RNA processing and/or repair in *D. discoideum*.

3.5. Structural Comparison of Thg1 and TLPs

The crystal structures of several members of the TLP clades of the Thg1 superfamily have been solved, and suggest many commonalities with Thg1, including an overall structural similarity to the *C. albicans* and *Homo sapiens* enzymes and the persistence of dimer-of-dimer quaternary structures [28, 29]. MaTLP assembles as a dimer-of-dimers similar to bona fide Thg1, yet uses a different mechanism of tRNA coordination. CaThg1 coordinates tRNA between both dimers of the tetramer, and binds Thg1–tRNA via 4:2 stoichiometry [54]. MaTLP independently binds one tRNA molecule per dimer, and does not seem to coordinate anticodon recognition by the opposing dimer, which is consistent with the notion of TLP repair activity (Figure 6) [23,40]. The present structures of the MaTLP and bona fide Thg1 variants differ most in tRNA coordination; the anticodon loop is not coordinated by the opposing dimer's fingers domain in TLPs [5,29]. MaTLP binds only the acceptor stem and T arm of its tRNA substrate; the flexible β-hairpin that coordinates the T arm has been speculated to enable the recognition of tRNA substrates with truncated acceptor stems [29]. The tRNA acceptor stem and T arm are coordinated by separate monomers in the dimer; the acceptor stem was hydrogen bonded to R136 and D137, and the triphosphate moiety at the 5′-end of the tRNA was bonded to the D21–K26 region and additionally coordinated by two Mg^{2+}, which is similar to tRNA coordination by CaThg1 [5,29]. Expectedly, divalent cations of the MaTLP active site are coordinated by the carboxylates of D21 and D69, which is similar to previously reviewed bona fide Thg1 structures, other polymerases, and BtTLP. The T arm is recognized by the opposing monomer in the MaTLP dimer; phosphate groups of U_{55} and G_{57} are involved in hydrogen-bond interactions with the protruding long arm or β-hairpin region of MaTLP [29]. Computational analysis and model structures of MaTLP–tRNA complexes have shown truncated but not full-length tRNA molecules binding to MaTLP, suggesting that TLP molecules recognize 5′-truncated tRNA molecules via the flexible β-hairpin, and terminate elongation by recognizing the acceptor stem length of its substrates [29].

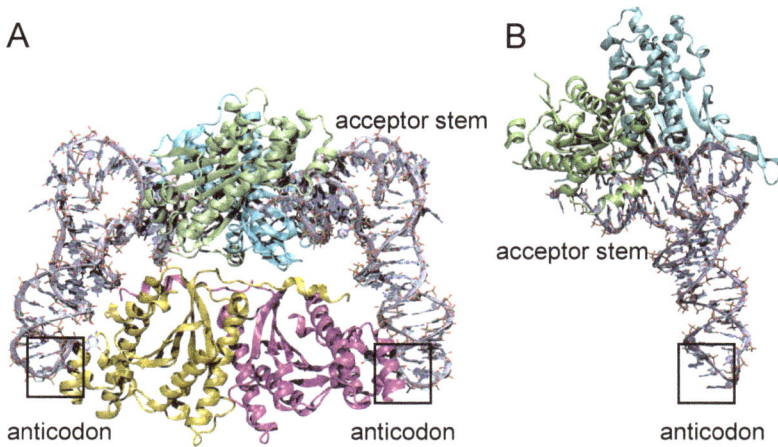

Figure 6. Structural comparison of Thg1 and TLPs. (**A**) The tetrameric eukaryotic Thg1 (monomers in green, teal, yellow and pink) coordinates two tRNAs molecules (grey) and binds both to the anticodon and the acceptor stem. (**B**) *Methanosarcina acetivorans* TLP forms a dimer (monomers in green and teal) coordinating on tRNA (grey). While the dimer binds the acceptor stem and T arm of its tRNA substrate, the anticodon loop is not coordinated by TLP enzymes.

The archaeon *I. hospitalis* encodes a minimalized TLP homolog (IhTLP) with a molecular weight of 18 kDa, which is smaller than its archaeal cousin PaTLP (25 kDa) and the bona fide Thg1 from *C. alibcans* (35 kDa). IhTLP encodes the three conserved carboxylates that Thg1 and other polymerases require to coordinate two metal ions for catalysis, but shares little overall homology to other TLP enzymes. Furthermore, the fingers domain, which are responsible for the anticodon loop and acceptor stem coordination, is significantly minimalized in IhTLP. IhTLP was found to be catalytically active in vitro, and catalyzes a significant tRNA repair reaction in vitro, adding up to 13 nucleotides to restore a truncated tRNA^{His}, yet the in vivo function of IhTLP remains to be investigated [35]. While this mode of recognition is suitable for rationalizing TLP roles in repairing 5′-truncated tRNA, the prospect of alternative non-tRNA substrates for at least some members of this enzyme family raises additional questions about how this mechanism could be adapted more broadly to control 3′ to 5′ polymerase reactions.

3.6. 5′-End Activation and Nucleotidyl Transfer in TLPs

Many features of the overall mechanism of 3′ to 5′ nucleotide addition appear to be shared between Thg1 and TLP members of the enzyme family. Distinct residues that participate in 5′-end activation (*ScThg1*: K44, S75, N161) and nucleotidyl transfer (*ScThg1*: R27, K96, R133) are found in similar positions in the BtTLP structure, except for the N161 residue, which is absent in TLPs. The differing effects on distinct catalytic steps of the reaction that are observed upon the mutagenesis of these residues led to the notion of partially separable active sites for 5′-end activation and nucleotidyl transfer, which appears to also be the case for TLPs [4,5,28–30,33]. However, several molecular details distinguish these two types of enzymes. In terms of the first 5′-end activation step of the reaction, BtTLP and the TLP from *M. xanthus* (MxTLP) can utilize GTP to activate the 5′-end of tRNA^{His} in vitro, whereas bona fide Thg1 enzymes are so far strictly ATP-dependent for catalyzing this reaction [6,23,28,40]. The crystal structures of TLP enzymes captured in the pre-activation conformation with either ATP or GTP nucleotides bound in the activation nucleotide-binding site have been determined, and confirm the roles for the residues implicated kinetically in the 5′-end activation reaction [28]. For the conserved lysine (K44 in ScThg1, K43 in BtTLP), this residue appears to be important specifically for the activation reaction with ATP, and not used for the GTP activation reaction, since the alteration of this residue in the context of the bifunctional BtTLP only affected 5′-adenylylation rates. Structural data support this ATP-specific role, since the side chain of K43 would predictably clash with the exocyclic 2′-amine of GTP bound in the active site. It is possible that ScThg1 K44 interacts with other yet-unidentified residues in the ScThg1 active site to ensure its ATP-dependent activity by preventing GTP binding in a catalytically competent conformation at this site.

Insight into the template-dependent nucleotidyl transfer step catalyzed by TLPs was provided by the structure of MaTLP bound to a GTP analog and a 5′-truncated triphosphorylated tRNA^{Phe} (ppptRNA$^{Phe}\Delta_1$), which mimics the activated tRNA ready for the nucleotidyl transfer. The incoming nucleotide uses Watson–Crick base-pairing and base-stacking interactions to facilitate its incorporation to the tRNA substrate [28]. In addition, binding of the incoming nucleotide seems to promote a significant structural change in the 5′-end of the tRNA substrate to accommodate the nucleotide incorporation across from the templating nucleotide. Also, no interactions between the enzyme and the base moiety of the GTP analog were observed here, supporting the idea of template-dependency exhibited by TLPs for nucleotide addition to their substrates, which is presumably distinct from the expected base-recognition capabilities for Thg1 homologs to catalyze template-independent GTP addition. In TLPs, coordination of the Mg^{2+} A to the 3′-OH of incoming NTP facilitates the nucleophilic attack on the α-phosphate of the activated tRNA substrate, which is analogous to the role of this residue in canonical polymerases. Meanwhile, the triphosphate of the incoming NTP coordinates to a third metal ion (Mg^{2+} C). It is arguable that the Mg^{2+} C seen in the active site of Thg1/TLP structures stabilizes the incoming NTP in the absence of metal coordination from other two Mg^{2+} ions (Mg^{2+} A and Mg^{2+} B), which are already coordinated to either the activating ATP that resides in the activation site or the activated tRNA substrate. Appearance of a third metal ion has also been observed in DNA pol β and DNA pol η [55,56]. Although these polymerases use the canonical two-metal ion mechanism, the third metal ion has been proposed to facilitate phosphoryl transfer to the nucleophilic 3′-OH group of the growing polynucleotide chain. Similarly, the use of a third Mg^{2+} ion (Mg^{2+} C) in TLPs seems to facilitate nucleophilic attack on the α-phosphate of the activated tRNA to promote PP$_i$ release upon subsequent nucleotidyl transfer.

4. Synthetic Biology Applications and Thg1/TLP Engineering

How TLP enzymes distinguish between truncated tRNA and full-length tRNA in the context of 5′-tRNA elongation or repair and how the elongation reaction is properly terminated in all cases remains to be fully solved. The proper termination of Thg1 enzymes after adding only the single G nucleotide to tRNA^{His} is intimately connected with the acquisition of the A$_{73}$ discriminator nucleotide of tRNA^{His} in eukaryotes, as opposed to the universally conserved (and likely ancestral) C$_{73}$ that is found in Bacteria and Archaea. Indeed, the molecular basis for the ability of the G$_{-1}$:A$_{73}$ base pair to trigger the termination of the addition in ScThg1 is demonstrated to result from ScThg1's highly efficient removal of the 5′-triphosphorylated end from tRNA species that terminate in this non-Watson–Crick base pair, as described above [36]. This mechanism for termination is unlikely to be conserved among TLPs, since these enzymes strictly add Watson–Crick base pairs that are not efficient substrates for the 5′-pyrophosphatase activity [36]. The previously demonstrated capability of TLPs to catalyze extended reverse RNA polymerization [23,29,35,40] makes these proteins promising candidates for protein engineering to harness their ability to extend RNAs in the 3′ to 5′ direction in a template-dependent manner.

4.1. TLPs Exhibit Broader RNA Recognition Properties Than Thg1 Homologs

The ability to act on a broader range of tRNA substrates than the tRNA^{His}-specific Thg1 homologs naturally positions TLP enzymes as candidates for engineering 3′ to 5′ polymerases for diverse functions. Understanding the basis for RNA recognition by various members of the Thg1 superfamily is a key element to these efforts. While recognition of the anticodon GUG is a major determinant for

Thg1-catalyzed G_{-1} addition to tRNA^{His}, the TLPs that have been tested to date do not depend on the presence of this sequence to be used as substrates for 3' to 5' polymerization [10,23]. This ability to act independently of the specific tRNA^{His} anticodon nucleotides is a necessary feature for TLPs to act outside of tRNA^{His} maturation, as has already been demonstrated by the physiological function of DdiTLP3 in mt-tRNA 5'-editing [42,44]. For this 5'-editing reaction, the ability to act on any tRNA is required, since many different tRNA species contain the genomically encoded 5'-mismatches that must be removed by 5'-editing and then repaired by the TLP prior to participation in translation [34,49,51–53]. Indeed, the pattern of tRNA species that have been demonstrated to undergo editing in the many diverse eukaryotes where this process has been investigated so far is quite different from organism to organism, and can involve from only a few species to nearly all of the mt-tRNA, thus suggesting that flexibility in tRNA recognition is important for this function [50]. The idea that some TLPs may act on other non-tRNA substrates also raises new questions for these enzymes and their ability to recognize various RNAs. Thus, TLP activity is not restricted to tRNA^{His} substrates, and may be engineered to further broaden their substrate specificity.

4.2. Steps Toward Utilizing TLPs for Targeted 3'-5' Addition Reactions

Although eukaryotic Thg1 is mostly restricted to acting on tRNA^{His}, bacterial and archaeal TLPs have been shown to exhibit broader RNA substrate specificity, suggesting that the 3' to 5' polymerase activity of TLPs could potentially be directed to site-specifically label diverse RNA substrates at their 5'-ends. TLPs are capable of adding all four nucleotides in a template-dependent fashion [23,29,35,40], which makes them true reverse polymerases. This could in principle also be extended to add labeled or modified nucleotides at specific positions of a given RNA substrate. This approach has previously been successful in 3'-biotinylation by the terminal nucleotidyltransferase Cid1 [38]. Other applications that we can envision include reverse polymerization on additional substrates, which would allow for applications such as the site-specific incorporation of nucleotides to the 5'-ends of tRNA transcripts that are difficult to incorporate during in vitro synthesis due to the limitations of the commonly used T7 RNA polymerase. Here, we describe several approaches that have been used successfully for this type of application.

One critical step in Thg1/TLP engineering is to broaden substrate specificity toward RNAs other than tRNA. While the tRNA^{His} GUG anticodon is not required for TLP activity, recognition of the overall tertiary tRNA structure is a conserved element for Thg1 and TLPs. Desai et al. developed a split tRNA approach, which was later successfully implemented by Nakamura et al., dividing the tRNA through the D-loop. Here, the tRNA structure is mostly provided by a guide RNA, complementary to the RNA of interest to be guanylylated [35,54]. The guided RNA approach has been successful to lead RNaseP to mRNA substrates and cleave pathogenic RNAs in cancer [57]. Thus, this approach could be used to direct Thg1 to any given RNA 5'-end when provided with a complementary sequence, resulting in a tRNA-like structure.

An alternative approach provides an external RNA-guide template, which is used to anneal to the full-length tRNA, ideally disrupting the structure in the aminoacyl acceptor and D-stems, and thus providing a template to add the desired nucleotide(s) to the tRNA. To demonstrate the feasibility of this approach, we exploited the known inability of TLP enzymes to efficiently add a non-Watson–Crick base-paired G_{-1} to wild-type tRNA^{His} with an A_{73} discriminator nucleotide [23,47]. Then, by adding an additional RNA oligonucleotide that is complementary to the tRNA sequence, we provided an alternative template that could possibly direct the addition of a nucleotide(s) to the tRNA 5'-end in the Watson–Crick-dependent manner that is preferred by TLPs. We reasoned that if TLP enzymes are capable of accommodating an external oligonucleotide template, 3'-5' polymerase activity could potentially be directed to add nucleotides to the 5'-end of any given RNA substrate [49].

To test this, we used labeled in vitro transcribed *S. cerevisiae* tRNA^{His} (Sc-tRNA^{His}) (Figure 7A). In the absence of any added oligonucleotide, ScThg1 adds G_{-1}, forming a non-Watson–Crick base pair, as evident from the $G_{-1}p*GpC$ product that is observed after G-addition (Figure 7B). BtTLP only weakly adds G_{-1}; however, as observed previously, it instead accumulates 5'-adenylylated and 5'-guanylylated intermediates that are the products of the first activation reaction (Figure 7B). Then, varied concentrations (1 μM and 10 μM) of an externally provided 38-mer oligonucleotide template that is complementary to 31 nucleotides on the 5'-end of Sc-tRNA^{His} were added to the assays. When this oligonucleotide anneals to Sc-tRNA^{His}, it will create a 7-nucleotide 3'-overhang that can serve as a template for nucleotide addition to the 5'-end of the Sc-tRNA^{His} (Figure 7A). Here, the 3'-CCCCCCA overhang has been chosen due to the preferential G–C base pairing activity exhibited by TLP enzymes.

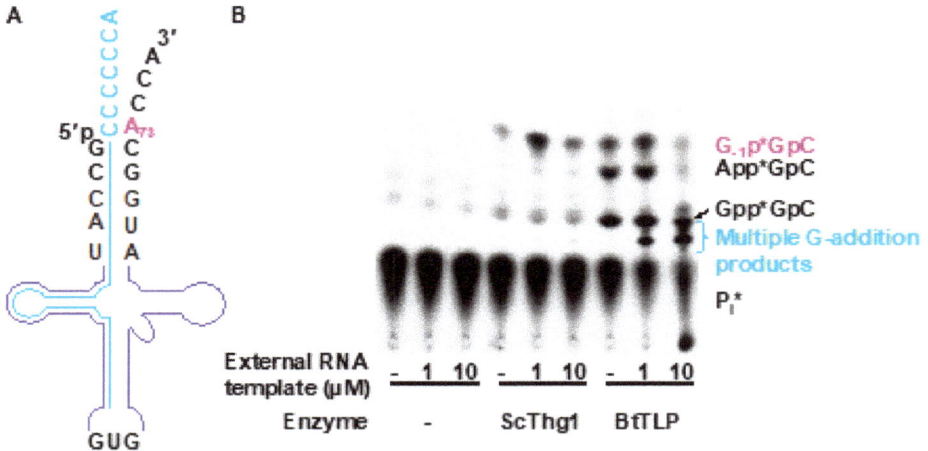

Figure 7. An external guide RNA oligonucleotide-mediated approach to manipulate nucleotide addition by *Bacillus thuringiensis* TLP homolog (BtTLP). (**A**) Schematic representation of tRNA^{His} (backbone in purple) hybridized to the external guide RNA oligonucleotide template (shown in blue). (**B**) Activity assay using the external guide RNA template and analyzed by thin-layer chromatography (TLC). Identities of the labeled reaction products are consistent with the known migration patterns of these species that have been validated by previous assays [23]. Nucleotide addition was tested using *Saccharomyces cerevisiae* Thg1 (ScThg1) and BtTLP with varying concentrations of the external guide RNA template (0 μM, 1 μM, and 10 μM, as indicated). In the absence of the guide RNA template, ScThg1 efficiently adds G_{-1} and in contrast, BtTLP only weakly adds G_{-1} across from A$_{73}$ (evident from the amount of the product G_{-1}p*GpC, which is labeled in purple). BtTLP instead accumulates 5'-adenylylated and 5'-guanylylated intermediates (evident from the products App*GpC and Gpp*GpC, which are labeled in black respectively). In the presence of the external guide RNA template, ScThg1 still maintains its ability to add a single G nucleotide across from A$_{73}$ in *S. cerevisiae* tRNA^{His} (Sc-tRNA^{His}, as evident from the product G_{-1}p*GpC shown in purple), whereas BtTLP adds multiple G nucleotides using the 3'-CCCCCCA of the external guide RNA oligonucleotide as a template. This multiple addition is evident from the appearance of several lower migrating products, which are labeled in blue, and are absent in ScThg1-containing reactions. The precise number of G-residues added to the tRNA cannot be determined due to the inability of this TLC system to resolve the distinct species that contain more than three added G-nucleotides [9,23,44], but the pattern of migration observed here is consistent with the migration described previously for these species.

Interestingly, the addition of the external templating oligonucleotide did not appreciably affect the products observed in the reactions with ScThg1. Only a single G_{-1} residue was still observed, suggesting that ScThg1 is not using the multiple C-containing template that would result in multiple nucleotide addition, as previously observed for variant tRNA with this 3'-end sequence [9]. Instead, ScThg1 must be utilizing the native Sc-tRNA^{His} 3'-end instead of the oligonucleotide-bound structure as a template. This is possibly due to changes in the overall shape of tRNA^{His} when bound to the oligonucleotide, which underscores this enzyme's stricter dependence on tRNA^{His} structure that has been demonstrated repeatedly, and also the unsuitability of Thg1 members of the 3 to 5′ polymerase family for this type of engineering approach.

In contrast, the addition of the external templating oligonucleotide to reactions with BtTLP results in significant changes in the observed reaction products. BtTLP efficiently created lower migrating products that have previously been identified as corresponding to multiple G-nucleotide additions [31], as expected for the use of the 3′-CCCCCCA sequence as a template. This observation provides proof for the first time that TLPs are capable of accommodating an externally provided oligonucleotide, and also implies that the 3′-5′ polymerase activity of TLPs can be guided to label an RNA substrate with a template of interest. Appearance of the products corresponding to multiple G-additions in the assays containing BtTLP supports the idea that BtTLP is an enzyme with more flexibility in accommodating an unnatural template and most importantly, BtTLP is capable of changing its nucleotide addition pattern to match the template that is available. The replacement of the tRNA^{His} acceptor stem by the external oligonucleotide alone was sufficient to guide BtTLP to change its nucleotide addition preferences as opposed to Thg1.

5. Conclusions

The tRNA^{His} guanylyltransferase family comprises a fascinating group of enzymes with novel catalytic activities. From the initial identification of these enzymes as the catalysts of adding a somewhat simple, albeit critical, single nucleotide to the 5′-ends of tRNA^{His} species, the mechanistic and biological complexity that is associated with the members of this family has grown significantly. New surprises about the functions of these enzymes, both engineered and on natural RNA substrates, are sure to continue to emerge, and will provide new opportunities to take advantage of 3′ to 5′ polymerization in the future.

Author Contributions: A.W.C., M.I.J., I.U.H., and J.E.J wrote the manuscript; M.I.J., C.Z.C., B.S.R., R.K., I.U.H., and J.E.J. prepared figures; B.S.R. carried out experiments; C.Z.C., M.I.J., I.U.H., and J.E.J. edited the manuscript.

Funding: This research was funded by Natural Sciences and Engineering Research Council of Canada grant number [RGPIN 04776-2014 to I.U.H.] and the J.P. Bickell Foundation (to I.U.H.), and the National Institutes of Health R01 GM087543 (to J.E.J.).

Acknowledgments: We are thankful to Patrick O'Donoghue for careful reading of the manuscript.

Conflicts of Interest: The authors declare no conflict of interest.

References

1. Joyce, C.M.; Steitz, T.A. Polymerase structures and function: Variations on a theme? *J. Bacteriol.* **1995**, *177*, 6321–6329. [CrossRef] [PubMed]
2. Lehman, I.R.; Richardson, C.C. The deoxyribonucleases of *Escherichia coli*. Iv. An exonuclease activity present in purified preparations of deoxyribonucleic acid polymerase. *J. Biol. Chem.* **1964**, *239*, 233–241. [PubMed]
3. Uptain, S.M.; Kane, C.M.; Chamberlin, M.J. Basic mechanisms of transcript elongation and its regulation. *Annu. Rev. Biochem.* **1997**, *66*, 117–172. [CrossRef] [PubMed]
4. Hyde, S.J.; Eckenroth, B.E.; Smith, B.A.; Eberley, W.A.; Heintz, N.H.; Jackman, J.E.; Doublie, S. tRNA^{His} guanylyltransferase (Thg1), a unique 3′-5′ nucleotidyl transferase, shares unexpected structural homology with canonical 5′-3′ DNA polymerases. *Proc. Natl. Acad. Sci. USA* **2010**, *107*, 20305–20310. [CrossRef] [PubMed]
5. Nakamura, A.; Nemoto, T.; Heinemann, I.U.; Yamashita, K.; Sonoda, T.; Komoda, K.; Tanaka, I.; Söll, D.; Yao, M. Structural basis of reverse nucleotide polymerization. *Proc. Natl. Acad. Sci. USA* **2013**, *110*, 20970–20975. [CrossRef]
6. Jahn, D.; Pande, S. Histidine tRNA guanylyltransferase from *Saccharomyces cerevisiae*. Ii. Catalytic mechanism. *J. Biol. Chem.* **1991**, *266*, 22832–22836.

7. Pande, S.; Jahn, D.; Söll, D. Histidine tRNA guanylyltransferase from *Saccharomyces cerevisiae*. I. Purification and physical properties. *J. Biol. Chem.* **1991**, *266*, 22826–22831. [PubMed]

8. Gu, W.; Hurto, R.L.; Hopper, A.K.; Grayhack, E.J.; Phizicky, E.M. Depletion of *Saccharomyces cerevisiae* tRNA^{His} guanylyltransferase Thg1p leads to uncharged tRNA^{His} with additional m(5)c. *Mol. Cell Biol.* **2005**, *25*, 8191–8201. [CrossRef] [PubMed]

9. Jackman, J.E.; Phizicky, E.M. tRNA^{His} guanylyltransferase catalyzes a $3'$-$5'$ polymerization reaction that is distinct from G$_{-1}$ addition. *Proc. Natl. Acad. Sci. USA* **2006**, *103*, 8640–8645. [CrossRef]

10. Jackman, J.E.; Phizicky, E.M. tRNA^{His} guanylyltransferase adds G$_{-1}$ to the $5'$ end of trnahis by recognition of the anticodon, one of several features unexpectedly shared with tRNA synthetases. *RNA* **2006**, *12*, 1007–1014. [CrossRef]

11. Ingle, R.A. Histidine biosynthesis. *Arabidopsis Book* **2011**, *9*, e0141. [CrossRef] [PubMed]

12. Lant, J.T.; Berg, M.D.; Heinemann, I.U.; Brandl, C.J.; O'Donoghue, P. Pathways to disease from natural variations in human cytoplasmic tRNAs. *J. Biol. Chem.* **2019**. [CrossRef]

13. Liao, S.M.; Du, Q.S.; Meng, J.Z.; Pang, Z.W.; Huang, R.B. The multiple roles of histidine in protein interactions. *Chem. Cent. J.* **2013**, *7*, 44. [CrossRef] [PubMed]

14. Sprinzl, M.; Vassilenko, K.S. Compilation of trna sequences and sequences of tRNA genes. *Nucleic Acids Res.* **2005**, *33*, D139–D140. [CrossRef]

15. Connolly, S.A.; Rosen, A.E.; Musier-Forsyth, K.; Francklyn, C.S. G-1:C73 recognition by an arginine cluster in the active site of *Escherichia coli* histidyl-tRNA synthetase. *Biochemistry* **2004**, *43*, 962–969. [CrossRef] [PubMed]

16. Nameki, N.; Asahara, H.; Shimizu, M.; Okada, N.; Himeno, H. Identity elements of *Saccharomyces cerevisiae* tRNA^{His}. *Nucleic Acids Res.* **1995**, *23*, 389–394. [CrossRef]

17. Rosen, A.E.; Musier-Forsyth, K. Recognition of G$_{-1}$:C$_{73}$ atomic groups by *Escherichia coli* histidyl-tRNA synthetase. *J. Am. Chem. Soc.* **2004**, *126*, 64–65. [CrossRef]

18. Burkard, U.; Willis, I.; Söll, D. Processing of histidine transfer RNA precursors. Abnormal cleavage site for RNase P. *J. Biol. Chem.* **1988**, *263*, 2447–2451. [PubMed]

19. Orellana, O.; Cooley, L.; Söll, D. The additional guanylate at the $5'$ terminus of *Escherichia coli* tRNA^{His} is the result of unusual processing by RNase P. *Mol. Cell Biol.* **1986**, *6*, 525–529. [CrossRef]

20. Cooley, L.; Appel, B.; Söll, D. Post-transcriptional nucleotide addition is responsible for the formation of the $5'$ terminus of histidine tRNA. *Proc. Natl. Acad. Sci. USA* **1982**, *79*, 6475–6479. [CrossRef]

21. Gu, W.; Jackman, J.E.; Lohan, A.J.; Gray, M.W.; Phizicky, E.M. tRNA^{His} maturation: An essential yeast protein catalyzes addition of a guanine nucleotide to the $5'$ end of tRNA^{His}. *Genes Dev.* **2003**, *17*, 2889–2901. [CrossRef] [PubMed]

22. Guo, D.; Hu, K.; Lei, Y.; Wang, Y.; Ma, T.; He, D. Identification and characterization of a novel cytoplasm protein Icf45 that is involved in cell cycle regulation. *J. Biol. Chem.* **2004**, *279*, 53498–53505. [CrossRef] [PubMed]

23. Rao, B.S.; Maris, E.L.; Jackman, J.E. tRNA $5'$-end repair activities of tRNA^{His} guanylyltransferase (Thg1)-like proteins from bacteria and archaea. *Nucleic Acids Res.* **2011**, *39*, 1833–1842. [CrossRef] [PubMed]

24. Yuan, J.; Gogakos, T.; Babina, A.M.; Söll, D.; Randau, L. Change of tRNA identity leads to a divergent orthogonal histidyl-tRNA synthetase/ tRNA^{His} pair. *Nucleic Acids Res.* **2011**, *39*, 2286–2293. [CrossRef]

25. Wang, C.; Sobral, B.W.; Williams, K.P. Loss of a universal tRNA feature. *J. Bacteriol.* **2007**, *189*, 1954–1962. [CrossRef]

26. Rao, B.S.; Mohammad, F.; Gray, M.W.; Jackman, J.E. Absence of a universal element for tRNA^{His} identity in *Acanthamoeba castellanii*. *Nucleic Acids Res.* **2013**, *41*, 1885–1894. [CrossRef] [PubMed]

27. Rao, B.S.; Jackman, J.E. Life without post-transcriptional addition of g-1: Two alternatives for trnahis identity in eukarya. *RNA* **2015**, *21*, 243–253. [CrossRef] [PubMed]

28. Hyde, S.J.; Rao, B.S.; Eckenroth, B.E.; Jackman, J.E.; Doublie, S. Structural studies of a bacterial tRNA^{His} guanylyltransferase (Thg1)-like protein, with nucleotide in the activation and nucleotidyl transfer sites. *PLoS ONE* **2013**, *8*, e67465. [CrossRef] [PubMed]

29. Kimura, S.; Suzuki, T.; Chen, M.; Kato, K.; Yu, J.; Nakamura, A.; Tanaka, I.; Yao, M. Template-dependent nucleotide addition in the reverse ($3'$-$5'$) direction by Thg1-like protein. *Sci. Adv.* **2016**, *2*, e1501397. [CrossRef] [PubMed]

30. Lee, K.; Lee, E.H.; Son, J.; Hwang, K.Y. Crystal structure of tRNA^{His} guanylyltransferase from *Saccharomyces cerevisiae*. *Biochem. Biophys. Res. Commun.* **2017**, *490*, 400–405. [CrossRef] [PubMed]

31. Jackman, J.E.; Phizicky, E.M. Identification of critical residues for G$_{-1}$ addition and substrate recognition by tRNA^{His} guanylyltransferase. *Biochemistry* **2008**, *47*, 4817–4825. [CrossRef] [PubMed]

32. Jackman, J.E.; Gott, J.M.; Gray, M.W. Doing it in reverse: 3′-to-5′ polymerization by the Thg1 superfamily. *RNA* **2012**, *18*, 886–899. [CrossRef] [PubMed]

33. Smith, B.A.; Jackman, J.E. Kinetic analysis of 3′-5′ nucleotide addition catalyzed by eukaryotic tRNA^{His} guanylyltransferase. *Biochemistry* **2012**, *51*, 453–465. [CrossRef] [PubMed]

34. Betat, H.; Long, Y.; Jackman, J.E.; Morl, M. From end to end: tRNA editing at 5′- and 3′-terminal positions. *Int. J. Mol. Sci.* **2014**, *15*, 23975–23998. [CrossRef] [PubMed]

35. Desai, R.; Kim, K.; Buchsenschutz, H.C.; Chen, A.W.; Bi, Y.; Mann, M.R.; Turk, M.A.; Chung, C.Z.; Heinemann, I.U. Minimal requirements for reverse polymerization and tRNA repair by tRNA^{His} guanylyltransferase. *RNA Biol.* **2018**, *15*, 614–622. [CrossRef]

36. Smith, B.A.; Jackman, J.E. *Saccharomyces cerevisiae* Thg1 uses 5′-pyrophosphate removal to control addition of nucleotides to tRNA^{His}. *Biochemistry* **2014**, *53*, 1380–1391. [CrossRef] [PubMed]

37. Lant, J.T.; Berg, M.D.; Sze, D.H.W.; Hoffman, K.S.; Akinpelu, I.C.; Turk, M.A.; Heinemann, I.U.; Duennwald, M.L.; Brandl, C.J.; O'Donoghue, P. Visualizing tRNA-dependent mistranslation in human cells. *RNA Biol.* **2018**, *15*, 567–575. [CrossRef] [PubMed]

38. Preston, M.A.; D'Silva, S.; Kon, Y.; Phizicky, E.M. tRNA^{His} 5-methylcytidine levels increase in response to several growth arrest conditions in *Saccharomyces cerevisiae*. *RNA* **2013**, *19*, 243–256. [CrossRef]

39. Rice, T.S.; Ding, M.; Pederson, D.S.; Heintz, N.H. The highly conserved tRNA^{His} guanylyltransferase Thg1p interacts with the origin recognition complex and is required for the G2/M phase transition in the yeast *Saccharomyces cerevisiae*. *Eukaryot. Cell* **2005**, *4*, 832–835. [CrossRef]

40. Heinemann, I.U.; Randau, L.; Tomko, R.J., Jr.; Söll, D. 3′-5′ tRNA^{His} guanylyltransferase in bacteria. *FEBS Lett.* **2010**, *584*, 3567–3572. [CrossRef] [PubMed]

41. Heinemann, I.U.; Söll, D.; Randau, L. Transfer RNA processing in archaea: Unusual pathways and enzymes. *FEBS Lett.* **2010**, *584*, 303–309. [CrossRef]

42. Abad, M.G.; Long, Y.; Willcox, A.; Gott, J.M.; Gray, M.W.; Jackman, J.E. A role for tRNA^{His} guanylyltransferase (Thg1)-like proteins from *Dictyostelium discoideum* in mitochondrial 5′-trna editing. *RNA* **2011**, *17*, 613–623. [CrossRef] [PubMed]

43. Heinemann, I.U.; Nakamura, A.; O'Donoghue, P.; Eiler, D.; Söll, D. tRNA^{His}-guanylyltransferase establishes tRNA^{His} identity. *Nucleic Acids Res.* **2012**, *40*, 333–344. [CrossRef]

44. Long, Y.; Abad, M.G.; Olson, E.D.; Carrillo, E.Y.; Jackman, J.E. Identification of distinct biological functions for four 3′-5′ RNA polymerases. *Nucleic Acids Res.* **2016**, *44*, 8395–8406. [CrossRef]

45. Heinemann, I.U.; O'Donoghue, P.; Madinger, C.; Benner, J.; Randau, L.; Noren, C.J.; Söll, D. The appearance of pyrrolysine in trnahis guanylyltransferase by neutral evolution. *Proc. Natl. Acad. Sci. USA* **2009**, *106*, 21103–21108. [CrossRef] [PubMed]

46. Prat, L.; Heinemann, I.U.; Aerni, H.R.; Rinehart, J.; O'Donoghue, P.; Söll, D. Carbon source-dependent expansion of the genetic code in bacteria. *Proc. Natl. Acad. Sci. USA* **2012**, *109*, 21070–21075. [CrossRef]

47. Abad, M.G.; Rao, B.S.; Jackman, J.E. Template-dependent 3′-5′ nucleotide addition is a shared feature of tRNA^{His} guanylyltransferase enzymes from multiple domains of life. *Proc. Natl. Acad. Sci. USA* **2010**, *107*, 674–679. [CrossRef]

48. Dodbele, S.; Moreland, B.; Gardner, S.M.; Bundschuh, R.; Jackman, J.E. 5′-end sequencing in *Saccharomyces cerevisiae* offers new insights into 5′-ends of tRNA^{His} and snoRNAs. *FEBS Lett.* **2019**, in press. [CrossRef]

49. Abad, M.G.; Long, Y.; Kinchen, R.D.; Schindel, E.T.; Gray, M.W.; Jackman, J.E. Mitochondrial tRNA 5′-editing in *Dictyostelium discoideum* and *Polysphondylium pallidum*. *J. Biol. Chem.* **2014**, *289*, 15155–15165. [CrossRef] [PubMed]

50. Dodbele, S.; Jackman, J.E.; Gray, M.W. Mechanisms and evolution of tRNA 5′-editing in mitochondria. In *RNA Metabolism in Mitochondria*; Jorge, C.-R., Gray, M.W., Eds.; Springer: Cham, Switzerland, 2018; Volume 34.

51. Gott, J.M.; Somerlot, B.H.; Gray, M.W. Two forms of RNA editing are required for tRNA maturation in Physarum mitochondria. *RNA* **2010**, *16*, 482–488. [CrossRef]

52. Lonergan, K.M.; Gray, M.W. Editing of transfer RNAs in *Acanthamoeba castellanii* mitochondria. *Science* **1993**, *259*, 812–816. [CrossRef] [PubMed]

53. Lonergan, K.M.; Gray, M.W. Predicted editing of additional transfer RNAs in *Acanthamoeba castellanii* mitochondria. *Nucleic Acids Res.* **1993**, *21*, 4402. [CrossRef] [PubMed]

54. Nakamura, A.; Wang, D.; Komatsu, Y. Molecular mechanism of substrate recognition and specificity of tRNA^{His} guanylyltransferase during nucleotide addition in the 3′-5′ direction. *RNA* **2018**, *24*, 1583–1593. [CrossRef] [PubMed]

55. Gao, Y.; Yang, W. Capture of a third Mg^{2+} is essential for catalyzing DNA synthesis. *Science* **2016**, *352*, 1334–1337. [CrossRef]

56. Yang, W.; Weng, P.J.; Gao, Y. A new paradigm of DNA synthesis: Three-metal-ion catalysis. *Cell Bioscience* **2016**, *6*, 51. [CrossRef]

57. Yuan, Y.; Altman, S. Selection of guide sequences that direct efficient cleavage of mRNA by human ribonuclease P. *Science* **1994**, *263*, 1269–1273. [CrossRef]

genes

MDPI

Article

Bacterial Aspartyl-tRNA Synthetase Has Glutamyl-tRNA Synthetase Activity

Udumbara M. Rathnayake and Tamara L. Hendrickson *

Department of Chemistry, Wayne State University, 5101 Cass Avenue, Detroit, MI 48202, USA;
udumbara.rathnayake@wayne.edu
* Correspondence: Tamara.Hendrickson@wayne.edu; Tel.: +1-313-577-6914

Received: 14 March 2019; Accepted: 27 March 2019; Published: 1 April 2019

Abstract: The aminoacyl-tRNA synthetases (aaRSs) are well established as the translators of the genetic code, because their products, the aminoacyl-tRNAs, read codons to translate messenger RNAs into proteins. Consequently, deleterious errors by the aaRSs can be transferred into the proteome via misacylated tRNAs. Nevertheless, many microorganisms use an indirect pathway to produce Asn-tRNA^{Asn} via Asp-tRNA^{Asn}. This intermediate is produced by a non-discriminating aspartyl-tRNA synthetase (ND-AspRS) that has retained its ability to also generate Asp-tRNA^{Asp}. Here we report the discovery that ND-AspRS and its discriminating counterpart, AspRS, are also capable of specifically producing Glu-tRNA^{Glu}, without producing misacylated tRNAs like Glu-tRNA^{Asn}, Glu-tRNA^{Asp}, or Asp-tRNA^{Glu}, thus maintaining the fidelity of the genetic code. Consequently, bacterial AspRSs have glutamyl-tRNA synthetase-like activity that does not contaminate the proteome via amino acid misincorporation.

Keywords: tRNA; misacylation; indirect tRNA aminoacylation; AspRS; GluRS-like

1. Introduction

The fidelity of protein translation depends on the accurate pairing of cognate tRNAs to their cognate amino acids. The ligation of the amino acid to its tRNA is catalyzed by a highly specific group of enzymes known as the aminoacyl-tRNA synthetases (aaRSs) [1–3]. Under normal conditions, these enzymes maintain high accuracy and specificity in selecting their cognate amino acid and tRNA substrates. In fact, decades of research have been dedicated to demonstrating that the aaRSs are exquisitely specific for their cognate substrates and are key players in defining the accuracy of the proteome [2,4–6]. An increasing body of work, however, is emerging that demonstrates that some aaRSs can alter their activity or relax their selectivity in response to stress, even promoting errors in translation [7–12].

Aspartyl-tRNA synthetase (AspRS) is an exception to the rule of one aaRS per amino acid/tRNA pair. This enzyme is found in two general forms, discriminating and non-discriminating, based on divergent tRNA specificities. Discriminating AspRS, the canonical enzyme, is found in eukaryotes, and some bacteria and archaea, and catalyzes the aspartylation of tRNA^{Asp} to produce Asp-tRNA^{Asp} (Figure 1A) [13]. The non-discriminating form, ND-AspRS, is found in many bacteria and archaea and some organelles; this enzyme cannot differentiate between tRNA^{Asp} and tRNA^{Asn} and aminoacylates both with aspartate to generate Asp-tRNA^{Asp} and the misacylated Asp-tRNA^{Asn}, respectively (Figure 1B) [14–17]. To maintain the fidelity of the genetic code, Asp-tRNA^{Asn} is then converted to Asn-tRNA^{Asn} by a glutamine-dependent amidotransferase (AdT) [14,15,18,19].

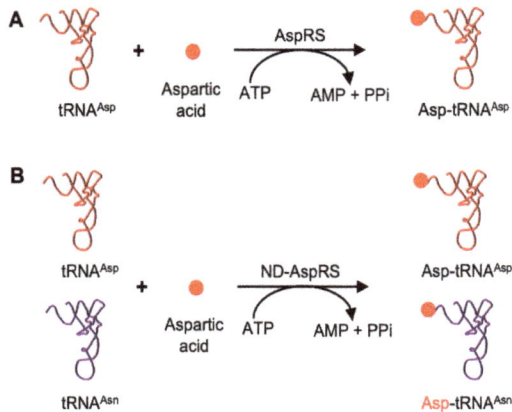

Figure 1. Canonical roles of AspRS and ND-AspRS. (**A**) AspRS aminoacylates tRNA^{Asp} with aspartic acid to produce Asp-tRNA^{Asp}. (**B**) ND-AspRS catalyzes the aspartylation of both tRNA^{Asp} and tRNA^{Asn} to produce Asp-tRNA^{Asp} and the misacylated Asp-tRNA^{Asn}.

The requirement for a discriminating AspRS versus an ND-AspRS differs from organism to organism; however, ND-AspRS is always accompanied by AdT [15,18–20]. ND-AspRS and AdT are typically found in organisms that lack asparaginyl-tRNA synthetase (AsnRS) and/or asparagine synthetase (AsnS). For example, *Pseudomonas aeruginosa* and *Helicobacter pylori* (*H. pylori* or *Hp*) lack both AsnRS and AsnS and consequently require ND-AspRS and AdT to produce both Asn-tRNA^{Asn} and asparagine [21,22]. *Mycobacterium tuberculosis* lacks AsnRS but has AsnS, so it only uses indirect tRNA aminoacylation to produce Asn-tRNA^{Asn} [12]. In contrast, *Staphylococcal aureus* (*S. aureus* or *Sa*) has a functioning AsnRS but lacks AsnS; thus, indirect aminoacylation is still required as the sole biosynthetic route to asparagine [23]. Some bacteria like *Thermus thermophilus* and *Deinococcus radiodurans* encode for more than one copy of AspRS [24,25]. In these cases, the longer, bacterial-type AspRS is discriminating and catalyzes the synthesis of Asp-tRNA^{Asp}. The second AspRS (AspRS2) is non-discriminating and produces Asp-tRNA^{Asn} along with Asp-tRNA^{Asp}.

While ND-AspRS clearly demonstrates relaxed tRNA substrate specificity, it remains fine-tuned as it only aspartylates tRNA^{Asp} and tRNA^{Asn}, only in organisms that have AdT, and it presumably discriminates against other tRNAs as potential substrates. In this way, the fidelity of the genetic code is maintained. Evidence of further relaxation in substrate specificity for either a discriminating AspRS or an ND-AspRS has not been reported to our knowledge. Here we demonstrate that these enzymes are capable of specifically producing Glu-tRNA^{Glu} while maintaining their canonical roles: The production of Asp-tRNA^{Asp} (AspRS and ND-AspRS) and Asp-tRNA^{Asn} (ND-AspRS only). To the best of our knowledge, this is the first example of an aaRS capable of producing a non-cognate, but correctly aminoacylated tRNA.

2. Materials and Methods

2.1. Materials

Potassium dihydrogen phosphate (KH$_2$PO$_4$), sodium dihydrogen phosphate (NaH$_2$PO$_4$), L-aspartic acid (catalog #A93100, batch #12002LC), L-glutamic acid (catalog #G1251, lot #SLBK8671V), magnesium chloride, and oligonucleotides were purchased from Sigma-Aldrich (St. Louis, MO, USA). The radiolabeled reagents aspartic acid, L-[2,3-3H] (catalog #ART 0211, lot #180606), glutamic acid, L-[2,3,4-3H] (catalog #ART 0132, lot #180710) and adenosine triphosphate (ATP) [γ-^{32}P] were purchased from American Radiolabeled Chemicals (St. Louis, MO, USA). Ampicillin, chloramphenicol, kanamycin, 4-(2-hydroxyethyl)-1-piperazineethanesulfonic acid (HEPES), ethylenediaminetetraacetic acid (EDTA), isopropyl β-ᴅ-thiogalactopyranoside (IPTG), phenylmethanesulfonyl fluoride (PMSF), and tris(hydroxymethyl) aminomethane (Tris) were from GoldBio Biotechnology, Inc., (St. Louis, MO, USA). Potassium chloride (KCl), sodium chloride (NaCl), trichloroacetic acid (TCA), and ammonium acetate (NH_4OAc) were purchased from Fisher Scientific (Hampton, NH, USA).

2.2. Overexpression and Purification of aaRSs

Hp ND-AspRS [22], the *Hp* tRNA^{Gln}-specific glutamyl-tRNA synthetase (GluRS2) [26], *Mycobacterium smegmatis* (*M. smegmatis* or *Ms*) ND-AspRS (vector provided by Dr. Babak Javid), and *Escherichia coli* (*E. coli* or *Ec*) AspRS and GluRS (vectors provided by Dr. Takuya Ueda [27]) were each overexpressed in *Ec* Bl21 (DE3) RIL in Luria-Bertani medium (LB, 1 L) inoculated with a saturated overnight culture grown from a single colony. The cultures were incubated at 37 °C, 200 rpm. IPTG (1 mM) was used to induce overexpression in each case. For *Hp* ND-AspRS and GluRS2 overexpression, the medium was supplemented with ampicillin (100 µg/mL), chloramphenicol (100 µg/mL), and glucose (0.5%). IPTG was added when the OD_{600} reached 0.8–1.0. Cells were collected after a 1 h induction period. For *Ms* ND-AspRS, the only difference was that kanamycin (25 µg/mL) was added instead of ampicillin. The two discriminating *Ec* enzymes were grown in the same medium as the two *Hp* enzymes. IPTG induction was initiated when the OD_{600} reached between 0.4 and 0.6 and continued for 4 h. Cells were harvested by centrifugation and frozen at −80 °C until ready for use.

The *Sa* ND-AspRS (vector provided by Dr. Kelly Sheppard) was overexpressed and purified as previously described [23]. In all other cases, cell pellets were suspended in lysis buffer (50 mM NaH_2PO_4, pH 7.4, 300 mM NaCl, 10 mM imidazole) and the cells were lysed with lysozyme (2 mg/mL) and sonication. Saturated PMSF (15 µL/mL) was added every 20 min to reduce proteolysis. Typically, a cell extract was added to a polyprep column that contained high-density cobalt agarose beads (~2 mL, Gold Biotechnology, Inc.) pre-washed with lysis buffer. The lysate was incubated with the resin by rotating at 4 °C for 1 h. The His$_6$-tagged aaRSs were eluted according to the manufacturer's protocol.

The eluents were dialyzed against dialysis buffer (20 mM KH_2PO_4, pH 7.2, 0.5 mM Na_2EDTA, 5 mM β-mercaptoethanol) at 4 °C for 1–2 h and then against fresh dialysis buffer overnight. The aaRSs were adsorbed onto a HiTrap^{TM} DEAE FF column (GE Healthcare Life Sciences, Pittsburgh, PA, USA). The proteins were eluted with a stepwise gradient of 20–300 mM KH_2PO_4, pH 7.2, supplemented with 0.5 mM Na_2EDTA, 5 mM β-mercaptoethanol. The proteins were concentrated with 30 kDa molecular weight spin filters (EMD Millipore, Burlington, MA, USA). After this two-column purification procedure, all enzymes were judged to have been purified to near homogeneity by SDS-PAGE (Figure S1). Final protein concentrations were determined by UV-Vis spectroscopy at 280 nm, using extinction coefficients calculated using the ExPASy ProtParam tool [28]. The aaRSs were immediately used in aminoacylation assays. All experiments were conducted using biological replicates in triplicate.

2.3. In Vivo Transcription and Purification of tRNAs

Hp tRNA^{Asn} [22] and tRNA^{Gln} [26] were overexpressed in *Ec* MV1184 and purified as previously described [29]. After deacylation, the tRNAs were electroeluted in a denaturing urea gel for further purification. The tRNA band was excised from the gel, crushed, and incubated overnight at 37 °C in crush and soak buffer (0.5 mM NH_4OAc, pH 5.2, 1 mM Na_2EDTA). The eluted tRNA sample was isopropanol precipitated. The resultant pellet was dissolved in water, folded, and quantified by aminoacylation assay, as previously described [29]. These purifications yield tRNA that is enriched with the overexpressed tRNA isoacceptor but also contains total *Ec* tRNA. Separately, total *Ec* tRNA was isolated from a saturated *Ec* MV1184 culture that had been grown at 37 °C in LB medium supplemented with glucose (0.5%) and purified as described [29]. The concentration of total *Ec* tRNA was calculated by UV-Vis spectroscopy at 260 nm assuming 1 OD = 1.6 µM.

2.4. Initial Rate Aminoacylation Assays

Each tRNA was pre-folded as previously described [29]. Aminoacylation assays were conducted in buffer containing 100 mM Na-HEPES, pH 7.5, 30 mM KCl, 12 mM $MgCl_2$, 2 mM ATP, 0.1 mg/mL BSA, 20 µM amino acid, 50 µCi/mL ^{3}H-labeled amino acid, and 10 µM *Hp* tRNA^{Asn} or tRNA^{Gln} as noted. Assays were conducted at 37 °C and were initiated with the addition of 200 nM *Hp* ND-AspRS or GluRS2. Aliquots were removed and quenched on Whatman filter pads containing TCA. The pads were washed 3 times for 15 min with chilled 5% TCA. Pads were dried and counted in 3 mL ECOLITE (+) scintillation fluid (MP Biomedicals, Solon, OH, USA). Aminoacylation assays with total *Ec* tRNA were carried out in the same buffer containing 50–100 µM total tRNA. *Hp* ND-AspRS (100 nM or 500 nM) or *Hp* GluRS2 (100 nM or 500 nM) were used in these assays as noted.

2.5. Extended Aminoacylation Assays (90 Min)

Hp ND-AspRS and GluRS2 used in these assays were only purified by cobalt affinity purification. The aminoacylation assays were conducted at 37 °C in buffer containing 20 mM HEPES-OH, pH 7.5, 4 mM $MgCl_2$, 2 mM ATP, 100 µM amino acid, and 25 µCi/mL ^{3}H-labeled amino acid. All aaRSs were added to a final concentration of 1 µM and overexpressed tRNA isoacceptor was added to a final concentration of 10 µM.

2.6. Acid Gel Electrophoresis and Northern Blot Analysis

Aminoacylation reactions were conducted in the same buffer used for initial rate aminoacylation assays only without radiolabeled amino acid for 90 min at 37 °C with 500 nM aaRS. NaOAc (0.3 mM, pH 4.5) was added to each reaction and they were quenched with phenol/chloroform (1:1 *v/v*, pH 4.5). The tRNAs were ethanol precipitated and the resultant pellets were dissolved in buffer containing 10 mM NaOAc, pH 4.5, 1 mM Na_2EDTA. Acid gel electrophoresis was used to separate deacylated tRNAs from their aminoacylated counterparts. The tRNA samples were quantified by UV-Vis spectroscopy at 260 nm, and 0.05 OD_{260} units were loaded per lane for overexpressed tRNAs. This amount was increased to 0.25 OD_{260} units for total *Ec* tRNA. Isoacceptor-specific oligonucleotides were used in Northern blots to visualize each tRNA (Table S1). The tRNA specific DNA probes were prepared as previously described [26]. The acid gel and Northern blot analyses were performed as previously described [26,30], with the exception that a non-specific DNA oligonucleotide (0.1 µM) was added during the first three washing steps after probe hybridization.

3. Results

3.1. H. pylori ND-AspRS Appears to Aminoacylate H. pylori tRNA^{Gln} with Aspartate and Glutamate

ND-AspRS has relaxed tRNA specificity and aspartylates both tRNA^{Asp} and tRNA^{Asn} to produce Asp-tRNA^{Asp} and Asp-tRNA^{Asn}, respectively [14,15]. We have previously demonstrated that the *Hp* ND-AspRS has this dual tRNA specificity, as expected [22]. Given the nature of the genetic code and decades of work characterizing the aaRSs, ND-AspRS should not demonstrate broader substrate specificity for non-cognate tRNAs or amino acids. We examined *Hp* ND-AspRS and GluRS2, a tRNA^{Gln}-specific GluRS [26,31], for possible cross-reactivity with overexpressed tRNA^{Asn} and tRNA^{Gln} with both aspartate and glutamate (Figure 2). We performed these assays with longer time points to look at plateau levels of aminoacylation and to detect any low levels of aminoacylation. As expected, ND-AspRS produced Asp-tRNA^{Asn} and did not produce Glu-tRNA^{Asn} (Figure 2A,B, respectively). Unexpectedly, this enzyme showed aminoacylation activity in both assays with tRNA^{Gln} (Figure 2C,D), suggesting that it was producing the non-cognate product *Hp* Asp-tRNA^{Gln} and Glu-tRNA^{Gln}. However, because these tRNAs were overexpressed in vivo, a practice necessary to introduce required post-transcriptional modifications like queuosine [32] and 2-thiouridine [33], the tRNAs used in these experiments were contaminated with total *Ec* tRNA. Thus, the possibility also exists that *Hp* ND-AspRS is aminoacylating one or more *Ec* tRNAs instead. For comparison, GluRS2 showed only its expected activity, producing Glu-tRNA^{Gln} (Figure 2D), but not Asp-tRNA^{Asn}, Glu-tRNA^{Asn}, or Asp-tRNA^{Gln} (Figure 2A–C). We conducted these same, longer aminoacylation assays using the *Ms* ND-AspRS and ND-GluRS (Figure S2): We observed the same behavior with glutamate, indicating that this phenomenon is found in bacteria beyond *H. pylori*.

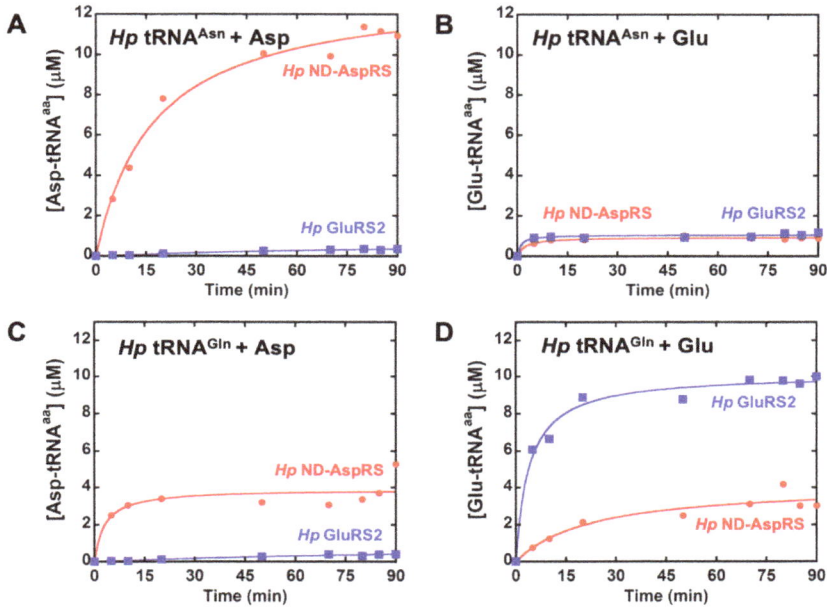

Figure 2. *H. pylori* ND-AspRS appears to attach aspartate and glutamate to tRNA^{Gln}. *Hp* ND-AspRS (●, 1 μM) and GluRS2 (■, 1 μM) were tested in cross-aminoacylation assays using *Hp* tRNA^{Asn} and tRNA^{Gln} with aspartate and glutamate. The tRNA isoacceptor concentration in each assay was 10 μM; each tRNA was contaminated with total *Ec* tRNA. (**A**) *Hp* tRNA^{Asn} aminoacylated with aspartate, (**B**) *Hp* tRNA^{Asn} aminoacylated with glutamate, (**C**) *Hp* tRNA^{Gln} aminoacylated with aspartate, and (**D**) *Hp* tRNA^{Gln} aminoacylated with glutamate.

We repeated the assays shown in Figure 2 with shorter time points and lower enzyme concentrations to look at initial rates (Figure S3). Under these more standard conditions, the ability of ND-AspRS to ligate glutamate onto either *Hp* tRNA^{Gln} or a contaminating *Ec* tRNA was masked (Figure S3D), offering an explanation for why this activity has remained undiscovered. Robust aspartylation by *Hp* ND-AspRS, presumably of contaminating *Ec* tRNA^{Asp} and/or tRNA^{Asn}, was observed as anticipated (Figure S3C).

Our use of extended time point assays (Figure 2) revealed an unexpected activity for ND-AspRS that was invisible under initial rate conditions (Figure S3). To our knowledge, this is the first evidence of an ND-AspRS utilizing glutamate instead of aspartate as its amino acid substrate. This activity makes some sense, however, given the close structural similarities between aspartate and glutamate. Nevertheless, the results presented in Figure 2 and Figures S2 and S3, still contained ambiguity with respect to the identity of the relevant tRNA substrate since each tRNA was overexpressed and purified with contaminating *Ec* tRNAs.

3.2. H. pylori ND-AspRS Aminoacylates E. coli tRNA^Glu with Glutamate to Produce Glu-tRNA^Glu

To more precisely examine the unexpected aminoacylation reaction(s) catalyzed by *Hp* ND-AspRS, we turned to acid gel electrophoresis and Northern blotting techniques [30]. We continued to use *Hp* GluRS2 as a control because of its high specificity for Glu-tRNA^Gln production [26,31]. The advantages of this approach are two-fold. First, the acidic pH of these gels enables the separation of aminoacyl-tRNAs from their deacylated counterparts due to the extra charge provided by the protonated amino terminus of the amino acid (30). Second, oligonucleotides can be designed for Northern blot visualizations that are specific for a desired tRNA isoacceptor. Thus, this approach allows us to unequivocally identify the tRNA(s) being aminoacylated in a given experiment, clearly resolving the ambiguity of the results presented above. Since *Hp* ND-AspRS uses glutamate as a substrate (Figure 2D), we hypothesized that it may be adding glutamate onto *Hp* tRNA^Gln or *Ec* tRNA^Glu. Consequently, we used this methodology to examine *Hp* ND-AspRS-catalyzed aminoacylation of *Hp* tRNA^Gln and *Ec* tRNA^Glu with aspartate versus glutamate as its amino acid substrates. Northern blots were conducted with overexpressed *Hp* tRNA^Gln contaminated with total *Ec* tRNA as a source of *Ec* tRNA^Glu. In both cases, oligonucleotides were designed to visualize only the tRNAs of interest: *Hp* tRNA^Gln and *Ec* tRNA^Glu.

Hp ND-AspRS does not aminoacylate *Hp* tRNA^Gln with either aspartate or glutamate (Figure 3, blot 1, lanes 2 and 3), suggesting that the activities observed above (Figure 2C,D) are the result of aminoacylation of one or more *Ec* tRNAs (compared to the production of Glu-tRNA^Gln by *Hp* GluRS2 in Figure 3, blot 1, lane 4.) In contrast, *Hp* ND-AspRS aminoacylates *Ec* tRNA^Glu with glutamate to produce Glu-tRNA^Glu (Figure 3, blot 2, lane 3), offering direct evidence that this ND-AspRS has non-canonical aminoacylation activity that goes beyond its ability to aspartylate tRNA^Asp and tRNA^Asn. We considered the possibility that *E. coli* glutamyl-tRNA synthetase (*Ec* GluRS) had inadvertently co-purified with *Hp* ND-AspRS, however, no evidence of this contamination was observed by SDS-PAGE (Figure S1). Remarkably, *Hp* ND-AspRS remains accurate by pairing glutamate with tRNA^Glu to produce correctly acylated Glu-tRNA^Glu but not the misacylated Asp-tRNA^Glu (Figure 3, blot 2, lanes 3 and 2 respectively). In other words, this enzyme is exhibiting activity consistent with a canonical GluRS, in that it is ligating glutamate specifically to tRNA^Glu.

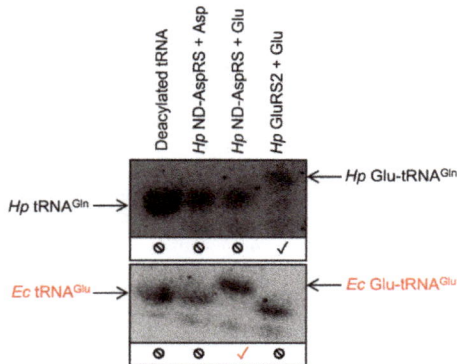

Figure 3. *H. pylori* ND-AspRS aminoacylates *E. coli* tRNA^Glu with glutamate producing Glu-tRNA^Glu. *Hp* tRNA^Gln, contaminated with total *Ec* tRNA, was aminoacylated with either *Hp* ND-AspRS or GluRS2 and with aspartate versus glutamate. Aminoacylated versus deacylated tRNAs were separated in an acid gel. Specific tRNAs and aminoacyl-tRNAs were imaged using ^32P-labeled oligonucleotides in Northern blots. The blots were visualized with a *Hp* tRNA^Gln-specific primer (blot 1) and an *Ec* tRNA^Glu-specific primer (blot 2). Expected tRNA aminoacylation activity is indicated with a black check mark (✓); unexpected aminoacylation activity is indicated in red (✓); the absence of aminoacylation activity with a given tRNA is indicated with a no symbol (⊘).

Next, we compared the initial rates and extents of glutamylation of total *Ec* tRNA by *Hp* GluRS2, an enzyme that naturally uses glutamate as its substrate [26,31], and *Hp* ND-AspRS, which unexpectedly also uses glutamate (Figure 4 and Figure S4). Here, to ensure that we could visualize this novel activity, we increased the concentration of each enzyme when looking at aminoacylation of a non-cognate tRNA. We also verified that *Ms* ND-AspRS has this ability to glutamylate total *Ec* tRNA (Figure S5). In these experiments, the ability of ND-AspRS to produce Glu-tRNA^{Glu} is clearly apparent.

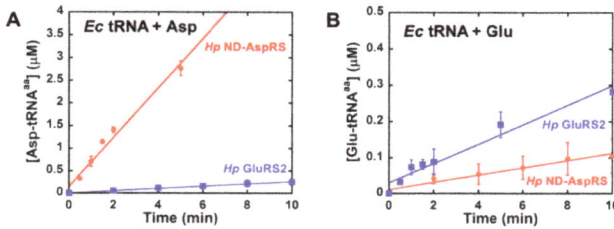

Figure 4. *H. pylori* ND-AspRS aminoacylates *E. coli* tRNA with glutamate. *Hp* ND-AspRS was tested for its activity to aminoacylate *Ec* tRNA (50–100 μM) with aspartate and glutamate. *Hp* GluRS2 was also assayed for comparison. (**A**) Aminoacylation of *Ec* tRNA with aspartate by *Hp* ND-AspRS (●, 100 nM) versus GluRS2 (■, 500 nM). (**B**) Aminoacylation of *Ec* tRNA with glutamate by *Hp* ND-AspRS (●, 500 nM) versus GluRS2 (■, 100 nM). Error bars represent standard deviation from biological replicates in triplicate.

3.3. Glu-tRNA^{Glu} Production Is Common among Bacterial Non-Discriminating and Discriminating AspRSs

All experiments described so far with *Hp* and *Ms* ND-AspRS demonstrate that these enzymes can attach glutamate to *Ec* tRNA and *Hp* ND-AspRS specifically produces *Ec* Glu-tRNA^{Glu}. Given that these are inter-species pairings in both cases, we decided that it was important to examine the *Ec* discriminating AspRS for this activity with its homologous *Ec* tRNA. We also included the ND-AspRS from *S. aureus* to gain additional perspective into the ubiquity of this unexpected activity. Acid gels and Northern blot analyses using oligonucleotide probes that are specific for four different *Ec* tRNAs (tRNA^{Glu}, tRNA^{Gln}, tRNA^{Asp}, and tRNA^{Asn}) were conducted with all four enzymes: The discriminating AspRS from *Ec* and the ND-AspRSs from *Hp, Ms,* and *Sa* (Figure 5). In many of the Northern blots shown below, the deacylated and the aminoacylated tRNAs appear as two bands, presumably due to differences in post-transcriptional modifications of the tRNAs, as previously reported [26].

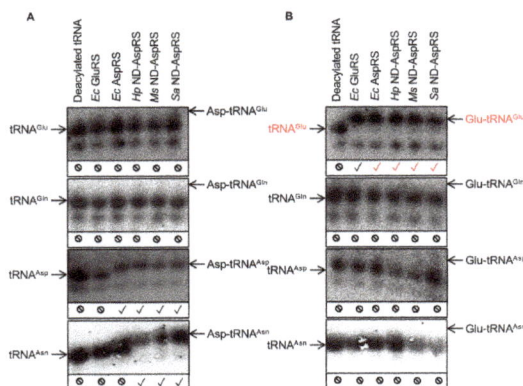

Figure 5. Some bacterial AspRSs aminoacylate *E. coli* tRNA^{Glu} with glutamate producing Glu-tRNA^{Glu}. Total *Ec* tRNA was aminoacylated with (**A**) aspartate and (**B**) glutamate by *Ec* GluRS, *Ec* AspRS, *Hp* ND-AspRS, *Ms* ND-AspRS, and *Sa* ND-AspRS. The aminoacylated versus deacylated tRNAs were separated in acid gels. ^{32}P-labeled oligonucleotides specific for *Ec* tRNA^{Glu}, tRNA^{Gln}, tRNA^{Asp}, and tRNA^{Asn} were used for each blot as indicated. Expected tRNA aminoacylation activities are indicated with a black check mark (✓); unexpected activities are indicated in red (✓); the absence of aminoacylation activity with a given tRNA is indicated with a no symbol (⊘).

Figure 5 shows the Northern blot results obtained from total *Ec* tRNA aminoacylated with aspartate (A) and glutamate (B) by different bacterial aaRSs: *Ec* GluRS was used as a control. The results of panel A reveal that each AspRS attaches aspartate to tRNA^{Asp} and the three ND-AspRSs also attach aspartate to tRNA^{Asn}, as expected (Figure 5A, blots 3 and 4). Furthermore, none of the enzymes tested were capable of attaching significant amounts of aspartate to either non-cognate tRNA substrates tRNA^{Glu} or tRNA^{Gln} (Figure 5A, blots 1 and 2).

The most interesting and critical results are revealed in blot 1 in Figure 5B and are highlighted with red checkmarks. This blot was probed with a ^{32}P-labeled oligonucleotide specific for *Ec* tRNA^{Glu} and Glu-tRNA^{Glu} formation with *Ec* GluRS was used as a positive control. A clear shift is observed between deacylated tRNA^{Glu} and Glu-tRNA^{Glu} produced by *Ec* GluRS (Figure 5B, blot 1, lanes 1 and 2). What is remarkable is that all four AspRSs showed this same shift, clearly indicative of ubiquitous Glu-tRNA^{Glu} production. The remaining blots in Figure 5B demonstrate that this glutamylation activity is specific for tRNA^{Glu} as none of the enzymes tested were capable of adding glutamate onto *Ec* tRNA^{Gln}, tRNA^{Asp}, or tRNA^{Asn}. These data demonstrate that *Ec* AspRS and the ND-AspRSs from *Hp*, *Ms*, and *Sa* aminoacylate tRNA^{Glu} with glutamate to produce Glu-tRNA^{Glu}. These reactions represent a new, non-canonical activity for AspRS that can be viewed as that of a discriminating GluRS.

4. Discussion

In this work, we demonstrate that bacterial discriminating and non-discriminating AspRSs have quantifiable GluRS activity as they are capable of specifically generating Glu-tRNA^{Glu}. Using total *Ec* tRNA, this activity was observed with four different AspRSs: The *Ec* discriminating AspRS and three non-discriminating AspRSs from *Hp*, *Ms*, and *Sa*. All four enzymes have canonical AspRS activity and produce Asp-tRNA^{Asp} (Figure 5A, blot 3). The three ND-AspRSs also produce misacylated Asp-tRNA^{Asn} (Figure 5A, blot 4); this intermediate would be converted to Asn-tRNA^{Asn} in vivo by AdT. All four enzymes were also capable of specifically producing Glu-tRNA^{Glu} (Figure 5B, blot 1) without adding glutamate to other related tRNAs.

We considered the possibility of the co-purification of contaminating *Ec* GluRS as an alternative explanation for this activity. However, each AspRS was purified in two steps: cobalt affinity and DEAE chromatography. *Ec* GluRS elutes from our DEAE column at a lower KH$_2$PO$_4$ concentration

(~50–100 mM) than the tested AspRSs (~200–300 mM), such that any GluRS that survived the affinity purification step would have been removed by DEAE chromatography. Furthermore, no evidence of *Ec* GluRS contamination was visible by using SDS-PAGE gel (Figure S1). Thus, this two-step purification strategy eliminates the possibility of contaminated *Ec* GluRS in the AspRS samples, confirming that this activity is due to AspRS alone.

This GluRS-like activity of AspRSs is remarkable and unexpected because it is accurate and yet outside the normal purview of an AspRS. It truly is GluRS activity as each AspRS specifically attaches glutamate only to tRNA^{Glu} producing Glu-tRNA^{Glu}. The fact that *Ec* AspRS demonstrates this activity is particularly important because of the homologous nature of this result: *Ec* AspRS produces *Ec* Glu-tRNA^{Glu}. All other results herein arose from heterologous pairings of an ND-AspRS with *Ec* tRNA. Our results with *Ec* AspRS demonstrate that this activity is not simply a result of cross-species recognition of a non-cognate tRNA. It is also remarkable that no evidence of misacylated Asp-tRNA^{Glu}, Glu-tRNA^{Asp}, or Glu-tRNA^{Asn} was observed, given the ability of these AspRSs to produce Glu-tRNA^{Glu}. This observation is important because it demonstrates that this GluRS-like activity does not threaten the fidelity of the genetic code. Nevertheless, it poses an unexpected problem with molecular recognition. How does AspRS know to specifically attach aspartate to tRNA^{Asp} (and tRNA^{Asn}) and glutamate to tRNA^{Glu} without cross misacylation of these tRNAs? Our results suggest that communication between the amino acid and tRNA binding sites must occur to achieve this specificity. Further research is needed to understand this apparent tRNA-induced conformational change.

AaRSs almost always recognize discrete molecular features in their tRNA substrates, termed identity elements. These determinants can be positive and favor recognition or negative and disfavor interactions. For *Ec* tRNA^{Asp}, the GUC anticodon, the G73 discriminator base, the G2:C71 base pair in the acceptor stem, C38 in the anticodon loop, and G10 in the D arm of the tRNA serve as the positive identity elements for *Ec* AspRS [34–36]. In contrast, *Ec* GluRS recognizes the G1:C72 and U2:A71 base pairs in the acceptor stem, A37 in the anticodon loop, U11:A24 base pair in D arm, and the U13:G22-A46 tertiary base pair in *Ec* tRNA^{Glu}. Further, the lack of a nucleotide at position 47 and the modified uridine (s^{2}U) in the UUC anticodon of tRNA^{Glu} are also known identity determinants [33,34,37–39]. *Ec* tRNA^{Glu} does contain several tRNA^{Asp} identity elements: The G73 discriminator base, G10 in the D arm, and C38 in the anticodon loop. But are these shared elements enough for recognition of *Ec* tRNA^{Glu} by *Ec* AspRS? Or is an alternative, expanded identity set recognized in such a way as to favor glutamylation of *Ec* tRNA^{Glu}? To answer these questions, the identity elements in *Ec* tRNA^{Glu} that allow recognition by *Ec* AspRS would have to be specifically evaluated.

In conclusion, we have demonstrated that bacterial AspRSs have GluRS activity and are capable of producing Glu-tRNA^{Glu}, at least in vitro. The relevance of this reaction in vivo is very difficult to demonstrate as it is unlikely that this activity is robust enough to enable deletion of the native *Ec* GluRS, especially in the presence of the cognate tRNA^{Asp} substrate. It is possible that this activity exists as a backup in case the cognate GluRS becomes damaged, for example. It is also possible that it is a remnant of early AspRS evolution. The early genetic code was likely sloppy with ancestral aaRSs capable of aminoacylating multiple tRNAs with different amino acids [5]. Evidence also suggests that Class I and II enzymes emerged in pairs [40,41], with early AspRSs and GluRSs perhaps as a co-evolving pair that recognized and acylated the same early tRNA or tRNA-like substrates. These enzymes would have recognized opposite faces of the tRNA, facilitating this co-evolution and protecting the tRNA from hydrolysis [41]. Given that the background GluRS activity exhibited by AspRS does not put the genetic code in jeopardy, it would not have been selected against as the genetic code evolved to be more selective and specific. The results presented here also suggest the possibility that other aaRSs retain similar background activities, a hypothesis that awaits further testing.

Supplementary Materials: The following are available online at http://www.mdpi.com/2073-4425/10/4/262/s1, Methodology for the overexpression and purification of *Ms* ND-GluRS, in vitro transcription of tRNAs, and extending tRNA aminoacylation assays. Figure S1. SDS-PAGE gels of purified aaRSs. Figure S2. Extended

Genes **2019**, *10*, 262

M. smegmatis tRNA^Asn and tRNA^Gln aminoacylation assays with *M. smegmatis* ND-AspRS and ND-GluRS with aspartate versus glutamate. Figure S3. *H. pylori* ND-AspRS shows unexpected aminoacylation activity with overexpressed *H. pylori* tRNA^Gln. Figure S4. Extended total *E. coli* tRNA aminoacylation assays with *H. pylori* ND-AspRS and GluRS2 with aspartate versus glutamate. Figure S5. Extended total *E. coli* tRNA aminoacylation assays with *M. smegmatis* ND-AspRS and ND-GluRS with aspartate versus glutamate. Table S1: The tRNA specific oligonucleotide sequences used in northern blot analysis

Author Contributions: Conceptualization, U.M.R. and T.L.H.; Methodology, U.M.R. and T.L.H.; Writing-Original Draft Preparation, U.M.R. and T.L.H.; Writing-Review & Editing, U.M.R. and T.L.H.; Visualization, U.M.R. and T.L.H.; Supervision, T.L.H.; Project Administration, T.L.H.; Funding Acquisition, T.L.H.

Funding: This research was supported by Wayne State University and private donors.

Acknowledgments: The authors thank Babak Javid, Takuya Ueda, and Kelly Sheppard for their kind gifts of plasmids. They also thank Whitney Wood for helpful discussions.

Conflicts of Interest: The authors declare no conflict of interest.

References

1. Ribas de Pouplana, L.; Schimmel, P. A view into the origin of life: aminoacyl-tRNA synthetases. *Cell. Mol. Life Sci.* **2000**, *57*, 865–870. [CrossRef] [PubMed]
2. Ibba, M.; Soll, D. Aminoacyl-tRNA synthesis. *Annu. Rev. Biochem.* **2000**, *69*, 617–650. [CrossRef]
3. Giege, R.; Springer, M. Aminoacyl-tRNA synthetases in the bacterial world. *EcoSal Plus* **2016**, *7*. [CrossRef]
4. Schimmel, P.; Giege, R.; Moras, D.; Yokoyama, S. An operational RNA code for amino acids and possible relationship to genetic code. *Proc. Natl. Acad. Sci. USA* **1993**, *90*, 8763–8768. [CrossRef]
5. Schimmel, P.; Ribas de Pouplana, L. Transfer RNA: From minihelix to genetic code. *Cell* **1995**, *81*, 983–986. [CrossRef]
6. Kubyshkin, V.; Acevedo-Rocha, C.G.; Budisa, N. On universal coding events in protein biogenesis. *Biosystems* **2018**, *164*, 16–25. [CrossRef] [PubMed]
7. Mohler, K.; Ibba, M. Translational fidelity and mistranslation in the cellular response to stress. *Nat. Microbiol.* **2017**, *2*, 17117. [CrossRef]
8. Ribas de Pouplana, L.; Santos, M.A.; Zhu, J.H.; Farabaugh, P.J.; Javid, B. Protein mistranslation: Friend or foe? *Trends Biochem. Sci.* **2014**, *39*, 355–362. [CrossRef]
9. Reynolds, N.M.; Lazazzera, B.A.; Ibba, M. Cellular mechanisms that control mistranslation. *Nat. Rev. Microbiol.* **2010**, *8*, 849–856. [CrossRef]
10. Wiltrout, E.; Goodenbour, J.M.; Frechin, M.; Pan, T. Misacylation of tRNA with methionine in *Saccharomyces cerevisiae*. *Nucleic Acids Res.* **2012**, *40*, 10494–10506. [CrossRef]
11. Schwartz, M.H.; Pan, T. Determining the fidelity of tRNA aminoacylation via microarrays. *Methods* **2017**, *113*, 27–33. [CrossRef] [PubMed]
12. Javid, B.; Sorrentino, F.; Toosky, M.; Zheng, W.; Pinkham, J.T.; Jain, N.; Pan, M.; Deighan, P.; Rubin, E.J. Mycobacterial mistranslation is necessary and sufficient for rifampicin phenotypic resistance. *Proc. Natl. Acad. Sci. USA* **2014**, *111*, 1132–1137. [CrossRef] [PubMed]
13. Woese, C.R.; Olsen, G.J.; Ibba, M.; Soll, D. Aminoacyl-tRNA synthetases, the genetic code, and the evolutionary process. *MMBR* **2000**, *64*, 202–236. [CrossRef] [PubMed]
14. Curnow, A.W.; Ibba, M.; Soll, D. tRNA-dependent asparagine formation. *Nature* **1996**, *382*, 589–590. [CrossRef] [PubMed]
15. Becker, H.D.; Kern, D. *Thermus thermophilus*: A link in evolution of the tRNA-dependent amino acid amidation pathways. *Proc. Natl. Acad. Sci. USA* **1998**, *95*, 12832–12837. [CrossRef]
16. Ibba, M.; Soll, D. Aminoacyl-tRNAs: Setting the limits of the genetic code. *Genes Dev.* **2004**, *18*, 731–738. [CrossRef] [PubMed]
17. Schon, A.; Kannangara, C.G.; Gough, S.; Soll, D. Protein biosynthesis in organelles requires misaminoacylation of tRNA. *Nature* **1988**, *331*, 187–190. [CrossRef]
18. Curnow, A.W.; Hong, K.; Yuan, R.; Kim, S.; Martins, O.; Winkler, W.; Henkin, T.M.; Soll, D. Glu-tRNAGln amidotransferase: A novel heterotrimeric enzyme required for correct decoding of glutamine codons during translation. *Proc. Natl. Acad. Sci. USA* **1997**, *94*, 11819–11826. [CrossRef]
19. Rathnayake, U.M.; Wood, W.N.; Hendrickson, T.L. Indirect tRNA aminoacylation during accurate translation and phenotypic mistranslation. *Curr. Opin. Chem. Biol.* **2017**, *41*, 114–122. [CrossRef]

20. Tumbula, D.L.; Becker, H.D.; Chang, W.Z.; Soll, D. Domain-specific recruitment of amide amino acids for protein synthesis. *Nature* **2000**, *407*, 106–110. [CrossRef] [PubMed]

21. Akochy, P.M.; Bernard, D.; Roy, P.H.; Lapointe, J. Direct glutaminyl-tRNA biosynthesis and indirect asparaginyl-tRNA biosynthesis in *Pseudomonas aeruginosa* PAO1. *J. Bacteriol.* **2004**, *186*, 767–776. [CrossRef] [PubMed]

22. Chuawong, P.; Hendrickson, T.L. The nondiscriminating aspartyl-tRNA synthetase from *Helicobacter pylori*: Anticodon-binding domain mutations that impact tRNA specificity and heterologous toxicity. *Biochemistry* **2006**, *45*, 8079–8087. [CrossRef]

23. Mladenova, S.R.; Stein, K.R.; Bartlett, L.; Sheppard, K. Relaxed tRNA specificity of the *Staphylococcus aureus* aspartyl-tRNA synthetase enables RNA-dependent asparagine biosynthesis. *FEBS Lett.* **2014**, *588*, 1808–1812. [CrossRef] [PubMed]

24. Becker, H.D.; Reinbolt, J.; Kreutzer, R.; Giege, R.; Kern, D. Existence of two distinct aspartyl-tRNA synthetases in *Thermus thermophilus*. Structural and biochemical properties of the two enzymes. *Biochemistry* **1997**, *36*, 8785–8797. [CrossRef]

25. Min, B.; Pelaschier, J.T.; Graham, D.E.; Tumbula-Hansen, D.; Soll, D. Transfer RNA-dependent amino acid biosynthesis: an essential route to asparagine formation. *Proc. Natl. Acad. Sci. USA* **2002**, *99*, 2678–2683. [CrossRef]

26. Skouloubris, S.; Ribas de Pouplana, L.; De Reuse, H.; Hendrickson, T.L. A noncognate aminoacyl-tRNA synthetase that may resolve a missing link in protein evolution. *Proc. Natl. Acad. Sci. USA* **2003**, *100*, 11297–11302. [CrossRef]

27. Shimizu, Y.; Inoue, A.; Tomari, Y.; Suzuki, T.; Yokogawa, T.; Nishikawa, K.; Ueda, T. Cell-free translation reconstituted with purified components. *Nat. Biotechnol.* **2001**, *19*, 751–755. [CrossRef]

28. Wilkins, M.R.; Gasteiger, E.; Bairoch, A.; Sanchez, J.C.; Williams, K.L.; Appel, R.D.; Hochstrasser, D.F. Protein identification and analysis tools in the ExPASy server. *Methods Mol. Biol.* **1999**, *112*, 531–552.

29. Zhao, L.; Rathnayake, U.M.; Dewage, S.W.; Wood, W.N.; Veltri, A.J.; Cisneros, G.A.; Hendrickson, T.L. Characterization of tunnel mutants reveals a catalytic step in ammonia delivery by an aminoacyl-tRNA amidotransferase. *FEBS Lett.* **2016**, *590*, 3122–3132. [CrossRef]

30. Varshney, U.; Lee, C.P.; RajBhandary, U.L. Direct analysis of aminoacylation levels of tRNAs in vivo. Application to studying recognition of *Escherichia coli* initiator tRNA mutants by glutaminyl-tRNA synthetase. *J. Biol. Chem.* **1991**, *266*, 24712–24718.

31. Salazar, J.C.; Ahel, I.; Orellana, O.; Tumbula-Hansen, D.; Krieger, R.; Daniels, L.; Soll, D. Coevolution of an aminoacyl-tRNA synthetase with its tRNA substrates. *Proc. Natl. Acad. Sci. USA* **2003**, *100*, 13863–13868. [CrossRef]

32. Kasai, H.; Oashi, Z.; Harada, F.; Nishimura, S.; Oppenheimer, N.J.; Crain, P.F.; Liehr, J.G.; von Minden, D.L.; McCloskey, J.A. Structure of the modified nucleoside Q isolated from *Escherichia coli* transfer ribonucleic acid. 7-(4,5-cis-Dihydroxy-1-cyclopenten-3-ylaminomethyl)-7-deazaguanosine. *Biochemistry* **1975**, *14*, 4198–4208. [CrossRef]

33. Sylvers, L.A.; Rogers, K.C.; Shimizu, M.; Ohtsuka, E.; Soll, D. A 2-thiouridine derivative in tRNAGlu is a positive determinant for aminoacylation by *Escherichia coli* glutamyl-tRNA synthetase. *Biochemistry* **1993**, *32*, 3836–3841. [CrossRef]

34. Giege, R.; Sissler, M.; Florentz, C. Universal rules and idiosyncratic features in tRNA identity. *Nucleic Acids Res.* **1998**, *26*, 5017–5035. [CrossRef]

35. Hasegawa, T.; Himeno, H.; Ishikura, H.; Shimizu, M. Discriminator base of tRNA[Asp] is involved in amino acid acceptor activity. *Biochem. Biophys. Res. Commun.* **1989**, *163*, 1534–1538. [CrossRef]

36. Nameki, N.; Tamura, K.; Himeno, H.; Asahara, H.; Hasegawa, T.; Shimizu, M. *Escherichia coli* tRNA[Asp] recognition mechanism differing from that of the yeast system. *Biochem. Biophys. Res. Commun.* **1992**, *189*, 856–862. [CrossRef]

37. Gregory, S.T.; Dahlberg, A.E. Effects of mutations at position 36 of tRNA[Glu] on missense and nonsense suppression in *Escherichia coli*. *FEBS Lett.* **1995**, *361*, 25–28. [CrossRef]

38. Sekine, S.; Nureki, O.; Sakamoto, K.; Niimi, T.; Tateno, M.; Go, M.; Kohno, T.; Brisson, A.; Lapointe, J.; Yokoyama, S. Major identity determinants in the "augmented D helix" of tRNA[Glu] from *Escherichia coli*. *J. Mol. Biol.* **1996**, *256*, 685–700. [CrossRef] [PubMed]

39. Sekine, S.; Nureki, O.; Tateno, M.; Yokoyama, S. The identity determinants required for the discrimination between tRNA^{Glu} and tRNA^{Asp} by glutamyl-tRNA synthetase from *Escherichia coli*. *Eur. J. Biochem.* **1999**, *261*, 354–360. [CrossRef]

40. Rodin, S.N.; Ohno, S. Four primordial modes of tRNA-synthetase recognition, determined by the (G,C) operational code. *Proc. Natl. Acad. Sci. USA* **1997**, *94*, 5183–5188. [CrossRef]

41. Ribas de Pouplana, L.; Schimmel, P. Two classes of tRNA synthetases suggested by sterically compatible dockings on tRNA acceptor stem. *Cell* **2001**, *104*, 191–193. [CrossRef]

MDPI

St. Alban-Anlage 66

4052 Basel

Switzerland

Tel. +41 61 683 77 34

Fax +41 61 302 89 18

www.mdpi.com

Genes Editorial Office

E-mail: genes@mdpi.com

www.mdpi.com/journal/genes

www.ingramcontent.com/pod-product-compliance
Lightning Source LLC
Chambersburg PA
CBHW051855210326
41597CB00033B/5905